Recent Titles in This Series

499 **Karl-Hermann Neeb,** Invariant subsemigroups of Lie groups, 1993
498 **J. Nikiel, H. M. Tuncali, and E. D. Tymchatyn,** Continuous images of arcs and inverse limit methods, 1993
497 **John Roe,** Coarse cohomology and index theory on complete Riemannian manifolds, 1993
496 **Stanley O. Kochman,** Symplectic cobordism and the computation of stable stems, 1993
495 **Min Ji and Guang Yin Wang,** Minimal surfaces in Riemannian manifolds, 1993
494 **Igor B. Frenkel, Yi-Zhi Huang, and James Lepowsky,** On axiomatic approaches to vertex operator algebras and modules, 1993
493 **Nigel J. Kalton,** Lattice structures on Banach spaces, 1993
492 **Theodore G. Faticoni,** Categories of modules over endomorphism rings, 1993
491 **Tom Farrell and Lowell Jones,** Markov cell structures near a hyperbolic set, 1993
490 **Melvin Hochster and Craig Huneke,** Phantom homology, 1993
489 **Jean-Pierre Gabardo,** Extension of positive-definite distributions and maximum entropy, 1993
488 **Chris Jantzen,** Degenerate principal series for symplectic groups, 1993
487 **Sagun Chanillo and Benjamin Muckenhoupt,** Weak type estimates for Cesaro sums of Jacobi polynomial series, 1993
486 **Brian D. Boe and David H. Collingwood,** Enright-Shelton theory and Vogan's problem for generalized principal series, 1993
485 **Paul Feit,** Axiomization of passage from "local" structure to "global" object, 1993
484 **Takehiko Yamanouchi,** Duality for actions and coactions of measured groupoids on von Neumann algebras, 1993
483 **Patrick Fitzpatrick and Jacobo Pejsachowicz,** Orientation and the Leray-Schauder theory for fully nonlinear elliptic boundary value problems, 1993
482 **Robert Gordon,** G-categories, 1993
481 **Jorge Ize, Ivar Massabo, and Alfonso Vignoli,** Degree theory for equivariant maps, the general S^1-action, 1992
480 **L. Š. Grinblat,** On sets not belonging to algebras of subsets, 1992
479 **Percy Deift, Luen-Chau Li, and Carlos Tomei,** Loop groups, discrete versions of some classical integrable systems, and rank 2 extensions, 1992
478 **Henry C. Wente,** Constant mean curvature immersions of Enneper type, 1992
477 **George E. Andrews, Bruce C. Berndt, Lisa Jacobsen, and Robert L. Lamphere,** The continued fractions found in the unorganized portions of Ramanujan's notebooks, 1992
476 **Thomas C. Hales,** The subregular germ of orbital integrals, 1992
475 **Kazuaki Taira,** On the existence of Feller semigroups with boundary conditions, 1992
474 **Francisco González-Acuña and Wilbur C. Whitten,** Imbeddings of three-manifold groups, 1992
473 **Ian Anderson and Gerard Thompson,** The inverse problem of the calculus of variations for ordinary differential equations, 1992
472 **Stephen W. Semmes,** A generalization of riemann mappings and geometric structures on a space of domains in $\mathbf{C}^n$, 1992
471 **Michael L. Mihalik and Steven T. Tschantz,** Semistability of amalgamated products and HNN-extensions, 1992
470 **Daniel K. Nakano,** Projective modules over Lie algebras of Cartan type, 1992
469 **Dennis A. Hejhal,** Eigenvalues of the Laplacian for Hecke triangle groups, 1992
468 **Roger Kraft,** Intersections of thick Cantor sets, 1992
467 **Randolph James Schilling,** Neumann systems for the algebraic AKNS problem, 1992

(*Continued in the back of this publication*)

Invariant Subsemigroups of Lie Groups

MEMOIRS
of the
American Mathematical Society

Number 499

Invariant Subsemigroups of Lie Groups

Karl-Hermann Neeb

July 1993 • Volume 104 • Number 499 (end of volume) • ISSN 0065-9266

American Mathematical Society
Providence, Rhode Island

1991 *Mathematics Subject Classification.*
Primary 22A15, 22E15, 22E60, 43A60.

Library of Congress Cataloging-in-Publication Data

Neeb, Karl-Hermann.
Invariant subsemigroups of Lie groups/Karl-Hermann Neeb.
p. cm. – (Memoirs of the American Mathematical Society, ISSN 0065-9266; no. 499)
Includes bibliographical references.
ISBN 0-8218-2562-3
1. Lie algebras. 2. Lie groups. 3. Semigroups. I. Title. II. Series.
QA3.A57 no. 499
[QA252.3]
510 s–dc20 93-17164
[512′.55] CIP

Memoirs of the American Mathematical Society

This journal is devoted entirely to research in pure and applied mathematics.

Subscription information. The 1993 subscription begins with Number 482 and consists of six mailings, each containing one or more numbers. Subscription prices for 1993 are $336 list, $269 institutional member. A late charge of 10% of the subscription price will be imposed on orders received from nonmembers after January 1 of the subscription year. Subscribers outside the United States and India must pay a postage surcharge of $25; subscribers in India must pay a postage surcharge of $43. Expedited delivery to destinations in North America $30; elsewhere $92. Each number may be ordered separately; *please specify number* when ordering an individual number. For prices and titles of recently released numbers, see the New Publications sections of the *Notices of the American Mathematical Society*.

Back number information. For back issues see the *AMS Catalog of Publications*.

Subscriptions and orders should be addressed to the American Mathematical Society, P. O. Box 1571, Annex Station, Providence, RI 02901-1571. *All orders must be accompanied by payment.* Other correspondence should be addressed to Box 6248, Providence, RI 02940-6248.

Memoirs of the American Mathematical Society is published bimonthly (each volume consisting usually of more than one number) by the American Mathematical Society at 201 Charles Street, Providence, RI 02904-2213. Second-class postage paid at Providence, Rhode Island. Postmaster: Send address changes to Memoirs, American Mathematical Society, P. O. Box 6248, Providence, RI 02940-6248.

Printed in the United States of America.
This volume was printed directly from author-prepared copy.
The paper used in this book is acid-free and falls within the guidelines
established to ensure permanence and durability. ♾
10 9 8 7 6 5 4 3 2 1 98 97 96 95 94 93

TABLE OF CONTENTS

ABSTRACT

We study closed invariant subsemigroups S of Lie groups G which are Lie semigroups, i.e., topologically generated by one-parameter semigroups. Such a semigroup S is determined by its cone $\mathbf{L}(S)$ of infinitesimal generators, a closed convex cone in the Lie algebra $\mathbf{L}(G)$ which is invariant under the adjoint action. First we investigate the structure of Lie algebras with invariant cones and give a characterization of those Lie algebras containing pointed and generating invariant cones. Then we study the global structure of invariant Lie semigroups, and how far Lie's third theorem remains true for invariant cones and Lie semigroups. Finally we describe the Bohr compactification S^b of an invariant Lie semigroup. Most remarkably, the lattice of idempotents of S^b is isomorphic to a certain lattice of faces of the cone dual to $\mathbf{L}(S)$.

Key words: invariant cones, Lie algebras, Lie semigroups, Bohr compactification, controllability, globality, causal structure

INTRODUCTION

This work is devoted to the study of closed invariant subsemigroups of Lie groups which are topologically generated by one-parameter semigroups. We call these semigroups *invariant Lie semigroups.* Dropping the invariance condition we speak of *Lie semigroups.* These semigroups are important for various reasons.

An *invariant order* on a Lie group G is a partial order on G which is invariant under both left and right shifts. Then the set

$$S := \{g \in G : \mathbf{1} \leq g\}$$

is an invariant submonoid of G with $H(S) = \{\mathbf{1}\}$. If, conversely, $S \subseteq G$ is an invariant submonoid with $H(S) = \{\mathbf{1}\}$ the prescription

$$g \leq g' \quad \text{if} \quad g' \in gS$$

defines an invariant order on G. We say that $\leq$ is a *continuous order* if the semigroup S is closed and topologically generated by every neighborhood of $\mathbf{1}$. According to a result of the author ([Ne91d]) a closed submonoid S of G with $H(S) = \{\mathbf{1}\}$ defines a continuous invariant order if and only if S is an invariant Lie semigroup. This is the connection of invariant Lie semigroups with these orders. Invariant orders on Lie groups are studied in [Vin80], [Ol82a], [Pa81] (in semi-simple groups), [Gi89] (in solvable groups) and by the author ([Ne91d], [Ne91e], [Ne88]). One of the most interesting questions in this context is the existence problem. When does a connected Lie group G admit a continuous invariant order?

If G is a connected Lie group and $\Delta \subseteq G \times G$ is the diagonal, then $M := (G \times G)/\Delta$ is a simple example of a symmetric space. If, in addition, $S \subseteq G$ is a Lie semigroup, then $S_1 := (S \times \{\mathbf{1}\})\Delta$ is a subsemigroup of $G \times G$ which defines a G-invariant partial order on M by

$$(g_1, g_2)\Delta \leq (g_1', g_2')\Delta \quad \text{if} \quad (g_1', g_2') \in (g_1, g_2)S_1.$$

So invariant Lie semigroups are closely related to ordered symmetric spaces. If, more generally, $S \subseteq G$ is a Lie semigroup, and $H(S) := S \cap S^{-1}$ its group of units, then the prescription

$$gH(S) \leq g'H(S) \quad \text{if} \quad g' \in gS$$

Received by the editors November 5, 1990.

Research supported by a postdoctoral grant of the Deutsche Forschungsgemeinschaft.

defines a partial order on the homogeneous space $M = G/H(S)$.

These spaces became increasingly important in recent years in such contexts as representation theory (Ol'shanskiĭ [Ol82a], [Ol82b], Howe [Fo89]) and harmonic analysis (Faraut [Fa87], [HOØ91]). Ordered homogeneous spaces have already been considered in [Ne91b], [DN92], [MN92], [HiHo90] and [La89]. They have also close relations to general relativity ([Gu76], [GL84], [HE73], [Ne91b], [MN92]).

We do not want to leave the reader with this bare idea of these applications, so we explain some of them in greater detail.

Representation Theory: Let G be a real semisimple Lie group. It is well known that the irreducible unitary representations of G fall into a finite number of series, each of which is associated with a conjugacy class of Cartan subgroups of G. Accordingly, the Hilbert space $L^2(G)$, with respect to right invariant Haar measure and the right regular representation of G, decomposes into a direct sum

$$L^2(G) = L_1 \oplus \ldots \oplus L_k$$

of invariant subspaces corresponding to the various series. In [GG77] Gelfand and Gindikin outlined a program to realize the function spaces L_j as spaces of boundary values of holomorphic functions on a complex manifold G_j with Shilov boundary G. It was Ol'shanskiĭ who observed that one may realize G_j, under certain additional conditions, as the interior of a Lie semigroup with G as group of units. This is where Lie semigroups appear on the scene. We give a short description of Ol'shanskiĭ's construction.

Let G be a real simple Lie group with Lie algebra $\mathfrak{g} = \mathbf{L}(G)$ and $C \subseteq \mathfrak{g}$ a pointed, generating invariant wedge. Such a cone exists if and only if the maximal compactly embedded subalgebras of $\mathfrak{g}$ have non-trivial center. We also assume that G is contained in a complex Lie group $G_{\mathbb{C}}$ with $\mathbf{L}(G_{\mathbb{C}}) = \mathfrak{g}_{\mathbb{C}}$, the complexification of $\mathfrak{g}$. The set

$$\Gamma = \Gamma(C) := G \exp(iC) \subseteq G_{\mathbb{C}}$$

is a closed subsemigroup of $G_{\mathbb{C}}$ ([Ol82a, p.278]) and its interior $\Gamma^0 = G \exp(iC^0)$ is a complex manifold with an antiholomorphic antiinvolution $g \mapsto g^\sharp$ corresponding to $X + iY \mapsto -X + iY$. These semigroups are the non-commutative analogs of the tube domains $V + iC \subseteq V_{\mathbb{C}}$, where C is a pointed generating wedge in the real vector space V.

Let H be a complex Hilbert space. We endow the set $\mathcal{C}$ of linear contractions of H with the weak operator topology. A representation of the semigroup Γ in H is a continuous homomorphism $\mathcal{T} : \Gamma \to \mathcal{C}$ such that

$$\mathcal{T}(\mathbf{1}) = \mathrm{id}_H \quad \text{and} \quad \mathcal{T}(\gamma^\sharp) = \mathcal{T}(\gamma)^* \quad \text{for all} \quad \gamma \in \Gamma.$$

Then the restriction $T := \mathcal{T}|_G$ is a strongly continuous unitary representation of G on H ([Lyu88, p.89]). $\mathcal{T}$ is said to be holomorphic if its restriction $\Gamma^0 \to B(H)$ is a holomorphic mapping.

For every function f on Γ^0 and $\gamma \in \Gamma^0$ we denote with γf the function $g \mapsto f(g\gamma)$ on G ($G\Gamma^0 \subseteq \Gamma\Gamma^0 \subseteq \Gamma^0$). The space of all holomorphic functions f on Γ^0 which satisfy

$$\|f\|_H := \sup_{\gamma \in \Gamma^0} \|\gamma f\|_2 < \infty$$

is denoted by $H^2(C)$ or simply by H. This space is called the *Hardy space associated with* C. We define $\mathcal{T}(\gamma)f(\gamma_1) := f(\gamma_1\gamma)$ for $\gamma \in \Gamma$ and $\gamma_1 \in \Gamma^0$. The following theorem is Ol'shanskiĭ's main result in [Ol82c]:

Theorem 1. 1) *H is a Hilbert space with respect to the norm $\|\cdot\|_H$.*

2) *There exists an isometry $I : H^2(C) \to L^2(G)$ such that for an arbitrary function $f \in H$ and an arbitrary sequence $\gamma_n \in \Gamma^0$ which converges to $\mathbf{1}$, the sequence $\gamma_n f$ converges to If with respect to the metric on $L^2(G)$.*
3) *I intertwines with right translations from G, i.e., $I\mathcal{T}(g) = R(g)I$.*
4) *$\mathcal{T}$ is a holomorphic representation of the semigroup Γ on H.*
5) *If $C = C_{\min}$ is a minimal invariant cone in L, then $I(H^2(C_{\min}))$ consists precisely of all the representations of the holomorphic discrete series, so $\Gamma(C_{\min})^0$ is one of the complex manifolds specified in the Gelfand Gindikin program.* ∎

There is a close connection between the semigroup $\Gamma = G\exp(iC)$ considered by Ol'shanskiĭ and Howe's oscillator semigroup (or metaplectic semigroup) ([Fo89], [Hi89]). To describe this connection, a few definitions are in order. First we set

$$S_{2n} := \{X \in \mathrm{Gl}(2n,\mathbb{C}) : X = X^\top, \mathrm{Re}\,X \text{ is positive definite }\}.$$

If $X \in S_{2n}$, then a function $K_X : \mathbb{R}^{2n} \to \mathbb{C}$ is called a *real Gauss kernel* if it is defined by $K_X(v) := e^{-\frac{1}{2}v^\top X v}$, and we write $GK_{\mathbb{R}} := \{K_X : X \in S_{2n}\}$. Real Gauss kernels are Schwartz functions on $\mathbb{R}^{2n}$ and $(GK_{\mathbb{R}}, *_{tw})$ is a semigroup, where *twisted convolution* $*_{tw}$ is defined by

$$F_1 *_{tw} F_2(v) = \int_{\mathbb{R}^{2n}} F_1(w)F_2(v-w)e^{-\pi i w^\top J v}\, dw$$

with

$$J = \begin{pmatrix} \mathbf{0} & \mathbf{1} \\ -\mathbf{1} & \mathbf{0} \end{pmatrix}.$$

The *oscillator* or *metaplectic semigroup* may be realized as the set

$$S_{tw} = \{C \cdot K_X \in GK_{\mathbb{R}} : C^2 = \frac{\det(X + i\pi J)}{(2\pi)^{2n}} \neq 0, X \in S_{2n}\}.$$

Let C be one of the two invariant cones in the Lie algebra $\mathrm{sp}(n,\mathbb{R})$ and let

$$\Gamma = \mathrm{Sp}(n,\mathbb{R})\exp(iC) \subseteq \mathrm{Sp}(n,\mathbb{C})$$

denote the associated semigroup. Then there exists a semigroup homomorphism $\gamma : S_{tw} \to \Gamma^0$ which is a double covering. The Shilov boundary of the semigroup S_{tw} is the *metaplectic group*, a double covering of $\mathrm{Sp}(n, \mathbb{R})$. The semigroup Γ agrees with the semigroup of contractions on $\mathbb{C}^n$ with respect to the hermitean form B given by the matrix iJ with J as above.

There are very recent generalizations of these concepts to affine symmetric spaces by J. Hilgert, G. Olafsson and B. Ørsted ([HOØ91]). J. Hilgert and G. Olafsson also have a generalization of the analytic extension of unitary representations to the solvable case ([HiOl90]).

Geometric Control Theory: Let M be a finite dimensional differentiable manifold and Ω be a system of complete smooth vector fields on M. A *steering function* is a mapping $t \mapsto f_t: [0, T] \to \Omega$ with $T \in \mathbb{R}^+$. The solution of the problem

$$\dot{x}(t) = f_t\big(x(t)\big), \quad x(0) = x_0$$

may be interpreted as the solution of a system steered by f. A typical question arising in this context is: What is the set of points reachable from x_0 by using steering functions in a given class? The closure of this set is denoted by $A(x_0)$ and is called the *attainable set*. To describe this problem in terms of Lie semigroups, we assume that the Lie algebra $\mathfrak{g}$ of smooth vector fields generated by Ω is finite dimensional. Then there exists a Lie group G with $\mathbf{L}(G) = \mathfrak{g}$ which acts as a group of diffeomorphisms on M such that the mapping $G \times M \to M, (g, m) \mapsto g.m$ is smooth ([Pal57, p.95]). It is no loss in generality to consider only the orbit $M_0 := G.x_0$ which is a homogeneous G-space. Let $H := \{g \in G : g.x_0 = x_0\}$. We may identify M_0 with G/H and determine the attainable set for the class of piecewise constant steering functions. The trajectories $\gamma : [0, T] \to G/H$ are concatenations of solutions of the autonomous systems

$$\dot{x}_k(t) = \mathcal{X}_k\big(x_k(t)\big), x_k(t_{k-1}) = x_{k-1}(t_{k-1}) \quad \text{with} \quad t_0 = 0 < t_1 < \ldots < t_n = T,$$

$k = 1, \ldots, n$, and $\mathcal{X}_k \in \Omega$. Consequently we may write

$$\gamma(t) = \exp\big((t - t_{k-1})\mathcal{X}_k\big) \exp\big((t_{k-1} - t_{k-2})\mathcal{X}_{k-1}\big) \ldots \exp\big(t_1 \mathcal{X}_1\big).x_0$$

for $t_{k-1} \leq t < t_k$. Let $S_0 := \langle \exp \mathbb{R}^+ \Omega \rangle$ be the subsemigroup of G generated by all one-parameter subsemigroups with generator in Ω. These semigroups are called *ray semigroups* because they are generated by rays, i.e., one-parameter semigroups. Now it is clear that $S_0.x_0$ is exactly the set of points reachable from x_0 and the exploration of this set reduces to the study of the semigroup $S_0 \subseteq G$ and the set $S_0 H = \{g \in G : g.x_0 \in \overline{S_0.x_0}\}$. This set is a semigroup if and only if $HS_0 \subseteq S_0 H$. The semigroup $S := \overline{S_0}$ is a Lie semigroup and $\overline{S.x_0} = \overline{S_0.x_0}$. Let us specialize further to the case where $H = \{\mathbf{1}\}$ and $M_0 = G$ is a Lie group, i.e., no vector field in $\mathfrak{g}$ has a zero on M_0. Then we see that the attainable set $A(\mathbf{1}) = S = \overline{S_0}$ is a Lie semigroup in G. It is interesting to ask if the system defined by $\Omega \subseteq \mathfrak{g}$ is *transitive*, i.e., $A(\mathbf{1}) = G$ which is equivalent to $S = G$ (Section VI). These systems are called *controllable* because we find for every $g \in G$ a piecewise constant steering function which steers a trajectory from $\mathbf{1}$ to g. Using semigroup theoretic notions, one may say that a set $\Omega \subseteq \mathfrak{g}$ is transitive if and only if $\exp \mathbb{R}^+\Omega$ is not contained in a Lie semigroup $S \neq G$ (cf. [La87], [Ne91a], [Hi92]).

Causality and Ordered Manifolds: Let M be a differentiable manifold. A *causal structure* on M is a function which assigns to each point $p \in M$ a wedge $\theta(p) \subseteq T_p(M)$, in the tangent space at p. We call such a function a *wedge field on* M. A *causal curve* (*conal curve*) is a piecewise smooth curve $\gamma : [t_0, t_1] \to M$ such that

$$\dot{\gamma}(t) \in \theta(\gamma(t))$$

whenever the derivative exists. We say that M is a *causal manifold* if there are no closed non-constant causal curves on M and *totally acausal* if every pair p, q of points in M can be joined by a causal curve. The set of points reachable from a point $p \in M$ is called its *causal future* $J^+(p)$ and the set of points $q \in M$ for which $p \in J^+(q)$ is called the *causal past* $J^-(p)$ of p.

Example 2. The events in space-time of general relativity form a 4-dimensional causal manifold M. In this special case, the cone $\theta(p)$ is called the *light-cone* in p. It is a four-dimensional Lorentzian cone, i.e.,

$$\theta(p) = \{(x, y, z, t) \in T_p(M) : t \geq 0, x^2 + y^2 + z^2 - t^2 \leq 0\}$$

with respect to a suitable basis of $T_p(M)$. The causal curves are called *timelike.* The causality of M is a reasonable physical assumption. For further information about general-relativity and causality of Lorentzian manifolds see [HE73], [HiHo90] and [Lev85]. ∎

Given a manifold M with a wedge field θ, we associate with θ an order on M defined by

$$p \prec q \quad \text{if} \quad q \in J^+(p).$$

Even in a causal manifold there may occur the pathology that there exist causal curves $\gamma : \mathbb{R}^+ \to M$ with $\gamma(0) = p$ such that $\gamma(t)$ comes arbitrary near to p and that the points $\gamma(t)$ which are near p cannot all be reached by short causal curves which stay in the vicinity of p. So we call a causal manifold M *strictly causal* (*globally orderable*) if every point $p \in M$ has a basis $\mathcal{U}$ of neighborhoods such that for every point $x \in U$ with $U \in \mathcal{U}$, every point $y \in J^+(x) \cap U$ can be reached from x on a causal curve which is contained in U.

Under still very general physical circumstances it is reasonable to assume that the manifold M is the homogeneous space of a Lie group G and that the wedge field θ is invariant, i.e.,

$$d\mu_g(\theta(p)) = \theta(\mu_g(p)) \quad \text{for} \quad p \in M, g \in G,$$

where $\mu_g : M \to M, p \mapsto g.p$ defines the action of G on M. We assume in addition that the wedges $\theta(p)$ are pointed. Let H denote the stabilizer of a point $p_0 \in M$ ($H = \{g \in G : g.p_0 = p_0\}$) and $\pi : G \to M, g \mapsto g.p_0$ the orbit mapping which induces a diffeomorphism of G/H and M. The striking idea which leads to Lie semigroups is to look at the wedge

$$W := d\pi(\mathbf{1})^{-1}\big(\theta(p_0)\big) \quad \text{and the set} \quad S := \pi^{-1}\big(J^+(p_0)\big).$$

We know from [La89, 4.9] that

$$(I) \quad H(W) := W \cap -W = \mathbf{L}(H) \quad \text{and} \quad \mathrm{Ad}(h)W = W \quad \text{for} \quad h \in H.$$

As a consequence of (I) we find that W is a *Lie wedge*, i.e., a wedge in a Lie algebra $\mathfrak{g}$ satisfying

$$e^{\mathrm{ad}\, h}W = W \quad \text{for all} \quad h \in H(W) = W \cap -W.$$

An easy computation shows that the mapping $\theta \mapsto W$ establishes a one-to-one correspondence between the G-invariant pointed wedge fields on M and the wedges $W \subseteq \mathbf{L}(G)$ satisfying condition (I). This is the correspondence between causal manifolds and Lie semigroups on the infinitesimal level. There is also a local one ([La89, 8]) which is too technical to present here. The next theorem describes the global correspondence. We "lift" the causal structure from the homogeneous space M to the group G by defining

$$\widetilde{\theta}(g) := d\lambda_g(\mathbf{1})W \quad \text{for} \quad g \in G,$$

where $\lambda_g : G \to G$ denotes left multiplication by g. Then $\widetilde{\theta}$ is a left invariant wedge field on G and $d\pi(g)\widetilde{\theta}(g) = \theta\big(\pi(g)\big)$. We write $S(W)$ for the causal future of the unit element $\mathbf{1}$ with respect to $\widetilde{\theta}$. Then we have:

Theorem 3. ([HiHo90, 1.3], [Ne91b, 1.29])

1) *S and $S(W)$ are semigroups and $S = S(W)H = HS(W)$.*
2) *M is causal if and only if H agrees with the unit group $H(S) = S \cap S^{-1}$ of the semigroup S.*
3) *W generates the Lie algebra $\mathbf{L}(G)$ if and only if $J^+(p_0)$ has non-empty (dense) interior.* ∎

At this point we recall what was said about geometric control theory and clarify the connections between causality and control theory. Let us take $\Omega := W \subseteq \mathbf{L}(G)$, where $H(W) = \mathbf{L}(H)$ and W generates $\mathbf{L}(G)$. To each element $x \in \Omega$ corresponds a vector field $\mathcal{X}$ on M defined by

$$\mathcal{X}(p) = \frac{d}{dt}\Big|_{t=0} \exp(tx).p \quad \text{for} \quad p \in M.$$

We consider the control system on M defined by Ω and set $S_W := \langle \exp W \rangle$. Then the set $S_W H = H S_W$ is a semigroup because

$$hS_W h^{-1} = \langle \exp \mathrm{Ad}(h) W \rangle = S_W \quad \text{for all} \quad h \in H$$

and the attainable set $A(p_0)$ equals $\pi(\overline{S_W H}) = \overline{\pi(S_W)}$. The relations between these semigroups can be described as

$$S_W \subseteq S(W) \quad \text{and} \quad S_W H \subseteq S.$$

Therefore $A(p_0)$ is contained in the closure of the causal future $\overline{J^+(p_0)} = \overline{\pi(S)} = \pi(\overline{S})$. Now it follows from [Ne91b, 1.29] that $A(p_0)$ contains the dense interior of $J^+(p_0)$ and therefore agrees with its closure:

$$(*) \qquad A(p_0) = \overline{J^+(p_0)}.$$

This points out that it is the same task to study the attainability set or the causal future of the point $p_0 = \pi(\mathbf{1})$. As an immediate consequence, we see that the control system defined above is controllable if and only if M is acausal. One should notice that the set $A(p)$ and the causal future $J^+(p)$ may differ substantially for $p \neq p_0$. We illustrate this with the following simple example.

Example 4. Let G denote $(\mathbb{R}^2, \star)$ with $(x,y) \star (x',y') = (x + e^y x', y + y')$, the non-abelian two dimensional Lie group, $H = \{0\} \times \mathbb{R}$ and $\Omega = W = \mathbb{R}^+ \times \mathbb{R}$. We identify $M = G/H$ with $\mathbb{R}$. Then we have $\pi(x,y) = x$. The action of G on $M = \mathbb{R}$ is given by $(x,y).z = x + e^y z$ and $\Omega = \mathbb{R}^+ \mathcal{X}_1 + \mathbb{R}\mathcal{X}_2$ with the vector fields

$$\mathcal{X}_1(z) = \frac{\partial}{\partial z}, \quad \mathcal{X}_2(z) = z\frac{\partial}{\partial z} \quad \text{with} \quad [\mathcal{X}_1, \mathcal{X}_2] = \mathcal{X}_1.$$

Therefore $J^+(z) = [z, \infty[$ for all $z \in \mathbb{R}$, $A(z) = \mathbb{R}^+$ for $z \geq 0$ and $A(z) = \mathbb{R}$ for $z < 0$. ■

The guiding idea in the theory of Lie groups is to study connected Lie groups G via their Lie algebra $\mathbf{L}(G)$ and the exponential function

$$\exp : \mathbf{L}(G) \to G.$$

Identifying $\mathbf{L}(G)$ with the set of one-parameter subgroups of G the exponential function maps a one-parameter subgroup $\gamma : \mathbb{R} \to G$ to its value $\exp(\gamma) = \gamma(1)$. Four of the main properties are:

1) The exponential function is a local diffeomorphism of a neighborhood B of 0 in $\mathbf{L}(G)$ onto the neighborhood $\exp(B)$ of $\mathbf{1}$ in G.
2) Every element of G is a finite product of exponentials, i.e.,

$$G = \langle \exp \mathbf{L}(G) \rangle := \bigcup_{n \in \mathbb{N}} \left(\exp \mathbf{L}(G)\right)^n.$$

This is a consequence of 1) and the connectedness of G.

3) The assignment $A \mapsto \langle \exp A \rangle$ defines a bijection between the set of subalgebras of $\mathbf{L}(G)$ and the set of analytic subgroups of G. These are exactly the subgroups which are connected analytic submanifolds of G, i.e., the images of Lie group morphisms into G.
4) (Lie's Third Theorem) The assignment $G \mapsto \mathbf{L}(G)$ establishes a one-to-one correspondence between the simply connected finite dimensional Lie groups and the finite dimensional Lie algebras. As a consequence the structure of a simply connected Lie group G and all its analytic properties are related to Lie algebraic properties of $\mathbf{L}(G)$.

These basic properties permit a translation mechanism between the group G and its Lie algebra $\mathbf{L}(G)$. Problems concerning the group G are translated into problems on the Lie algebra $\mathbf{L}(G)$. This translation turns analytic problems which are often very unhandy and difficult to cope with into algebraic problems. Thus it extends the amount of available methods essentially.

As an example for the above method one may consider the classification of the compact simple Lie groups which is translated to the classification of root systems and certain discrete subgroups of vector spaces associated to root systems. The characteristic data are an abelian subalgebra $\mathfrak{h} \subseteq \mathbf{L}(G)$, called a Cartan algebra, a set Ω of linear functionals on $\mathfrak{h}$, called the roots, and a discrete subgroup $D \subseteq \mathfrak{h}$. Following this philosophy E. Cartan even classified irreducible symmetric spaces by classifying real semisimple Lie algebras with involutive automorphisms. Again the characteristic data are associated to a set of linear functions on a small abelian subspace of a certain Lie algebra.

Another example is the theory of unitary representations of connected Lie groups. A differentiation process associates to a given unitary representation $\pi : G \to \mathrm{U}(\mathcal{H})$ on the hilbert space $\mathcal{H}$ a representation of its Lie algebra on a dense subspace $\mathcal{H}^\omega$. This was the fundamental idea which was used by Harish-Chandra and others to develop the representation theory of semisimple Lie groups up to its present state. Again the characteristic data is closely related to Cartan algebras $\mathfrak{h}$ in $\mathbf{L}(G)$ and the representations of the associated subgroups $\exp \mathfrak{h} \subseteq G$.

Having these ideas in mind we want to study subsemigroups of Lie groups. So we have to associate a tangent object $\mathbf{L}(S)$ to a subsemigroup S of a Lie group G. Let us assume for simplicity that S is closed (cf. Definition IV.4). Then we define the tangent object

$$\mathbf{L}(S) := \{x \in \mathbf{L}(G) : \exp(\mathbb{R}^+ x) \subseteq S\}.$$

This is the set of all continuous one-parameter semigroups $\gamma : \mathbb{R}^+ \to S$. To make the translation mechanism work for semigroups we have to consider only those which are reconstructable from their tangent object. Thus we say that S is a *Lie semigroup* if

$$S = \overline{\langle \exp \mathbf{L}(S) \rangle}.$$

We say that S is an *invariant Lie semigroup* if it is, in addition, invariant under all inner automorphisms of G. These are the objects we will be concerned with in the following.

The assignment $S \mapsto \mathbf{L}(S)$ assigns to an invariant Lie semigroup a closed convex cone $\mathbf{L}(S)$ in the Lie algebra $\mathbf{L}(G)$ which is invariant under all inner automorphisms of $\mathbf{L}(G)$ (Proposition IV.7). So it is natural to start with a consideration of closed convex cones, called *wedges*, W in a Lie algebra $\mathfrak{g}$ which are *invariant*, i.e., invariant under all inner automorphisms of $\mathbf{L}(G)$. This is what we do in Sections I, II and III. We call it the infinitesimal theory because it deals with the sets of infinitesimal generators of invariant Lie semigroups.

In Section I we recall some material on the geometry of wedges W in finite dimensional vector spaces. The main concepts here are the faces $\mathcal{F}(W)$ and the exposed faces $\mathcal{F}_e(W)$ of W. These two sets are complete lattices with respect to set inclusion. We will generalize these concepts to Lie semigroups in Section IV. A wedge W in $\mathfrak{g}$ is said to be *pointed* if the *edge* $H(W) := W \cap (-W)$ consists only of $\{0\}$ and *generating* if $W - W = \mathfrak{g}$. It will turn out in Section II that a Lie algebra $\mathfrak{g}$ containing a pointed invariant wedge W also contains a Cartan algebra $\mathfrak{h}$ such that $K := \overline{e^{\operatorname{ad}\mathfrak{h}}}$ is a compact group of automorphisms of $\mathfrak{g}$. Moreover the wedge W is determined completely by its intersection with the fixed point module $\mathfrak{g}_{\text{fix}}$ with respect to the action of K. To understand this situation properly we consider in the second half of Section I a compact group K acting on $\mathfrak{g}$ and a wedge $W \subseteq \mathfrak{g}$ which is invariant under the action of K. The mapping $W \mapsto W \cap \mathfrak{g}_{\text{fix}}$ has very nice properties concerning interior points, the edge and the faces.

As we have already mentioned above, a Lie algebra $\mathfrak{g}$ containing a pointed generating invariant wedge contains also a Cartan algebra $\mathfrak{h}$ such that the group $e^{\operatorname{ad}\mathfrak{h}}$ is relatively compact. We call $\mathfrak{h}$ a *compactly embedded Cartan algebra.* We fix such a Cartan algebra. Then a decomposition of the $\mathfrak{h}$-module $\mathfrak{g}$ into isotypical components leads to a real roots decomposition of the Lie algebra $\mathfrak{g}$. Moreover, it has the two additional properties of having (strong) cone potential and of being quasihermitean. Section II is devoted to a structure analysis of Lie algebras with these properties. We clarify the interdependence of these properties and give a general construction for Lie algebras with cone potential which is due to Spindler ([Sp88]). This construction principle permits us to find counterexamples which illuminate the necessity of the assumptions in our theorems. It also shows that there is a great variety of quasihermitean Lie algebras with strong cone potential. As we will see in Section III these are exactly the conditions needed to guarantee the existence of a pointed generating invarint wedge.

Having done all this structure theory in Section II we study invariant wedges in quasihermitean Lie algebras with a compactly embedded Cartan algebra in Section III. Since $\mathfrak{h}$ agrees with the fixed point submodule $\mathfrak{g}_{\text{fix}}$ of the $\overline{e^{\operatorname{ad}\mathfrak{h}}}$-module $\mathfrak{g}$, we assign to each invariant wedge $W \subseteq \mathfrak{g}$ its intersection $C := \mathfrak{h} \cap W$ with $\mathfrak{h}$ (cf. Section I). The main problem is to find conditions for wedges $C \subseteq \mathfrak{h}$ which guarantee the existence of an invariant wedge $W \subseteq \mathfrak{g}$ such that $W \cap \mathfrak{h} = C$. To find such conditions we define the *Weyl group* $\mathcal{W}$ and a wedge $\mathcal{C} \subseteq \operatorname{End}(\mathfrak{h})$ which is spanned by a set of rank-one operators defined in terms of the real root decomposition (Definitions III.2, III.4). It is not difficult to prove that the invariance of C under $\mathcal{W}$ and $\mathcal{C}$ is necessary (Proposition III.7). If C is pointed and generating in $\mathfrak{h}$ the sufficiency of this invariance condition is proved in [HHL89]. The principal new achievements are the following:

1) A vector space $\widetilde{\mathfrak{h}} \subseteq \mathfrak{h}$ is the intersection $I \cap \mathfrak{h}$ of an ideal $I \subseteq \mathfrak{g}$ if and only if it is invariant under $\mathcal{W}$ and $\mathcal{C}$ (Theorem III.23). There exists an additional condition (CA) which may be formulated with $\widetilde{\mathfrak{h}}$ and the real root system such that I may be chosen to have a compactly embedded Cartan algebra if and only if $\widetilde{\mathfrak{h}}$ satisfies (CA).
2) Assume that $C \subseteq \mathfrak{h}$ is a pointed wedge. Then $C = W \cap \mathfrak{h}$ for an invariant wedge $W \subseteq \mathfrak{g}$ if and only if C is invariant under $\mathcal{W}$ and $\mathcal{C}$ and $C - C$ satisfies the condition (CA) (Theorem III.31).
3) Assume that $C \subseteq \mathfrak{h}$ is a generating wedge. Then $C = W \cap \mathfrak{h}$ for an invariant wedge $W \subseteq \mathfrak{g}$ if and only if C is invariant under $\mathcal{W}$ and $\mathcal{C}$ and the edge $H(C)$ satisfies an additional condition (Rec) (Theorem III.33).

These are the general reconstruction theorems. In most of the cases where we need those theorems we do not have to take care of the additional conditions (Rec) and (CA) because we have Theorem III.35 (The Special Reconstruction Theorem): Suppose that $W \subseteq \mathfrak{g}$ is pointed and generating. Then the following assertions hold:

1) A wedge $C \subseteq W \cap \mathfrak{h}$ is the trace $W' \cap \mathfrak{h}$ of an invariant wedge $W' \subseteq \mathfrak{g}$ if and only if C is invariant under $\mathcal{W}$ and $\mathcal{C}$.
2) Write $D^\star$ for the dual wedge in $\widehat{\mathfrak{h}}$ of a wedge $D \subseteq \mathfrak{h}$. Then a wedge $F \subseteq \mathfrak{h}$ with $F^\star \in \mathcal{F}(C^\star)$ is the trace $F = W' \cap \mathfrak{h}$ of an invariant wedge $W' \subseteq \mathfrak{g}$ if and only if F is invariant under $\mathcal{W}$ and $\mathcal{C}$.

We conclude Section III with the Characterization Theorem of those Lie algebras which contain pointed generating invariant wedges.

In the first three sections we have described an infinitesimal theory of invariant subsemigroups of Lie groups, the theory of invariant wedges in Lie algebras. Now we turn to the global theory. Some of the results we shall need later in the special case of invariant subsemigroups are true for general Lie semigroups. These results are contained in Section IV. The crucial concept is that of a face of a Lie semigroup.

Let S be a closed submonoid of the Lie group G and $F \subseteq S$ a closed subsemigroup. Then we say that F is a

1) *a face of* S if $S \setminus F$ is an ideal in S.
2) *an exposed face of* S if $F = S \cap H(\overline{\langle SF^{-1} \rangle})$.
3) *a compact exposed face of* S if F is an exposed face and if there exists a continuous homomorphism $\varphi : S \to K$, where K is a compact monoid, such that $F = \varphi^{-1}(H(K))$.
4) *a normal exposed face of* S if $F = S \cap H(S_1)$, where $S_1 \subseteq G$ is a closed subsemigroup with $S \subseteq S_1$ and which has a normal group of units.

We denote the set of faces (exposed faces, compact exposed faces, normal exposed faces) of S with $\mathcal{F}(S)\big(\mathcal{F}_e(S),\ \mathcal{F}_c(S),\ \mathcal{F}_n(S)\big)$. Then we prove the hierarchy of faces, namely that a normal exposed face is compact exposed and that an exposed face is a face and discuss how faces of Lie semigroups generalize faces of wedges in vector spaces. We relate faces of the semigroup S to faces of the tangent wedge $\mathbf{L}(S)$ of S. Some of the most important results are:

1) $\mathbf{L}\left(\mathcal{F}_e(S)\right) \subseteq \mathcal{F}_e\left(\mathbf{L}(S)\right)$.
2) $\mathbf{L}\left(\mathcal{F}(S)\right) \subseteq \mathcal{F}\left(\mathbf{L}(S)\right)$ whenever S is invariant.
3) If $F \in \mathcal{F}_e(S)$ and S is a Lie semigroup, then the semigroup $\overline{\langle SF^{-1}\rangle}$ is a Lie semigroup.

The fifth section may be counted to the theory of compact semigroups. We construct an interesting compactification of a Lie semigroup S with compact group of units. We also give a criterion for the Alexandrov compactification to be a topological semigroup and a more general construction if this criterion fails. The results of this section will be used to describe the Bohr compactification of invariant Lie semigroups but they are interesting in their own right.

In Section VI we turn our attention to invariant Lie semigroups and work out the special tools which are available for this class of Lie semigroups. A crucial property is the *semialgebra property* of invariant wedges W in a Lie algebra $\mathfrak{g}$, i.e., that

$$(W \cap B) * (W \cap B) \subseteq W$$

for every Baker-Campbell-Hausdorff neighborhood B in $\mathfrak{g}$.

In Lemma VI.1 we use the semialgebra property of $\mathbf{L}(S)$. We also often use the fact that the order $\leq_S$ is directed and that it satisfies

$$x, y \leq_S xy \quad \text{for} \quad x, y \in S.$$

To prove that $\mathbf{L}\left(\overline{SF^{-1}}\right)^*$ is a face of the dual wedge $\mathbf{L}(S)^*$ for every face of S (the Face Theorem) we need results on the closedness of normal subgroups of G whenever the fundamental group $\pi_1(G)$ has at most rank 1. All this information is not available for general Lie semigroups. At the end of this section we present the two most essential low dimensional examples of non-abelian invariant Lie semigroups: the invariant Lie semigroup in the group $(2,\mathbb{R})^{\sim}$ and an invariant Lie semigroup in the Oscillator group. These examples are very instructive because we know that at least one of them is contained in every invariant Lie semigroup $S \subseteq G$ if $\mathbf{L}(G)$ is not a compact Lie algebra and $H(S) = \{1\}$ (Theorem VI.25). This permits us to conclude that exactly the invariant Lie semigroups with $\mathbf{L}(G)/\mathbf{L}\left(H(S)\right)$ compact and $H(S)$ compact have no non-compact order intervals.

The global theory of invariant Lie semigroups is divided into two main parts. The existence theory and the structure theory. The following two sections, Section VII and VIII, are devoted to the existence theory. In Section VIII we consider the problem of deciding whether a given invariant wedge is the tangent wedge of a Lie semigroup or not (the globality problem) and in Section VII we consider a weaker version of this problem, the controllability problem. When does the exponential image $\exp W$ of an invariant wedge generate the whole group? Recall that a Lie wedge $W \subseteq \mathbf{L}(G)$, where G is a connected Lie group, is said to be *controllable in* G if $\langle \exp W \rangle = G$. We will characterize those invariant wedges which are controllable in the associated simply connected group. Moreover, one can easily reduce the problem to the case where W is pointed and generating and then give a criterion for its intersection $C = W \cap \mathfrak{h}$ with a compactly embedded Cartan algebra $\mathfrak{h}$ which is checkable by consideration of the root data $(\mathfrak{h}, \Omega)$.

Also for the globality problem in Section VIII the main idea is to reduce the problem to the reductive case. This is much harder than it was for the controllability problem. The main result in this direction is the Reduction Theorem VIII.7, which gives only sufficient conditions for globality but which permits proving the globality of all pointed generating invariant wedges in some large classes of Lie algebras. Another main result is the existence of at least one pointed generating invariant wedge which is global in a simply connected group G whenever $\mathbf{L}(G)$ contains pointed generating invariant wedges. This leads to a characterization of those simply connected Lie groups which possess non-degenerate group orders in terms of algebraic properties of the Lie algebra.

In the last part of this work (Section IX to XII) we describe the Bohr compactification S^b of an invariant Lie semigroup S. The *Bohr compactification* of a topological semigroup S is a compact semigroup S^b together with a continuous morphism of topological semigroups $i_S : S \to S^b$ such that the following universal property holds. If $\varphi : S \to K$ is a morhism into a compact topological semigroup K, then there exists a morphism $\varphi^b : S^b \to K$ such that $\varphi^b \circ i_S = \varphi$. The Bohr compactification is a functor from the category $\underline{TSg}$ of topological semigroups to the category $\underline{CTSg}$ of compact topological semigroups. It is the adjoint of the forgetful functor from $\underline{CTSg}$ to $\underline{TSg}$. We endow S^b with the quasi order

$$s \prec s' \quad \text{if } s' \in sS^b$$

and we write $E(S^b)$ for the set of idempotents in the semigroup S^b.

The motivation for this work is the following. First it is interesting to see how such an abstract universal object as the Bohr compactifiction can be described by using the data at hand. Another motivation was to find interesting examples of compact topological semigroups which are not too far from Lie semigroups. This means that the subsemigroup generated by the images of all one-parameter subsemigroups is dense and the set of all one-parameter semigroups is a finite dimensional wedge. We call these semigroups *generalized Lie semigroups*. They show how the nowadays classical theory of compact semigroups and the very recent Lie theory of semigroups combine to yield new interesting results.

In Section IX we analyze the global structure of S^b. The main results are the following:

1) $E(S^b)$ is an abelian zero dimensional compact semigroup.
2) $\mathcal{F}_n(S) = \mathcal{F}_c(S)$ for all generating invariant Lie semigroups S.
3) For every idempotent $e \in E(S^b)$ there exists a generating invariant Lie semigroup $S_e \supseteq S$ such that $S_e^b \cong eS^b$ and this isomorphism is given by the mapping $\varphi_e^b : S_e^b \to eS^b$ induced by the unique continuation $\varphi_e : S_e \to eS^b$ of the mapping $\lambda_e \circ i_S : S \to eS^b$.

4) $S^\flat = \bigcup_{e \in E(S^\flat)} S(e)$, where

$$S(e) := \varphi_e^\flat(S_e(\mathbf{1})), \quad S_e(\mathbf{1}) = i_{S_e}(\widetilde{\mathrm{comp}}(S_e))H(S_e^\flat),$$

and $\widetilde{\mathrm{comp}}(S_e)$ is the set of all elements s in S_e for which the order interval $[H(S_e), sH(S_e)]$ in $G/H(S_e)$ is compact. The sets $S(e)$ are a disjoint cover of $S^\flat$ and the set $S(e)$ is open in $eS^\flat$.

Having obtained these results it is easy to see that equivalence classes

$$[s] := [s,s] = \{t \in S^\flat : s \prec t \prec s\}$$

of the elements in the set $S(e)$ are exactly the translates of the groups $H(e)$ (Proposition IX.16). So we are interested in these groups. By the isomorphism $S(e) \cong S_e(\mathbf{1})$ it is clear that it suffices to assume that $e = \mathbf{1}$ and to consider $H(\mathbf{1}) = H(S^\flat)$. We know already that $i_S(H(S))$ is dense in $H(S^\flat)$ (Proposition IX.16). Hence $H(S^\flat)$ is a homomorphic image of $H(S)^\flat$. As we will see below it may be substantially smaller. We determine the ideal corresponding to the kernel of the mapping $i_S\,|_{H(S)}$ in the Lie algebra $\mathbf{L}(G)$. Then we consider a reduced situation and describe how to get $H(S^\flat)$ from $H(S)$.

In this Section XI we consider the lattice $E(S^\flat)$ of idempotents of the Bohr compactification of a generating invariant Lie semigroup S. Since this lattice is the same for $S_0 := S/H(S)$ we assume throughout this section that $H(S) = \{\mathbf{1}\}$ (Proposition IX.14). Moreover we show that, provided $\operatorname{rank} \pi_1(G) \leq 1$, $E(S^\flat)$ can be embedded into the lattice of faces of the wedge $C^\star$, where $C := \mathbf{L}(S) \cap \mathfrak{h}$ is the intersection of $\mathbf{L}(S)$ with a compactly embedded Cartan algebra of $\mathbf{L}(G)$. Such a compactly embedded Cartan algebra exists since $\mathbf{L}(S)$ is a pointed generating invariant cone in $\mathbf{L}(G)$ (Proposition III.15). One should note that this result rests substantially on almost everything that we did in the previous sections. First we need the result on the geometry of cones from Section I, then the Special Reconstruction Theorem from Section III to get the correspondence between the faces of $C^\star$ and the faces of $\mathbf{L}(S)^*$. We also need the crucial Theorem IX.24 on the existence of the semigroup S_e for every idempotent and its properties. The assumption on the fundamental group of G is needed to apply the Face Theorem from Section VI (cf. Theorem IX.24). Finally the results on globality from Section VIII are needed to characterize those faces of $C^\star$ which correspond to idempotents of $S^\flat$.

In the last section we consider interesting examples of Lie semigroups and their compactifications. We use these examples to study finite dimensional representations of G which map S into a relatively compact subset of endomorphism of $\mathbb{R}^n$.

Acknowledgement

The author would like to express his appreciation to Karl Heinrich Hofmann under whose direction he prepared this work, to J. D.

Lawson for many useful discussions on Section VII, to W. A. F. Ruppert for teaching him so much about semigroup compactifications, to U. Zimmermann and A. Eggert for proof reading and to the whole "Seminar Sophus Lie" at the "Technische Hochschule Darmstadt" for providing such a pleasant atmosphere. The author also greatfully acknowledges the financial support of the "Deutsche Forschungsgemeinschaft" during the time while he was preparing this manuscript.

I. INVARIANT CONES IN K-MODULES

In this first section we collect some material concerning the geometry of wedges in vector spaces and invariant wedges in finite dimensional modules of compact groups. A rather complete reference for the geometry of wedges is the first chapter in [HHL89]. We recall the definitions and propositions which are needed later and sketch some of the elementary proofs. We will see in Section III that it is possible to classify invariant wedges in Lie algebras by their intersections with certain subalgebras which are the set of fixed points of an action of a compact group of automorphisms of the Lie algebra. This is where the theory of wedges in K-modules is needed. Throughout this section L denotes a finite dimensional vector space.

Definition I.1. Let L be a finite dimensional vector space. A subset W is called a *wedge* if it is a closed convex cone. The vector space $H(W) := W \cap -W$ is called the *edge of the wedge.* We say that W is *pointed* if the edge of W is trivial and that W is *generating* if $W - W = L$. We denote the dual of L with $\widehat{L}$. The *dual wedge* $W^* \subseteq \widehat{L}$ is the set of all functionals which are non-negative on W. We set

$$\operatorname{algint} W := \{x \in W : \omega(x) > 0 \text{ for all } \omega \in W^* \setminus H(W^*)\}.$$

■

Proposition I.2. *We identify the dual of $\widehat{L}$ with L. Then the following assertions hold for a wedge $W \subseteq L$:*

(1) $(W^*)^* = W$.

(2) *W is generating iff W^* is pointed and conversely W is pointed iff W^* is generating.*

(3) *$\omega \in \operatorname{algint} W^*$ iff $\omega(x) > 0$ for all $x \in W \setminus H(W)$ and $\operatorname{algint} W$ is the interior of W with respect to the subspace $W - W$ of L.*

(4) *For a family $(W_i)_{i \in I}$ of wedges in L we have that*

$$(\bigcap_{i\in I} W_i)^* = \overline{\sum_{i\in I} W_i^*} \quad \text{and} \quad (\sum_{i\in I} W_i)^* = \bigcap_{i\in I} W_i^*.$$

Proof. (1) [HHL89, I.1.4]
(2) [HHL89, I.1.7]
(3) [HHL89, I.2.21]
(4) [HHL89, I.1.6] ■

Definition I.3. Let F, $W \subseteq L$ be wedges. Then we set

$$L_F(W) := \overline{W + F - F} \quad \text{and} \quad T_F(W) := H\big(L_F(W)\big) = L_F(W) \cap -L_F(W).$$

The fact that $W + F - F$ is stable under multiplication with non-negative scalars and convex shows that $L_F(W)$ is a wedge. Note that $L_F(W) = \overline{W - F}$ if $F \subseteq W$. We say that a wedge $F \subseteq W$ is an *exposed face* of W if

$$F = W \cap T_F(W)$$

and a *face* of W if its complement $W \setminus F$ is an ideal in the additive semigroup W. We write $\mathcal{F}(W)$ for the set of faces of W and $\mathcal{F}_e(W)$ for the set of exposed faces of W. We will see later (Section IV) how to generalize these concepts to arbitrary subsemigroups of Lie groups. ∎

The following proposition describes how the faces of W and its dual wedge are related.

Proposition I.4. (Exposed Faces) *The set $\mathcal{F}_e(W)$ is stable under arbitrary intersections and therefore a complete lattice with $H(W)$ as minimal and W as maximal element. Moreover, the following assertions hold.*

(1) *The mappings*

$$\mathrm{op} : \mathcal{F}_e(W^*) \to \mathcal{F}_e(W), \ E \mapsto W \cap E^\perp$$

and

$$\mathrm{op} : \mathcal{F}_e(W) \to \mathcal{F}_e(W^*), \ F \mapsto W^* \cap F^\perp$$

are order reversing bijections. Moreover, for every subset $E \subseteq W^*$ the set

$$\mathrm{op}(E) := E^\perp \cap W$$

is an exposed face of W and for every exposed face there exists $\omega \in W^*$ with $F = \ker \omega \cap W = \mathrm{op}(\{\omega\})$.

(2) *For a wedge $F \subseteq W$ we have that*

$$L_F(W)^* = W^* \cap F^\perp = \mathrm{op}(F).$$

Proof. Let $(F_i)_{i \in I}$ be a family of exposed faces of W and $F := \bigcap_{i \in I} F_i$. Then F is a wedge. The relation

$$F \subseteq W \cap T_F(W) \subseteq W \cap T_{F_i}(W) = F_i \quad \text{for all} \quad i \in I$$

shows that $F = W \cap T_F(W) \in \mathcal{F}_e(W)$. We conclude that every non-empty subset in $\mathcal{F}_e(W)$ has an infimum. Thus this poset is a complete lattice.

(1) [HHL89, I.2.2-4]

(2) This follows from the definition of $L_F(W)$ and from Proposition I.2(4). ∎

Proposition I.5. (Faces) *The set $\mathcal{F}(W)$ of faces of W is stable under arbitrary intersections and therefore a complete lattice with $H(W)$ as minimal and W as maximal element. Moreover, the following assertions hold.*

(1)
$$\mathcal{F}(V) = \{F \in \mathcal{F}(W) : F \subseteq V\} \quad \textit{for every} \quad V \in \mathcal{F}(W),$$
i.e., the faces of a face V are exactly the faces of W which are contained in V.

(2) *For every element $f \in \operatorname{algint} W$ the whole wedge W is the only face containing f.*

(3) *A subset $F \subseteq W$ is a face iff there exists a finite chain*
$$F_0 = F \subseteq F_1 \subseteq \ldots \subseteq F_n = W$$
of wedges such that $F_i \in \mathcal{F}_e(F_{i+1})$ for $i = 0, \ldots, n-1$. In particular every face $F \neq W$ is contained in $\ker \omega \cap W$ for a suitable $\omega \in W^* \setminus H(W^*)$.

(4) *A subset $E \subseteq W^*$ is a face iff there exists a finite chain*
$$W = W_0 \subseteq W_1 \subseteq \ldots \subseteq W_n = E^*$$
of wedges such that $W_{i+1} = L_{F_i}(W_i)$ for an exposed face $F_i \in \mathcal{F}_e(W_i)$. In particular $H(E^*) \cap W \neq H(W)$ if $E \neq W^*$.

(5) *For every face $E \in \mathcal{F}(W^*)$ we have that*
$$E = H(E^*)^\perp \cap W^* = (E - E) \cap W^* \quad \textit{and} \quad E^* = L_{H(E^*)}(W).$$

(6) *For faces $E_1, E_2 \in \mathcal{F}(W^*)$ the relations $E_1 \subseteq E_2$ and $E_2^\perp = H(E_2^*) \subseteq E_1^\perp = H(E_1^*)$ are equivalent.*

Proof. Let $(F_i)_{i \in I}$ be a family of faces of W and $F := \bigcap_{i \in I} F_i$. Then F is a wedge and $x + y \in F, x, y \in W$ implies that $x, y \in F_i$ for all $i \in I$. Therefore $x, y \in F$. Hence $F \in \mathcal{F}(W)$.

(1) Let $F \in \mathcal{F}(V)$ and suppose that $x, y \in W$ with $x + y \in F \subseteq V$. Then $x, y \in V$ since V is a face of W, and therefore $x, y \in F$. Hence F is a face of W which is contained in V. Conversely, assume that F is a face of W which is contained in V. Then, since $x + y \in F$ and $x, y \in V$ imply that $x, y \in F$, F is is a face of V. We conclude that every non-empty subset in $\mathcal{F}(W)$ has an infimum. Thus this poset is a complete lattice.

(2) Suppose that $f \in F \in \mathcal{F}(W)$ and let $x \in W$. Then $f - W$ is an open neighborhood of 0 in $W - W$ (Proposition I.2). Hence we find an $n \in \mathbb{N}$ such that $\frac{1}{n}x \in f - W$. Therefore we find $y \in W$ with $\frac{1}{n}x + y = f \in F$. Thus $\frac{1}{n}x \in F$. This implies that $x \in F$.

(3) An easy induction shows that F_i is a face of W for each subscript $i = n, n-1, \ldots, 0$. Hence the sufficiency of the condition follows from (1) above.

To see that this condition is necessary, we use induction on $\dim(W - W)$. If $F \cap \operatorname{algint} W \neq \emptyset$ then, in view of (2) above, we have that $F = W$. Suppose that $W \neq F$. Then we find a linear functional ω such that $\omega(\text{algint } W) \subseteq]0, \infty[$ and $\omega(F) = \{0\}$ (Hahn-Banach). Thus $\omega \in W^* \setminus H(W^*)$. Then $F' := \text{op}(\{\omega\})$ is an exposed face of W with $\dim(F' - F') < \dim W - W$. Now the induction hypothesis applies and shows that there are wedges

$$F_0 = F \subseteq F_1 \subseteq \ldots \subseteq F_{n-1} = F'$$

such that $F_i \in \mathcal{F}_e(F_{i+1})$. If we set $F_n := W$ the proof is complete.
(4) Necessity: Let $E \in \mathcal{F}(W^*)$. Then, in view of (3), we find a sequence of wedges

$$E = E_0 \subseteq E_1 \subseteq \ldots \subseteq E_n = W^*$$

such that $E_i \in \mathcal{F}_e(E_{i+1})$. We set $W_i := E_{n-i}^*$ for $i = 0, \ldots, n$. Then $W_0 = (W^*)^* = W \subseteq W_1 \subseteq \ldots \subseteq W_n = E^*$ and, according to Proposition I.4(1), we find exposed faces $F_i \subseteq W_i = E_{n-i}^*$ such that

$$E_{n-i-1} = E_{n-i} \cap F_i^{\perp} = L_{F_i}(W_i)^*.$$

Thus $W_{i+1} = E_{n-i-1}^* = L_{F_i}(W_i)$ with $F_i \in \mathcal{F}_e(W_i)$.
Sufficieny: For every sequence satisfying the conditions we know from Proposition I.4 that

$$W_{i+1}^* = L_{F_i}(W_i)^* = F_i^{\perp} \cap W_i^* \in \mathcal{F}_e(W_i^*),$$

and that

$$E = W_n^* \subseteq W_{n-1}^* \subseteq \ldots \subseteq W_0^* = W^*.$$

Now (3) implies that $F \in \mathcal{F}(W^*)$.
(5) The second statement follows from the first one by duality (Proposition I.2). It is clear that $E - E = \text{span } E = H(E^*)^{\perp} = (E^{\perp})^{\perp}$. Let $f = e - e' \in W^*$. Then $e = f + e' \in E$ and therefore $f, e' \in E$. This proves that $(E - E) \cap W^* = E$.
(6) This is a direct consequence of (5). ∎

Example I.6. To visualize the difference between faces and exposed faces we give an example whith $\mathcal{F}_e(W) \neq \mathcal{F}(W)$.

$$W = \{(x, y, z) : x \leq 0, z \geq 0, x^2 + y^2 \leq z^2 \ \text{ or } \ x \geq 0, |y| \leq z\}.$$

Then $F := \mathbb{R}^+(0,1,1)$ is a face and the smallest exposed face containing F is $\widetilde{F} := \mathbb{R}^+(0,1,1) + \mathbb{R}^+(1,0,0)$ (cf. [Rup87, 3.7]). The dual wedge is

$$W^* = \{(x, y, z) : x \geq 0, z \geq 0, x^2 + y^2 \leq z^2\}.$$

Using Proposition I.5 or geometric visualization it is easy to see that

$$\mathcal{F}_e(W^*) = \mathcal{F}(W^*)$$

but the above face F shows that $\mathcal{F}(W) \neq \mathcal{F}_e(W)$. The lattices $\mathcal{F}(W)$ and $\mathcal{F}(W^*)$ are not antiisomorphic as it is the case for the lattices of exposed faces (Proposition I.4) because $\mathcal{F}(W^*)$ contains only two chains of length 3 but $\mathcal{F}(W)$ contains 4 chains of length 3. ∎

Proposition I.7. *Let W, W' be wedges in the finite dimensional vector space L and suppose that*

$$W \cap -W' \subseteq H(W) \cap H(W'),$$

i.e., $W \cap -W'$ is a vector space. Then $V := W + W'$ is closed and hence a wedge. The edge of V is $H(V) = H(W) + H(W')$. If, in addition, W' is a vector space, then

$$\operatorname{algint}(W + W') = \operatorname{algint} W + W'.$$

Proof. It is clear that the above condition implies that $W \cap -W' = H(W) \cap H(W')$ is a vector space. Then [HHL89, I.2.32] implies that V is a wedge. Let $v = w + w' \in H(V)$. Then $-v = -w - w' = w_1 + w_1' \in V$ with $w_1 \in W$ and $w_1' \in W'$. Therefore

$$w + w_1 = -w' - w_1' \in W \cap -W' = H(W) \cap H(W')$$

and consequently $w, w_1 \in H(W)$, $w', w_1' \in H(W')$. Hence $v \in H(W) + H(W')$. The converse, that $H(W) + H(W') \subseteq H(V)$, is clear. To prove the assertion on the algebraic interior of W we assume that W' is a vector space. According to Proposition I.2.4) the dual wedge of V is

$$V^* = W^* \cap W'^* = W^* \cap W'^{\perp}.$$

Therefore

$$V^* \setminus H(V^*) = W^* \cap W'^{\perp} \setminus W^{\perp}.$$

Let $\omega \in V^* \setminus H(V^*)$ and $x \in \operatorname{algint} W + W'$. Then $\omega \in W^* \setminus H(W^*)$ and consequently $\omega(x) > 0$ since $\omega(W') = \{0\}$ (Proposition I.2). From $V - V = W - W + W'$ we conclude that

$$\operatorname{algint} W + W' = \operatorname{int}_{W-W}(W) + W' \subseteq \operatorname{int}_{V-V}(V) = \operatorname{algint} V.$$

Let $x \in \operatorname{algint} V$. Suppose that $x \notin \operatorname{algint} W + W'$. Then, using the Separation Theorem of Hahn-Banach, we find $\nu \in (\operatorname{algint} W + W')^*$ such that $\nu(x) = 0$ and $\nu(V) \neq \{0\}$. This contradicts the above fact that $\omega(x) > 0$ for all $\omega \in V^* \setminus H(V^*)$. ∎

Corollary I.8. *Let $W \subseteq L$ be a wedge and $F_1, F_2 \in \mathcal{F}(W)$. Then $F_1 + F_2$ is closed and hence a wedge.*

Proof. Since every face contains the edge $H(W)$ (Proposition I.4) we find that

$$H(W) \subseteq F_1 \cap -F_2 \subseteq W \cap -W = H(W)$$

is a vector space. An application of Proposition I.7 completes the proof. ∎

Corollary I.9. *Let $F_1, F_2 \in \mathcal{F}_e(W)$. Then their supremum in the lattice $\mathcal{F}_e(W)$ is*

$$F_1 \vee_e F_2 = T_{F_1+F_2}(W) \cap W.$$

The supremum of two faces $F_1, F_2 \in \mathcal{F}(W)$ may be obtained by setting

$$W_0 := F_1 \vee_e F_2 \text{ in } \mathcal{F}_e(W) \quad \text{and} \quad W_{i+1} := F_1 \vee_e F_2 \text{ in } \mathcal{F}_e(W_i) \text{ for } i \in \mathbb{N}.$$

Then there exists an $i \in \mathbb{N}$ such that $W_{i+1} = W_i$ and

$$F_1 \vee F_2 = W_i = \bigcap_{j \in \mathbb{N}} W_j.$$

Proof. It is clear that $F := T_{F_1+F_2}(W) \cap W$ is an exposed face of W since

$$F \subseteq T_F(W) \cap W \subseteq T_{F_1+F_2}(W) \cap W = F.$$

For every exposed face F' containing F_1 and F_2 we have

$$F = T_{F_1+F_2}(W) \cap W \subseteq T_{F'}(W) \cap W = F'$$

and therefore F is the supremum of F_1 and F_2 in $\mathcal{F}_e(W)$. To prove the second assertion, let $F := F_1 \vee F_2$ in $\mathcal{F}(W)$. It follows from Proposition I.5 that the wedges F_1 and F_2 are faces of W_i for all $i \in \mathbb{N}$ because $W_{i+1} \in \mathcal{F}_e(W_i)$. Hence $F \subseteq \bigcap_{i \in \mathbb{N}} W_i$. If $F_1 + F_2$ does not intersect the algebraic interior of W_i, then $W_{i+1} \neq W_i$ has smaller dimension than W_i since it is contained in its boundary (Proposition I.5). We conclude that there exists an index i such that $W_i = W_j$ for $j \geq i$. Then $F_1 + F_2$ intersects the interior of W_i and therefore $F_1 + F_2 \subseteq F \subseteq W_i$ implies that $F = W_i$ (Proposition I.5). ■

Next we consider wedges in a finite dimensional vector space L which has the additional structure of a module of a compact group K.

Theorem I.10. *Let K be a compact group, m normalized Haar measure on K, L a finite dimensional K-module and $\widehat{L}$ the dual module. Then L decomposes into a direct sum*

$$L = L_{\text{fix}} \oplus L_{\text{eff}},$$

where $L_{\text{fix}} = \{x \in L : k.x = x \text{ for all } k \in K\}$ and $L_{\text{eff}} = \operatorname{span}\{k.x - x : x \in L, k \in K\}$. The averaging operator $p : L \to L, x \mapsto \int_K k.x \, dm(k)$ is a projection onto L_{fix} and $\ker p = L_{\text{eff}}$. The dual module has the direct decomposition

$$\widehat{L} = \widehat{L}_{\text{fix}} \oplus \widehat{L}_{\text{eff}} = L_{\text{eff}}^{\perp} \oplus L_{\text{fix}}^{\perp}$$

and the adjoint of p, $\widehat{p} : \widehat{L} \to \widehat{L}, \omega \mapsto \omega \circ p$, is the projection onto $\widehat{L}_{\text{fix}}$ along $\widehat{L}_{\text{eff}}$. Suppose that $W \subseteq L$ is a K-invariant wedge, $W^* \subseteq \widehat{L}$ its dual, and

$C := W \cap L_{\text{fix}}$. *For a wedge* $D \subseteq L_{\text{fix}}$ *we set* $D^\star := D^* \cap L_{\text{eff}}^\perp$. *Then the following assertions hold:*

(1) $p(W) = C$, $\widehat{p}(W^*) = W^* \cap L_{\text{eff}}^\perp = C^\star$.

(2) $\operatorname{algint}(W \cap L_{\text{fix}}) = p(\operatorname{algint} W) = \operatorname{algint} W \cap L_{\text{fix}} \neq \emptyset$, $\operatorname{algint} W^* \cap L_{\text{eff}}^\perp \neq \emptyset$.

(3) $W \cap L_{\text{eff}} \subseteq H(W)$, $W^* \cap L_{\text{fix}}^\perp \subseteq H(W^*)$.

(4) $p(H(W)) = H(C)$.

Suppose that $F \subseteq W$, W' *are also* K*-invariant wedges. Then we have that:*

(5) $p(W \cap W') = p(W) \cap p(W')$.

(6) $p(L_F(W)) = L_{p(F)}(C)$.

(7) $p(T_F(W)) = T_{p(F)}(C)$.

Proof. Using Weyl's unitary trick we may assume that L is a euclidean vector space and the elements of K act on V as isometries. Moreover we may identify L with $\widehat{L}$ by the duality with respect to the euclidean scalar product $\langle \cdot, \cdot \rangle$.

Let $z \in L_{\text{fix}}$, $x \in L$ and $k \in K$. Then

$$\langle z, k.x - x \rangle = \langle k^{-1}.z, x \rangle - \langle z, x \rangle = \langle k^{-1}.z - z, x \rangle = \langle z - z, x \rangle = 0.$$

Thus L_{fix} is orthogonal to L_{eff}. Suppose that x is orthogonal to both subspaces. Then, for every $k \in K$ and $z \in L$, we have that

$$\langle z, k.x - x \rangle = \langle k^{-1}.z - z, x \rangle \in \langle L_{\text{eff}}, x \rangle = \{0\}.$$

This implies that $k.x = x$ and therefore $x \in L_{\text{fix}} \cap L_{\text{fix}}^\perp = \{0\}$. This proves the decomposition of L as a direct sum of the K-modules L_{fix} and L_{eff}. Moreover, we have seen that $L_{\text{fix}}^\perp = L_{\text{eff}}$ and that $L_{\text{eff}}^\perp = L_{\text{fix}}$. Hence the second assertion of (1) to (3) follows from the first one by duality. It is easy to verify that p is a K-equivariant projection onto L_{fix} since Haar measure is a left and right invariant probability measure on K. We conclude that $p(k.x - x) = k.p(x) - p(x) = 0$ for $k \in K$ and $x \in L$. Hence $L_{\text{eff}} = \ker p$.

(1) It is clear that $W \cap L_{\text{fix}} \subseteq p(W)$. But W is closed and convex and m is a probability measure on K. Hence $p(x)$ is contained in the closed convex hull of $K.x \subseteq W$. Thus $p(W) \subseteq W \cap L_{\text{fix}}$.

(2) Let $x \in \operatorname{algint} W$. First we show that $p(x) \in \operatorname{algint} W$. To see this, let $\omega \in W^* \setminus H(W^*)$. Then

$$\langle \omega, p(x) \rangle = \int_K \langle \omega, k.x \rangle \, dm(k) > 0$$

since $\langle \omega, \mathbf{1}.x \rangle = \langle \omega, x \rangle > 0$ and each open subset of K has non-zero Haar measure. Now Proposition I.2 implies that $p(x) \in \operatorname{algint} W$. So $p(\operatorname{algint} W) = \operatorname{algint} W \cap L_{\text{fix}}$. Before we prove the last assertion we have to prove (3).

(3) We consider the K-module $L' := L_{\text{eff}}$ and $W' := W \cap L'$. Then $L'_{\text{fix}} = \{0\}$ and (2) above show that $0 \in \operatorname{algint} W'$. Hence W' is a vector space and therefore $W \cap L_{\text{eff}} \subseteq H(W)$.

(2) (continued) Using Proposition I.7 we find that

$$\begin{aligned} p(\operatorname{algint} W) &= p(\operatorname{algint} W + L_{\text{eff}}) = p\big(\operatorname{algint}(W + L_{\text{eff}})\big) \\ &= \operatorname{algint}(W + L_{\text{eff}}) \cap L_{\text{fix}} \\ &= \operatorname{algint}\big((W + L_{\text{eff}}) \cap L_{\text{fix}}\big) = \operatorname{algint}(W \cap L_{\text{fix}}). \end{aligned}$$

(4) The inclusion $H(C) \subseteq p\big(H(W)\big)$ is trivial. But $p\big(H(W)\big)$ is a vector space and therefore contained in $H(C)$.

(5) This follows from

$$p(W \cap W') = W \cap W' \cap L_{\text{fix}} = p(W) \cap p(W').$$

(6) The inclusion $p\big(L_F(W)\big) \subseteq L_{p(F)}(C)$ follows from

$$p\big(L_F(W)\big) = p\big(\overline{W - F}\big) \subseteq \overline{p(W) - p(F)} = L_{p(F)}\big(p(W)\big).$$

But

$$p(W) - p(F) = W \cap L_{\text{fix}} - F \cap L_{\text{fix}} \subseteq L_F(W) \cap L_{\text{fix}} = p\big(L_F(W)\big)$$

and therefore $L_{p(F)}\big(p(W)\big) \subseteq p\big(L_F(W)\big)$.

(7) This is a consequence of (6) and (4). ■

Lemma I.11. *Let L be a finite dimensional module of the compact group K,*

$$C_0 \subseteq C \subseteq C_1 \subseteq L_{\text{fix}}$$

wedges such that there exist K-invariant wedges $W_0, W_1 \subseteq L$ with $C_0 = W_0 \cap L_{\text{fix}}$, $C_1 = W_1 \cap L_{\text{fix}}$, $W_0 \subseteq W_1$, W_0 is pointed, W_1 is generating and C is pointed and generating in L_{fix}. Then there exists a pointed and generating K-invariant wedge $W \subseteq L$ with

$$W_0 \subseteq W \subseteq W_1 \quad \text{and} \quad W \cap L_{\text{fix}} = C.$$

Proof. Let $\omega \in \operatorname{algint} C^\star \subseteq \widehat{L}_{\text{fix}}$ and set $B := \omega^{-1}(1) \cap C$. For a compact convex K-invariant neighborhood $B_1 \subseteq L_{\text{eff}}$ we set

$$W := \big(W_0 + \mathbb{R}^+(B + B_1)\big) \cap W_1.$$

W is closed and pointed: It is clear that $\mathbb{R}^+(B + B_1)$ is pointed, generating and K-invariant. Thus

$$\big(\mathbb{R}^+(B + B_1) \cap -W_0\big) \cap L_{\text{fix}} = \mathbb{R}^+ B \cap -C_0 = C \cap -C_0 \subseteq H(C) = \{0\}.$$

Therefore $\mathbb{R}^+(B + B_1) \cap -W_0 \subseteq L_{\text{eff}}$ is a K-invariant vector space (Theorem I.10). Thus W is closed and pointed (Proposition I.7).

The relations $W_0 \subseteq W \subseteq W_1$ are consequences of the definition.

W is generating: According to Theorem I.10 we have that

$$\operatorname{int}(C) \subseteq \operatorname{int}(C_1) = \operatorname{int}(W_1) \cap L_{\text{fix}}.$$

Thus $\mathbb{R}^+ B \cap \operatorname{int} W_1 \neq \emptyset$ and therefore $\operatorname{int} W \neq \emptyset$.

The last assertion follows from

$$\begin{aligned} C = \mathbb{R}^+ B &\subseteq \big(W_0 + \mathbb{R}^+(B + B_1)\big) \cap C_1 \\ &\subseteq W \cap L_{\text{fix}} \subseteq C_1 \cap (C_0 + \mathbb{R}^+ B) = C_1 \cap C = C. \end{aligned}$$

■

Definition I.12. We write $\mathcal{F}_i(W)$ respectively $\mathcal{F}_{ei}(W)$ for the sets of K-invariant faces respectively exposed faces of W. These sets are stable under intersections (Propositions I.4, I.5) and therefore complete lattices in their own right. ∎

Proposition I.13. *Assume the notation of* Theorem I.10. *Then the mapping*

$$\Psi : F \mapsto F \cap L_{\text{fix}} = p(F)$$

maps invariant faces of W into faces of C, i.e.,

$$\Psi(\mathcal{F}_i(W)) \subseteq \mathcal{F}(C) \tag{1.1}$$

and

$$\Psi(\mathcal{F}_{ei}(W)) \subseteq \mathcal{F}_e(C). \tag{1.2}$$

It preserves arbitrary intersections and it is a surjective lattice-homomorphism in both cases.

Proof. (1.1) Let $F \in \mathcal{F}_i(W)$ and $x, y \in p(W)$ with $x + y \in p(F)$. Then $x, y \in W$ and $x + y \in F$. Hence

$$x, y \in F \cap p(W) \subseteq F \cap L_{\text{fix}} = p(F).$$

This proves that $p(F) \in \mathcal{F}(p(W))$.

(1.2) Let $F \in \mathcal{F}_{ei}(W)$. Using Proposition I.4(1) we find $\omega \in W^*$ such that $F = \ker \omega \cap W$. Set $\omega' := \omega | L_{\text{fix}}$. Then $\omega' \in C^\star$ and

$$p(F) = F \cap L_{\text{fix}} = \ker \omega' \cap C.$$

We conclude that $p(F) \in \mathcal{F}_e(C)$.

The fact that Ψ preserves arbitrary intersections follows from $\Psi(F) = F \cap L_{\text{fix}}$.

Let $E \in \mathcal{F}_e(C)$ and choose $\omega' \in C^\star$ with $\ker \omega' \cap C = E$. Set $\omega := \omega' \circ p$. Then $\omega \in W^*$ beause $\omega(W) = \omega'(p(W)) \subseteq \mathrm{I\!R}^+$ and $F := \ker \omega \cap W$ is an exposed face of W with $F \cap L_{\text{fix}} = E$. It is invariant since $\ker \omega$ is invariant. This shows that $\Psi(\mathcal{F}_{ei}(W)) = \mathcal{F}_e(C)$. The supremum of two faces $F_1, F_2 \subseteq \mathcal{F}_{ei}(W)$ is

$$F_1 \vee_e F_2 = W \cap T_{F_1+F_2}(W)$$

since $F_1 + F_2$ is closed (Corollary I.8). Consequently

$$p(F_1 \vee_e F_2) = C \cap T_{p(F_1+F_2)}(C) = C \cap T_{p(F_1)+p(F_2)}(C) = p(F_1) \vee_e p(F_2).$$

Now assume that $F_1, F_2 \in \mathcal{F}_i(W)$ and let

$$W_0 = F_1 \vee_e F_2 \supseteq W_1 \supseteq \ldots \supseteq W_i = \ldots = F_1 \vee F_2$$

be the sequence constructed in Corollary I.9. Then $W_{i+1} = F_1 \vee_e F_2$ with respect to the lattice $\mathcal{F}_e(W_i)$. Inductively we find that $p(W_{k+1}) \in \mathcal{F}_e(p(W_k)) \subseteq \mathcal{F}(C)$ for $k = 0, \ldots, i-1$ and that $p(F_1) \vee p(F_2) \subseteq p(W_k)$ because $p(W_k)$ is a face of C. Thus

$$p(W_i) = p(F_1 \vee F_2) = p(F_1) \vee p(F_2)$$

because $F_1 + F_2$ intersects the algebraic interior of W_i (Proposition I.5) and consequently $p(F_1) + p(F_2)$ intersect the algebraic interior of $p(W_i)$ which leads to $p(W_i) = p(F_1) \vee p(F_2)$. Since $W_i \in \mathcal{F}(W)$ these considerations prove that $\Psi : \mathcal{F}_i(W) \to \mathcal{F}(C)$ is surjective and a lattice homomorphism. ∎

II. LIE ALGEBRAS WITH CONE POTENTIAL

The next two sections are dedicated to the infinitesimal theory of invariant subsemigroups of Lie groups, the theory of invariant cones in Lie algebras. This section is purely Lie algebraic. Invariant cones firstly occur in Section III. Every Lie algebra containing a pointed generating invariant cone has a property called cone potential. In this section we study Lie algebras having this property and additional conditions which are needed to guarantee the existence of invariant cones. We give a detailed description of the Lie algebras with cone potential. Moreover, we provide a method to construct Lie algebras with cone potential from prescribed data. There exist already some structure theorems for Lie algebras with cone potential which are proved in [HHL89]. We restate them in a condensed form which contains the exact assumptions and assertions to which we will refer. Our way to the subject is somewhat different from that one described in [HHL89] because it was one of our objectives to clarify the interdependence of the new notions which we will introduce. This goal will be reached in Section III.

Definition II.1. Let $\mathfrak{h}$ be a finite dimensional vector space and $\Omega \subseteq \widehat{\mathfrak{h}}$ a finite symmetric subset, i.e., $\Omega = -\Omega$. We say that $\Omega^+ \subseteq \Omega$ is a positive system if there exists $h \in \mathfrak{h}$ such that

$$\Omega^+ = \{\omega \in \Omega : \omega(h) > 0\} \quad \text{and} \quad \Omega = \Omega^+ \cup -\Omega^+ \cup \{0\}.$$

■

Definition II.2. Let $\mathfrak{g}$ be a finite dimensional Lie algebra. For a subset $A \subseteq \mathfrak{g}$ we define

$$\mathrm{Inn}_{\mathfrak{g}}(A) := \langle e^{\operatorname{ad} A} \rangle \quad \text{and} \quad \mathrm{INN}_{\mathfrak{g}}(A) := \overline{\mathrm{Inn}_{\mathfrak{g}}(A)}.$$

We usually omit the subscript if no confusion is possible. We call an element $x \in \mathfrak{g}$ *compactly embedded* if $\mathrm{INN}_{\mathfrak{g}}(\mathbb{R}x)$ is compact and write

$$\operatorname{comp}(\mathfrak{g}) := \{x \in \mathfrak{g} : \mathrm{INN}_{\mathfrak{g}}(\mathbb{R}x) \text{ is compact}\}$$

for the set of compactly embedded elements of $\mathfrak{g}$. ■

Theorem II.3. (The Real Root Decomposition) *Let $\mathfrak{g}$ be a finite dimensional Lie algebra with compactly embedded Cartan algebra $\mathfrak{h}$ and Λ be the set of roots of $\mathfrak{g}_{\mathbb{C}}$ with respect to $\mathfrak{h}_{\mathbb{C}}$. These are all purely imaginary on $\mathfrak{h}$. We set*

$$\Omega := \{-i\lambda|_{\mathfrak{h}} : \lambda \in \Lambda\} \quad \textit{and} \quad \mathfrak{g}^{\omega} = \mathfrak{g}^{-\omega} := \mathfrak{g} \cap (\mathfrak{g}_{\mathbb{C}}^{\lambda} \oplus \mathfrak{g}_{\mathbb{C}}^{-\lambda}) \quad \textit{for} \quad \omega = -i\lambda|_{\mathfrak{h}}.$$

For any choice of a positive sytem $\Omega^+ \subseteq \Omega$ there is a unique complex structure $I : \mathfrak{g}_{\mathrm{eff}} \to \mathfrak{g}_{\mathrm{eff}}$ with $I^2 = -\mathrm{id}_{\mathfrak{g}_{\mathrm{eff}}}$ and a direct decomposition of $\mathfrak{g}$ into isotypic $\mathfrak{h}$-submodules under the adjoint action

$$\mathfrak{g} = \mathfrak{h} \oplus \mathfrak{g}_{\mathrm{eff}}, \qquad \mathfrak{g}_{\mathrm{eff}} = [\mathfrak{h}, \mathfrak{g}] = \bigoplus_{\omega \in \Omega +} \mathfrak{g}^{\omega}, \tag{2.1}$$

where the action of $\mathfrak{h}$ is described by

$$[h, x] = \omega(h) I x \qquad \text{for all} \qquad h \in \mathfrak{h}, x \in \mathfrak{g}^{\omega}. \tag{2.2}$$

The complexification of $\mathfrak{g}^{\omega}$ is $\mathfrak{g}_{\mathbb{C}}^{\lambda} \oplus \mathfrak{g}_{\mathbb{C}}^{-\lambda}$, where λ is the unique complex extension of $i\omega$. We have

$$[\mathfrak{g}^{\omega}, \mathfrak{g}^{\omega'}] \subseteq \mathfrak{g}^{\omega+\omega'} + \mathfrak{g}^{\omega-\omega'}, \tag{2.3}$$

and if q is any invariant symmetric bilinear form on $\mathfrak{g} \times \mathfrak{g}$, then

$$q(x, Ix) = 0 \qquad \text{and} \qquad q(x) = q(Ix) \qquad \text{for all} \qquad x \in \mathfrak{g}^{\omega}. \tag{2.4}$$

Proof. In view of [HHL89, III.6.5/8], it only remains to show (2.3). This follows easily:

$$\begin{aligned}
[\mathfrak{g}^{\omega}, \mathfrak{g}^{\omega'}] &= [\mathfrak{g} \cap (\mathfrak{g}_{\mathbb{C}}^{\lambda} \oplus \mathfrak{g}_{\mathbb{C}}^{-\lambda}), \mathfrak{g} \cap (\mathfrak{g}_{\mathbb{C}}^{\lambda'} \oplus \mathfrak{g}_{\mathbb{C}}^{-\lambda'})] \\
&\subseteq \mathfrak{g} \cap [\mathfrak{g}_{\mathbb{C}}^{\lambda} \oplus \mathfrak{g}_{\mathbb{C}}^{-\lambda}, \mathfrak{g}_{\mathbb{C}}^{\lambda'} \oplus \mathfrak{g}_{\mathbb{C}}^{-\lambda'}] \\
&\subseteq \mathfrak{g} \cap (\mathfrak{g}_{\mathbb{C}}^{\lambda+\lambda'} \oplus \mathfrak{g}_{\mathbb{C}}^{-\lambda-\lambda'} \oplus \mathfrak{g}_{\mathbb{C}}^{\lambda-\lambda'} \oplus \mathfrak{g}_{\mathbb{C}}^{\lambda'-\lambda}) \\
&= \mathfrak{g} \cap (\mathfrak{g}^{\omega+\omega'} \oplus i\mathfrak{g}^{\omega+\omega'} \oplus \mathfrak{g}^{\omega-\omega'} \oplus i\mathfrak{g}^{\omega-\omega'}) \\
&= \mathfrak{g}^{\omega+\omega'} \oplus \mathfrak{g}^{\omega-\omega'}.
\end{aligned}$$

■

Definition II.4. We set $Q(x, y) := p([Ix, y])$ for $x, y \in \mathfrak{g}_{\mathrm{eff}}$, where $p : \mathfrak{g} \to \mathfrak{h}$ is the projection along $\mathfrak{g}_{\mathrm{eff}}$ and $Q(x) := Q(x, x)$. For every $x \in \mathfrak{g}^{\omega}$ we write $\langle x \rangle := \mathrm{span}\{x, Ix, Q(x)\}$. A finite dimensional Lie algebra with a compactly embedded Cartan algebra is said to have *cone potential* if $Q(x) \neq 0$ for every $x \in \mathfrak{g}^{\omega}$ and $\omega \in \Omega^+$. Note that

$$Q(x, y) = \sum_{\omega \in \Omega +} Q(x_\omega, y_\omega) = \sum_{\omega \in \Omega +} p([Ix_\omega, y_\omega]) \qquad \text{for} \qquad x, y \in \mathfrak{g}_{\mathrm{eff}}$$

with $x = \sum_{\omega \in \Omega +} x_\omega$, $y = \sum_{\omega \in \Omega +} y_\omega$ and $x_\omega, y_\omega \in \mathfrak{g}^{\omega}$. ■

Proposition II.5. *Let $\mathfrak{h}$ be a compactly embedded Cartan algebra of $\mathfrak{g}$. Then there exists a unique maximal compactly embedded subalgebra $\mathfrak{k}_{\mathfrak{h}} \subseteq \mathfrak{g}$ with $\mathfrak{h} \subseteq \mathfrak{k}_{\mathfrak{h}}$.*

Proof. [HHL89, A.2.40] ■

Definition II.6. Let $\mathfrak{g}$ be a Lie algebra with a compactly embedded Cartan algebra $\mathfrak{h}$. Set $\mathfrak{z}_{\mathfrak{k}} := Z(\mathfrak{k}_{\mathfrak{h}})$. Then $\mathfrak{g}$ is called *quasihermitean* if

$$\mathfrak{z}_{\mathfrak{k}} \cap \operatorname{int} \operatorname{comp}(\mathfrak{g}) \neq \emptyset.$$

Note that, in view of [HHL89, A.2.25], this conditon is equivalent to the existence of $x \in \mathfrak{z}_{\mathfrak{k}}$ such that $\ker \operatorname{ad} x = \mathfrak{k}_{\mathfrak{h}}$. ■

The notions quasihermitean and cone potential are motivated by the fact that a Lie algebra which contains a pointed generating invariant cone has these properties. Later in this section we will need a third notion called strong cone potential and clarify the interdependence of these properties of a Lie algebra. ■

Definition II.7. Let $V := \mathbb{C}^n$ and $\langle \cdot, \cdot \rangle$ the standard hermitean product on V. We define a Lie algebra structure on $\mathfrak{g} := V \times \mathbb{R} \times \mathbb{R}$ by setting

$$[(v, z, h), (v', z', h')] := (ihv' - ih'v, \operatorname{Im}\langle v, v' \rangle, 0).$$

This turns $\mathfrak{g}$ into a Lie algebra (cf. Proposition II.21) as one checks easily by verifying the Jacobi identity. This Lie algebra is called the $(2n+2)$-dimensional *oscillator algebra* A_{2n+2}. The subalgebra $\mathbb{C}^n \times \mathbb{R} \times \{0\} \subseteq A_{2n+2}$ is called the $(2n+1)$-dimensional *Heisenberg algebra* $\mathfrak{h}_n$. ■

Proposition II.8. *Let $\mathfrak{g}$ be a finite dimensional Lie algebra with a compactly embedded Cartan algebra. For any $\omega \in \Omega^+$ and $x \in \mathfrak{g}^\omega \setminus \{0\}$ there are four mutually exclusive possibilities:*

(1) $\omega(Q(x)) < 0$. *Then* $\langle x \rangle \cong \operatorname{su}(2) \cong \operatorname{so}(3)$.

(2) $\omega(Q(x)) > 0$. *Then* $\langle x \rangle \cong \operatorname{sl}(2, \mathbb{R}) \cong \operatorname{su}(1,1)$.

(3) $\omega(Q(x)) = 0$ *and* $Q(x) \neq 0$. *Then $\langle x \rangle$ is isomorphic to the three dimensional Heisenberg algebra. For every $h \in \mathfrak{h}$ with $\omega(h) \neq 0$ the algebra $A(x) := \mathbb{R}h \oplus \langle x \rangle$ is isomorphic to the four dimensional Oscillator algebra A_4.*

(4) $Q(x) = 0$. *Then* $\langle x \rangle \cong \mathbb{R}^2$.

If $\mathfrak{g}$ has cone potential, then (4) is impossible, and if $\mathfrak{g}$ is semisimple, then either (1) or (2) holds.

Proof. [HHL89, III.6.12, 15, 16, 18] and [Ne89c, II.3]. ■

Corollary II.9. *Every semisimple Lie algebra with a compactly embedded Cartan algebra has cone potential.* ■

Proposition II.10.

(a) *If $\mathfrak{g}$ has cone potential and $\mathfrak{a} \subseteq \mathfrak{g}$ is a subalgebra with $[\mathfrak{a}, \mathfrak{h}] \subseteq \mathfrak{a}$, then the subalgebra $\mathfrak{a} + \mathfrak{h}$ has cone potential and $\mathfrak{h}$ is a compactly embedded Cartan algebra in $\mathfrak{a} + \mathfrak{h}$.*

(b) *Let $\mathfrak{g}$ be a Lie algebra with cone potential and $\mathfrak{r}$ its radical, $\mathfrak{n}$ its nilradical and $\mathfrak{z}$ its center. Then*

(1) $\mathfrak{z} = Z(\mathfrak{n}) = \mathfrak{n} \cap \mathfrak{h}$,
(2) $\mathfrak{r}^\omega = \mathfrak{n}^\omega$ *for* $\omega \in \Omega^+$,
(3) $[\mathfrak{r}_{\text{eff}}, \mathfrak{r}_{\text{eff}}] = [\mathfrak{n}_{\text{eff}}, \mathfrak{n}_{\text{eff}}] \subseteq \mathfrak{z}$, *and*
(4) $[\mathfrak{r}^\omega, \mathfrak{r}^{\omega'}] \neq \{0\}$ *iff* $\omega = \omega'$ *for* $\omega, \omega' \in \Omega^+$.
(5) $Q(x,y) = [Ix, y]$ *for* $x, y \in \mathfrak{r}_{\text{eff}}$.

(c) *If* $\mathfrak{g} = \mathfrak{r}$ *is a solvable Lie algebra with cone potential, then*

$$\mathfrak{r} = \mathfrak{h} \oplus \mathfrak{r}_{\text{eff}}, \qquad \mathfrak{n} = \mathfrak{z} \oplus \mathfrak{r}_{\text{eff}}, \qquad \mathfrak{n}/\mathfrak{z} \quad \textit{is abelian},$$

and

(1) $\mathfrak{r}' = [\mathfrak{r}, \mathfrak{n}] = \mathfrak{n}' \oplus \mathfrak{r}^+$, *and*
(2) $\mathfrak{r}'' = \mathfrak{n}' = [\mathfrak{r}_{\text{eff}}, \mathfrak{r}_{\text{eff}}] \subseteq \mathfrak{z} = \mathfrak{h} \cap \mathfrak{n}$.

Proof. (a) This follows from

$$\mathfrak{a} = \mathfrak{a} \cap \mathfrak{h} \oplus \bigoplus_{\omega \in \Omega^+} \mathfrak{a} \cap \mathfrak{g}^\omega$$

which is the decomposition of the $\mathfrak{h}$-module $\mathfrak{a}$ into isotypical summands.
(b) In view of [HHL89, III.6.23] only (5) remains to be proved, but this follows from (3).
(c) [HHL89, III.6.24, 25]. ∎

The previous proposition implies in particular that a solvable Lie algebra $\mathfrak{r}$ with cone potential which is not abelian has solvable length 3, i.e., $\mathfrak{r}^{(3)} = \{0\}$ and $\mathfrak{r}^{(2)} \neq \{0\}$. To see this, the fact that $\mathfrak{r}$ has merely a compactly embedded Cartan algebra has no consequences for the solvable length of $\mathfrak{r}$, we define $\mathfrak{n} \subseteq \mathrm{gl}(n,\mathbb{C})$ to be the set of all upper triangular matrices with zero diagonal. We take a diagonal matrix $D \in \mathrm{gl}(n,\mathbb{C})$ such that all differences $d_i - d_j$ of the diagonal entries d_i are different from 0 and $\mathrm{Re}(d_i) = 0$. We set $\mathfrak{g} := \mathfrak{n} \rtimes \mathbb{R}D \subseteq \mathrm{gl}(n,\mathbb{C})$. This is a subalgebra since $[D, \mathfrak{n}] \subseteq \mathfrak{n}$. Moreover

$$[D, E_{ij}] = (d_i - d_j)E_{ij} \quad \text{for} \quad i < j,$$

where E_{ij} is the matrix with only one entry 1 in the i-th row and j-th column. Since all the numbers $d_i - d_j$ are non-zero, the subspace $\mathbb{R}D \subseteq \mathfrak{g}$ is a compactly embedded Cartan algebra of $\mathfrak{g}$. The solvable length of $\mathfrak{r}$ is n because $\mathfrak{r}^{(n-1)} \neq \{0\}$ and $\mathfrak{r}^{(n)} = \{0\}$. ∎

Proposition II.11. *Let* $\mathfrak{g}$ *be a finite dimensional Lie algebra with compactly embedded Cartan algebra* $\mathfrak{h}$ *and* $\mathfrak{r}$ *its radical. Then there exists an* $\mathfrak{h}$*-invariant Levi subalgebra* $\mathfrak{s} \subseteq \mathfrak{g}$. *This Levi algebra satisfies*

$$(S) \qquad \mathfrak{h} = (\mathfrak{h} \cap \mathfrak{r}) \oplus (\mathfrak{h} \cap \mathfrak{s}) \quad \textit{and} \quad [\mathfrak{h} \cap \mathfrak{r}, \mathfrak{s}] = \{0\}.$$

For every Levi algebra $\mathfrak{s} \subseteq \mathfrak{g}$ *satisfying* (S) *we have*

(1) $\mathfrak{h} \cap \mathfrak{s}$ *is a compactly embedded Cartan algebra of* $\mathfrak{s}$.

(2) $\mathfrak{h} \subseteq Z_{\mathfrak{r}}(\mathfrak{s}) \oplus \mathfrak{s}$, *where* $Z_{\mathfrak{r}}(\mathfrak{s}) = \{x \in \mathfrak{r} : [x, \mathfrak{s}] = \{0\}\}$.

(3) $[\mathfrak{h}, \mathfrak{s}] \subseteq \mathfrak{s}$ *and* $\mathfrak{g}_1 := \mathfrak{h} + \mathfrak{s} = (\mathfrak{h} \cap \mathfrak{r}) \oplus \mathfrak{s}$ *is a reductive subalgebra with the compactly embedded Cartan algebra* $\mathfrak{h}$.

(4) $\mathfrak{g}$ *is a semisimple* $\mathfrak{g}_1$*-module and* $Z_{\mathfrak{r}}(\mathfrak{s})$ *is a* $\mathfrak{g}_1$*-submodule. In particular, there is a direct* $\mathfrak{g}_1$*-module decomposition*

$$\mathfrak{g} = \mathfrak{s} \oplus Z_{\mathfrak{r}}(\mathfrak{s}) \oplus M_1$$

with $\mathfrak{r} = Z_{\mathfrak{r}}(\mathfrak{s}) \oplus M_1$ *and*

$$M_1 = \bigoplus_{\omega \in \Omega^+} M_1 \cap \mathfrak{g}^\omega \subseteq \mathfrak{r}_{\text{eff}}.$$

(5) *For each* $\omega \in \Omega^+$ *we have*

$$\mathfrak{g}^\omega = \mathfrak{r}^\omega \oplus \mathfrak{s}^\omega,$$

where $\mathfrak{r}^\omega = \mathfrak{r} \cap \mathfrak{g}^\omega$ *and* $\mathfrak{s}^\omega = \mathfrak{s} \cap \mathfrak{g}^\omega$ *are the isotypic components of the* $\mathfrak{h}$*-modules* $\mathfrak{r}$ *and* $\mathfrak{s}$ *indexed by* $\omega \in \Omega^+$.

(6) $\mathfrak{g}$ *has cone potential iff* $\mathfrak{h} + \mathfrak{r}$ *has cone potential.*

Proof. It is clear that (S) implies that $\mathfrak{s}$ is $\mathfrak{h}$-invariant. If, conversely, $\mathfrak{s}$ is $\mathfrak{h}$-invariant, then

$$\mathfrak{s} = \mathfrak{s} \cap \mathfrak{h} \oplus \bigoplus_{\omega \in \Omega^+} \mathfrak{s} \cap \mathfrak{g}^\omega \quad \text{and} \quad \mathfrak{r} = \mathfrak{r} \cap \mathfrak{h} \oplus \bigoplus_{\omega \in \Omega^+} \mathfrak{r} \cap \mathfrak{g}^\omega$$

Thus $\mathfrak{h} = (\mathfrak{h} \cap \mathfrak{r}) \oplus (\mathfrak{h} \cap \mathfrak{s})$ and $[\mathfrak{h} \cap \mathfrak{r}, \mathfrak{s}] \subseteq \mathfrak{s} \cap \mathfrak{r} = \{0\}$.

(1)-(5) Using [HHL89, III.6.28] and [Sp88, p.67] we only have to prove (4) which follows from the fact that M_1 is $\mathfrak{h}$-invariant, and

$$\mathfrak{h} \cap M_1 \subseteq Z_{\mathfrak{r}}(\mathfrak{s}) \cap M_1 = \{0\}.$$

(6) That the cone potential of $\mathfrak{g}$ implies that $\mathfrak{h} + \mathfrak{r}$ has cone potential follows from Proposition II.10(a). To see that the converse is true, we consider a root $\omega \in \Omega^+$. Then $\mathfrak{g}^\omega = \mathfrak{r}^\omega + \mathfrak{s}^\omega$. Let $x = x_R + x_S \in \mathfrak{g}^\omega$ with $x_R \in \mathfrak{r}^\omega$ and $x_S \in \mathfrak{s}^\omega$. Now

$$Q(x) = Q(x_R + x_S) = Q(x_S) + Q(x_R) + 2Q(x_R, x_S) \in Q(x_S) + \mathfrak{r}$$

since Q is symmetric ([HHL89, III.6.7]). Moreover $Q(x_S) \in \mathfrak{s}$. If $Q(x) = 0$, this implies that $Q(x_S) = 0$ and therefore $x_S = 0$ since $\mathfrak{s}$ has cone potential (Corollary II.9). This leads to $Q(x) = Q(x_R) = 0$. Hence $x_R = 0$ since $\mathfrak{r} + \mathfrak{h}$ has cone potential. This proves that $\mathfrak{g}$ has cone potential if $\mathfrak{r} + \mathfrak{h}$ has. ∎

Note that (4) implies that $[\mathfrak{s}, \mathfrak{r}^+] \subseteq \mathfrak{r}_{\text{eff}}$ because

$$[\mathfrak{s}, \mathfrak{r}_{\text{eff}}] \subseteq [\mathfrak{s}, \mathfrak{r}] \subseteq [\mathfrak{s}, M_1] \subseteq M_1 \subseteq \mathfrak{r}_{\text{eff}}.$$

Lemma II.12. *Let* $\mathfrak{g} = \mathrm{sl}(2, \mathbb{R})$ *or* $\mathrm{so}(3)$, $\mathfrak{h} = \mathbb{R}U \subseteq \mathfrak{g}$ *a compactly embedded Cartan algebra with* $\operatorname{Spec}(\operatorname{ad} U) \subseteq \{\pm 2i, 0\}$ *and* V *a simple* $\mathfrak{g}$*-module. Suppose that*

$$Z_V(U) = \{x \in V : U.x = 0\} = \{0\}.$$

Then $\dim V \in 2\mathbb{Z}$ *and* $\operatorname{Spec}\big(\pi(U)\big) \subseteq (2\mathbb{Z}+1)i$, *where* $\pi : \mathfrak{g} \to \mathrm{gl}(V)$ *defines the action of* $\mathfrak{g}$ *on* V.

Proof. If $Z_V(U) = \{0\}$, then, by definition, U acts on V without non-zero fixed points. Hence the fact that $\operatorname{Spec}\big(\pi(U)\big) \subseteq i\mathbb{R}$ implies that $\dim V \in 2\mathbb{Z}$. We consider the complexification $V_{\mathbb{C}} = \mathbb{C} \otimes V$. Then we get a representation

$$\pi_{\mathbb{C}} : \mathfrak{g}_{\mathbb{C}} = \mathrm{sl}(2,\mathbb{C}) \to \mathrm{gl}(V_{\mathbb{C}})$$

which extends the given representation by

$$\pi_{\mathbb{C}}(z \otimes x)(z' \otimes v') := zz' \otimes \pi(x)v'.$$

It is clear that $\operatorname{Spec}\big(\pi_{\mathbb{C}}(U)\big) = \operatorname{Spec}\big(\pi(U)\big)$ and therefore that $0 \notin \operatorname{Spec}\big(\pi_{\mathbb{C}}(U)\big)$. There are two cases:

Case 1: The $\mathfrak{g}_{\mathbb{C}}$-module $V_{\mathbb{C}}$ is simple. Then it follows from the classification of the simple complex $\mathfrak{g}_{\mathbb{C}}$-modules ([Bou75, p.70]) that

$$\operatorname{Spec}\big(\pi_{\mathbb{C}}(U)\big) = \{-\frac{1}{2}\dim V, \ldots, -1, 1, \ldots, \frac{1}{2}\dim V\}i \subseteq (2\mathbb{Z}+1)i.$$

Case 2: The $\mathfrak{g}_{\mathbb{C}}$-module $V_{\mathbb{C}}$ is semisimple. Then there exists a non-trivial $\mathfrak{g}_{\mathbb{C}}$-submodule $V_1 \subseteq V_{\mathbb{C}}$. But, as a real $\mathfrak{g}$-module, $V_{\mathbb{C}} \cong V \oplus V$. Thus $V_1 \cong V$ as $\mathfrak{g}$-modules. Therefore we find a complex structure I on V which commutes with $\pi(\mathfrak{g})$ and we get a continuation of π to $\mathrm{sl}(2,\mathbb{C})$ on V by

$$\pi_{\mathbb{C}}\big((a+ib) \otimes x\big)v' := (a + Ib)\pi(x)v'.$$

Now V is a simple complex $\mathrm{sl}(2,\mathbb{C})$-module and again the classification of these modules ([Bou75, p.70]) shows that

$$\begin{aligned}\operatorname{Spec}\big(\pi(U)\big) &= \operatorname{Spec}\big(\pi_{\mathbb{C}}(1 \otimes U)\big) \\ &= \{-\frac{1}{2}\dim V, \ldots, -1, 1, \ldots, \frac{1}{2}\dim V\}i \subseteq (2\mathbb{Z}+1)i\end{aligned}$$

on V. ∎

Proposition II.13. *Suppose that* $\mathfrak{g}$ *has cone potential and* $\nu \in \Omega^+$ *with* $\mathfrak{s}^\nu \neq \{0\}$. *If* $\omega \in \mathbb{R}\nu$ *with* $\mathfrak{r}^\omega \neq \{0\}$, *then there exists* $x \in \mathfrak{s}^\nu$ *with* $|\nu(Q(x))| = 2$ *and* $\omega(Q(x)) \in 2\mathbb{Z} + 1$.

Proof. The Levi subalgebra $\mathfrak{s}$ is semisimple and contains the compactly embedded Cartan algebra $\mathfrak{h} \cap \mathfrak{s}$ (Proposition II.11). Therefore Proposition II.8 implies that $\nu(Q(x)) \neq 0$ for $x \in \mathfrak{s}^\nu$ and we may assume that $\nu(Q(x)) \in \{\pm 2\}$. We have $\nu(Q(x)) = 2$ iff $\langle x \rangle \cong \mathrm{sl}(2, \mathbb{R})$ and $\nu(Q(x)) = -2$ iff $\langle x \rangle \cong \mathrm{su}(2)$ (Proposition II.8). We consider the subspace

$$\mathfrak{r}_1 := \bigoplus_{\omega \in \mathbb{R}\nu \cap \Omega^+} \mathfrak{r}^\omega.$$

According to our assumption

$$[\langle x \rangle, \mathfrak{r}_1] \subseteq \bigoplus_{\omega \in \mathbb{R}\nu \cap \Omega^+} (\mathfrak{r}^{\omega+\nu} + \mathfrak{r}^{\omega-\nu}) \subseteq \mathfrak{r}_1$$

since $\omega \pm \nu \in \mathbb{R}\nu$. Therefore we find a simple $\langle x \rangle$-module $V \subseteq \mathfrak{r}_1$. For every $y = \sum y_\omega \in V$ we get

$$[Q(x), y] = \sum \omega(Q(x)) I y_\omega = 0$$

if and only if $\omega(Q(x)) = 0$ for $y_\omega \neq 0$. But $\omega \in \mathbb{R}\nu$ and therefore $\omega(Q(x)) \neq 0$ for all such $\omega \in \Omega^+$. Now Lemma II.12 applies and shows that $\dim V \in 2\mathbb{Z}$ and $\omega(Q(x)) \in 2\mathbb{Z} + 1$. ∎

Definition II.14. We set

$$\Omega_R := \{\omega \in \Omega : \mathfrak{g}^\omega \subseteq \mathfrak{r}\} \quad \text{and} \quad \Omega_S := \{\omega \in \Omega : \mathfrak{g}^\omega \subseteq \mathfrak{s}\},$$

where $\mathfrak{s}$ satisfies (S) from Proposition II.11. ∎

Corollary II.15. *Let* $\mathfrak{g}$ *be a Lie algebra with cone potential. Then the following assertions hold:*

(1) $\Omega^+ = \Omega_R^+ \cup \Omega_S^+$ *and* $\Omega_R \cap \Omega_S = \emptyset$.

(2) $\mathfrak{r} = (\mathfrak{h} \cap \mathfrak{r}) \oplus \bigoplus_{\omega \in \Omega_R^+} \mathfrak{g}^\omega$ *and* $\mathfrak{s} = (\mathfrak{h} \cap \mathfrak{s}) \oplus \bigoplus_{\omega \in \Omega_S^+} \mathfrak{g}^\omega$.

(3) *The property* (S) *from* Proposition II.11 *determines* $\mathfrak{s}$ *uniquely.*

(4) $\mathfrak{h} \cap \mathfrak{r} = \bigcap_{\omega \in \Omega_S^+} \ker \omega$.

(5) $\Omega_R^+ = \{\omega \in \Omega^+ : \omega(Q(\mathfrak{g}^\omega)) = \{0\}\}$ *and* $\Omega_S^+ = \{\omega \in \Omega^+ : \omega(Q(\mathfrak{g}^\omega)) \neq \{0\}\}$.

Proof. (cf. [Sp89]) (1) Suppose that $\mathfrak{s}^\nu = \mathfrak{g}^\nu \cap \mathfrak{s} \neq \{0\}$ and $0 \neq x \in \mathfrak{s}^\nu$. Then $\nu(Q(x)) \neq 0$ and therefore $[\langle x \rangle, \mathfrak{r}^\nu] \neq \{0\}$ whenever $\mathfrak{r}^\nu \neq \{0\}$. This is a contradiction to Proposition II.13 because $\omega(Q(x)) \neq \nu(Q(x))$ for all

$\omega \in \Omega_R^+ \cap \mathbb{R}\nu$. Hence $\mathfrak{r}^\nu = \{0\}$ if $\mathfrak{s}^\nu \neq \{0\}$. Now $\mathfrak{g}^\omega = (\mathfrak{g}^\omega \cap \mathfrak{r}) \oplus (\mathfrak{g}^\omega \cap \mathfrak{s})$ implies (1).
(2) This is a consequence of (1).
(3) The set $\Omega_S := \Omega \setminus \Omega_R$ is independent of $\mathfrak{s}$ and

$$\mathfrak{s} = [\mathfrak{h} \oplus \bigoplus_{\omega\in\Omega_S^+} \mathfrak{g}^\omega, \mathfrak{h} \oplus \bigoplus_{\omega\in\Omega_S^+} \mathfrak{g}^\omega]$$

is therefore uniquely determined by property (S).
(4) It follows from Proposition II.11 that $[\mathfrak{h} \cap \mathfrak{r}, \mathfrak{s}] = \{0\}$, hence

$$\mathfrak{h} \cap \mathfrak{r} \subseteq \bigcap_{\omega\in\Omega_S^+} \ker\omega = Z_{\mathfrak{h}}(\mathfrak{s}).$$

Then $\mathfrak{h} \cap \mathfrak{s} \cap Z_{\mathfrak{h}}(\mathfrak{s}) \subseteq Z(\mathfrak{s}) = \{0\}$ implies that $\mathfrak{h} \cap \mathfrak{r} = Z_{\mathfrak{h}}(\mathfrak{s})$.
(5) It is proved in (1) above that $\omega\big(Q(\mathfrak{g}^\omega)\big) \neq \{0\}$ for $\omega \in \Omega_S^+$. Let $\omega \in \Omega_R^+$. Then $Q(\mathfrak{g}^\omega) \subseteq [\mathfrak{r}^\omega, \mathfrak{r}^\omega] \subseteq Z(\mathfrak{g})$ (Proposition II.10). Therefore $\omega\big(Q(\mathfrak{r}^\omega)\big) \subseteq \omega\big(Q(\mathfrak{g}^\omega)\big) = \{0\}$. ∎

Let $\mathfrak{k}_{\mathfrak{h}}$ denote the uniquely determined maximal compactly embedded subalgebra of $\mathfrak{g}$ containing $\mathfrak{h}$ (Proposition II.5).

Lemma II.16. *Let $\omega \in \Omega_S^+$. Then $\dim \mathfrak{g}^\omega = 2$ and the following are equivalent:*

(1) $\mathfrak{g}^\omega \cap \mathfrak{k}_{\mathfrak{h}} \neq \emptyset$.
(2) *There exists $x \in \mathfrak{g}^\omega \cap \mathfrak{k}_{\mathfrak{h}}$ such that $\langle x \rangle \cong \mathrm{so}(3)$.*
(3) *There exists $x \in \mathfrak{g}^\omega \cap \mathfrak{k}_{\mathfrak{h}}$ such that $\omega\big(Q(x)\big) < 0$.*
(4) $\mathfrak{g}^\omega \subseteq \mathfrak{k}_{\mathfrak{h}}$.

Proof. The complexification $\mathfrak{s}_{\mathbb{C}}$ of $\mathfrak{s}$ is a complex semisimple Lie algebra, hence

$$\dim \mathfrak{g}^\omega = \dim_{\mathbb{C}} \mathfrak{g}_{\mathbb{C}}^\omega = \dim_{\mathbb{C}}(\mathfrak{g}_{\mathbb{C}}^\lambda \oplus \mathfrak{g}_{\mathbb{C}}^{-\lambda}) = 2$$

for $\omega = -i\lambda|_{\mathfrak{h}}$ ([Hum72, p.37]). We have $\mathfrak{h} \subseteq \mathfrak{k}_{\mathfrak{h}}$ and therefore

$$\mathfrak{k}_{\mathfrak{h}} = \mathfrak{h} \oplus \bigoplus_{\omega\in\Omega+} \mathfrak{g}^\omega \cap \mathfrak{k}_{\mathfrak{h}}.$$

(1) $\Rightarrow$ (2): Let $0 \neq x \in \mathfrak{k}_{\mathfrak{h}} \cap \mathfrak{g}^\omega$. Then $\langle x \rangle \subseteq \mathfrak{k}_{\mathfrak{h}}$ is a compact Lie algebra, hence $\langle x \rangle \cong \mathrm{so}(3)$ (Proposition II.8).
(2) $\Leftrightarrow$ (3): Proposition II.8.
(2) $\Rightarrow$ (4): We use Proposition II.8 to see that $\omega\big(Q(x)\big) < 0$. Therefore $\mathfrak{g}^\omega \subseteq \mathfrak{s}$ (Corollary II.15) and $\mathfrak{g}^\omega = \mathrm{span}\{x, Ix\}$ because $\dim \mathfrak{g}^\omega = 2$. We conclude that $Ix \in [\mathfrak{h}, x] \subseteq \mathfrak{k}_{\mathfrak{h}}$ and this implies that $\mathfrak{g}^\omega \subseteq \mathfrak{k}_{\mathfrak{h}}$.
(4) $\Rightarrow$ (1): This is trivial. ∎

Definition II.17. Set $\Omega_K^+ := \{\omega \in \Omega^+ : \mathfrak{g}^\omega \subseteq \mathfrak{k}_\mathfrak{h}\}, \Omega_P^+ := \Omega^+ \setminus \Omega_K^+$ and

$$\mathfrak{p}_\mathfrak{h} := \bigoplus_{\omega \in \Omega_P^+} \mathfrak{g}^\omega.$$

We call Ω_K the set of *compact roots* and Ω_P the set of *non-compact roots.* ■

Proposition II.18. *Let $\mathfrak{g}$ be a Lie algebra with cone potential. Then $\mathfrak{k}_\mathfrak{h} := \mathfrak{h} \oplus \bigoplus_{\omega \in \Omega_K^+} \mathfrak{g}^\omega$ is a maximal compactly embedded subalgebra of $\mathfrak{g}$ and $\mathfrak{p}_\mathfrak{h}$ satisfies*

(1) $\mathfrak{g} = \mathfrak{k}_\mathfrak{h} \oplus \mathfrak{p}_\mathfrak{h}$ *and* $[\mathfrak{k}_\mathfrak{h}, \mathfrak{p}_\mathfrak{h}] \subseteq \mathfrak{p}_\mathfrak{h}$.

(2) *If $\mathfrak{g}$ is semisimple then $\mathfrak{g} = \mathfrak{k}_\mathfrak{h} \oplus \mathfrak{p}_\mathfrak{h}$ is a Cartan decomposition of $\mathfrak{g}$.*

(3) $\mathfrak{z}_\mathfrak{k} = \bigcap_{\omega \in \Omega_K^+} \ker \omega = (\mathfrak{h} \cap \mathfrak{r}) \oplus (\mathfrak{z}_\mathfrak{k} \cap \mathfrak{s})$.

(4)

$$\operatorname{int} \operatorname{comp}(\mathfrak{g}) \cap \mathfrak{z}_\mathfrak{k} = \mathfrak{z}_\mathfrak{k} \setminus \bigcup_{\omega \in \Omega_P^+} \ker \omega.$$

Proof. The definition of Ω_K^+ and Lemma II.16 imply that

$$\mathfrak{k}_\mathfrak{h} = \mathfrak{h} \oplus \bigoplus_{\omega \in \Omega_K^+} \mathfrak{g}^\omega.$$

(1) The $\mathfrak{k}_\mathfrak{h}$-module $\mathfrak{g}$ is semisimple because $\mathfrak{k}_\mathfrak{h}$ is compactly embedded and we find a submodule $M \subseteq \mathfrak{g}$ with $\mathfrak{g} = M \oplus \mathfrak{k}_\mathfrak{h}$. Now $\mathfrak{h} \subseteq \mathfrak{k}_\mathfrak{h}$ implies that $M = \bigoplus_{\omega \in \Omega^+} M \cap \mathfrak{g}^\omega$. Whence $M \cap \mathfrak{g}^\omega = \mathfrak{g}^\omega$ for $\omega \in \Omega_P^+$ and $M \cap \mathfrak{g}^\omega = \{0\}$ for $\omega \in \Omega_K^+$. This proves that $M = \mathfrak{p}_\mathfrak{h}$ and $[\mathfrak{k}_\mathfrak{h}, \mathfrak{p}_\mathfrak{h}] \subseteq \mathfrak{p}_\mathfrak{h}$.
(2) This follows from (1) and [Hel78, III.7].
(3) This is clear.
(4) This follows from the fact that an element $h \in \mathfrak{z}_\mathfrak{k}$ is contained in $\operatorname{int} \operatorname{comp} \mathfrak{g}$ if and only if $\ker \operatorname{ad} h = \mathfrak{k}_\mathfrak{h}$ ([HHL89, A.2.25]) and this is equivalent to $\omega(h) \neq 0$ for all $\omega \in \Omega_P^+$ and $\omega(h) = 0$ for all $\omega \in \Omega_K^+$. ■

Proposition II.19. *Let $\mathfrak{g}$ be a simple Lie algebra and $\mathfrak{g} = \mathfrak{k} \oplus \mathfrak{p}$ a Cartan decomposition of $\mathfrak{g}$. Then the following assertions hold:*

(1) *$\mathfrak{p}$ is a simple $\mathfrak{k}$-module and $\mathfrak{k}$ is a maximal subalgebra of $\mathfrak{g}$,*

(2) $\dim Z(\mathfrak{k}) \leq 1$,

(3) $\dim Z(\mathfrak{k}) = 1$ *iff* $Z\big(Z(\mathfrak{k}), \mathfrak{g}\big) = \mathfrak{k}$, *and*

(4) *$\mathfrak{g}$ is quasihermitean iff $\mathfrak{g} = \mathfrak{k}$ or $\dim Z(\mathfrak{k}) = 1$, i.e., the quasihermitean simple Lie algebras are exactly the compact and hermitean simple Lie algebras.*

Proof. (1)- (3) [Sp88. p.115]
4) " $\Rightarrow$ ": Suppose that $\mathfrak{g}$ is quasihermitean. Then $\operatorname{int} \operatorname{comp}(\mathfrak{g}) \cap Z(\mathfrak{k}) \neq \emptyset$. There are two cases:
Case 1) $0 \in \operatorname{int} \operatorname{comp}(\mathfrak{g})$. Then $\mathfrak{g} = \mathfrak{k}$ is compact simple.

Case 2) $0 \notin \operatorname{int}\operatorname{comp}(\mathfrak{g})$. Then $Z(\mathfrak{k}) \neq \{0\}$ and therefore $\dim Z(\mathfrak{k}) = 1$ (cf. (2) above).
" $\Leftarrow$ ": It is clear that $\mathfrak{g}$ is quasihermitean if $\mathfrak{g} = \mathfrak{k}$ is compact. Suppose that $\dim Z(\mathfrak{k}) = 1$ and choose $z \in Z(\mathfrak{k}) \setminus \{0\}$. Then, in view of (1), $\ker \operatorname{ad} z = \mathfrak{k}$ and therefore $\mathfrak{k}$ contains a Cartan algebra $\mathfrak{h}$ of $\mathfrak{g}$ ([Bou75, p.25]). We conclude that $\mathfrak{h}$ is a compactly embedded Cartan algebra and $\mathfrak{k}_{\mathfrak{h}} = \mathfrak{k}$. Then [HHL89, III.6.34] applies and shows that $z \in Z(\mathfrak{k}_{\mathfrak{h}}) \cap \operatorname{int}\operatorname{comp}(\mathfrak{g})$. ■

Proposition II.20. *A Lie algebra $\mathfrak{g}$ with cone potential is quasihermitean if and only if no non-compact root vanishes on $\mathfrak{z}_{\mathfrak{k}}$. Suppose that this holds. Then the Levi algebra $\mathfrak{s}$ contains only quasihermitean simple ideals.*

Proof. We know from Proposition II.18 that

$$\operatorname{int}\operatorname{comp}\mathfrak{g} \cap \mathfrak{z}_{\mathfrak{k}} = \mathfrak{z}_{\mathfrak{k}} \setminus \bigcup_{\omega \in \Omega_P^+} \ker \omega$$

and this set is non-empty if and only if no non-compact root vanishes on $\mathfrak{z}_{\mathfrak{k}}$. Suppose that this condition is satisfied. We assume that $\mathfrak{s}$ contains a non-quasihermitean ideal $\mathfrak{s}_0$ and choose a complementary ideal $\mathfrak{s}_1$. Then we find a non-compact root $\nu \in \Omega_S^+ \cap \Omega_P^+$ such that $\mathfrak{g}^\nu \subseteq \mathfrak{s}_0$. Then $\mathfrak{z}_{\mathfrak{k}} = (\mathfrak{h} \cap \mathfrak{r}) \oplus (\mathfrak{z}_{\mathfrak{k}} \cap \mathfrak{s}) \subseteq (\mathfrak{h} \cap \mathfrak{r}) + (\mathfrak{z}_{\mathfrak{k}} \cap \mathfrak{s}_1)$ because $Z(\mathfrak{k}_{\mathfrak{h}} \cap \mathfrak{s}_0) = \{0\}$ (Proposition II.19) and therefore

$$\nu(\mathfrak{z}_{\mathfrak{k}}) \subseteq \nu(\mathfrak{h} \cap \mathfrak{r}) + \nu(\mathfrak{h} \cap \mathfrak{s}_1) = \{0\}$$

since $[\mathfrak{h} \cap \mathfrak{r}, \mathfrak{s}_0] = \{0\}$ and $[\mathfrak{s}_1, \mathfrak{s}_0] = \{0\}$. This is impossible if $\mathfrak{g}$ is quasihermitean. ■

Now we have all tools which are necessary to give a description of the Lie algebras with cone potential. We use the universal construction for Lie algebras with cone potential from [Sp88, pp.73-82].

Proposition II.21. *Let $\mathfrak{g}_1 = \mathfrak{z}_1 \oplus \mathfrak{s}$ be a reductive Lie algebra with compactly embedded Cartan algebra $\mathfrak{h}_1 = \mathfrak{z}_1 \oplus \mathfrak{h}_1 \cap \mathfrak{s}$, commutator algebra $\mathfrak{s}$ and M a $\mathfrak{g}_1$-module with the associated representation $\varphi : \mathfrak{g}_1 \to \operatorname{End}(M)$ such that*

(1) $\ker \varphi \cap \mathfrak{z}_1 = \{0\}$,
(2) *$\varphi(h)$ is semisimple with purely imaginary spectrum for every $h \in \mathfrak{h}_1$,*
(3) $M_0 := \{x \in M : \varphi(\mathfrak{h}_1).x = \{0\}\} = \{0\}$
(4) *$q : M \times M \to V$ is an $\mathfrak{g}_1$-invariant skew-symmetric bilinear mapping.* *Then the vector space $\mathfrak{g} := M \oplus V \oplus \mathfrak{g}_1$ with the bracket*

$$[(m, x, a), (m', x', a')] = \Big((\varphi(a)m' - \varphi(a')m), q(m, m'), [a, a']\Big)$$

is a Lie algebra such that

$$\begin{aligned} Z(\mathfrak{g}) &= \{0\} \oplus V \oplus \{0\}, \\ \mathfrak{n} &= M \oplus V \oplus \{0\} \quad \textit{is the nilradical,} \\ \mathfrak{r} &= M \oplus V \oplus \mathfrak{z}_1 \quad \textit{is the radical, and} \\ \mathfrak{h} &= \{0\} \oplus V \oplus \mathfrak{h}_1 \end{aligned}$$

is a compactly embedded Cartan algebra of $\mathfrak{g}$.

Proof. First we note that $[V,\mathfrak{g}] = \{0\}$ and that the bracket is skew symmetric. Let $\mathfrak{n} := M + V$. Then

$$[(m,x,0),(m',x',0)] = \big(0, q(m,m'),0\big).$$

Thus $\big[[\mathfrak{n},\mathfrak{n}],\mathfrak{n}\big] = \{0\}$ and the skew symmetry of $[\cdot,\cdot]$ implies that $\mathfrak{n}$ is a nilpotent Lie algebra. If we show that the mapping

$$\widetilde{\varphi}(a) : \mathfrak{n} \to \mathfrak{n}, \quad (m,x) \mapsto \big(\varphi(a)m, 0\big)$$

is a derivation, it follows from the above formula for the bracket that

$$\mathfrak{g} \cong \mathfrak{n} \rtimes \mathfrak{g}_1$$

is the semidirect product of the Lie algebras $\mathfrak{n}$ and $\mathfrak{g}_1$ and therefore a Lie algebra. We show that this is true:

$$\begin{aligned}
&\big[\widetilde{\varphi}(a)(m,x),(m',x')\big] + \big[(m,x),\widetilde{\varphi}(a)(m',x')\big] \\
&= \Big(0, q\big(\varphi(a)m,m'\big) + q\big(m,\varphi(a)m'\big)\Big) \\
&= (0,0) \\
&= \widetilde{\varphi}(a)\big(0,q(m,m')\big) \\
&= \widetilde{\varphi}(a)\big[(m,x),(m',x')\big].
\end{aligned}$$

We know already that $\mathfrak{n}$ is a nilpotent ideal of $\mathfrak{g}$. It is also clear that $\mathfrak{r} := \mathfrak{n} + \mathfrak{z}_1$ is a solvable ideal. But $\mathfrak{g} = \mathfrak{r} \rtimes \mathfrak{s}$ and $\mathfrak{s}$ is semisimple. Thus $\mathfrak{r}$ is the radical of $\mathfrak{g}$. The ideal $\mathfrak{n}$ is the nilradical because the mappings $\varphi(\operatorname{ad} h)$, $h \in \mathfrak{z}_1 \setminus \{0\}$ are not nilpotent since, in view of (2), $\varphi(\operatorname{ad} h) \neq 0$ (1) and $\varphi(h)$ is semisimple. We conclude that $\mathfrak{n}$ is the largest nilpotent ideal and consequenty the nilradical. Assertion (3) shows that $Z(\mathfrak{g}) = V$ since $[V,\mathfrak{g}] = \{0\}$ and $Z(\mathfrak{g}) \cap M = \{0\}$. To prove the last assertion we firstly note that $V + \mathfrak{h}_1$ is an abelian compactly embedded subalgebra of $\mathfrak{g}$. If an element $(m,0,x)$ normalizes $\mathfrak{h}$, then x normalizes $\mathfrak{h}_1$ in S and so $x \in \mathfrak{h}_1$ because $\mathfrak{h}_1$ is a Cartan algebra of $\mathfrak{g}_1$ and $m = 0$ (3). We conclude that $\mathfrak{h}$ is self-normalizing, hence a Cartan algebra. ∎

Definition II.22. With the notations and assumptions of Proposition II.21 we write

$$\operatorname{Lie}(\mathfrak{g}_1, M, q, V)$$

for the Lie algebra constructed from the $\mathfrak{g}_1$-module M and the invariant skew symmetric bilinear form $q : M \times M \to V$. ∎

Lemma II.23. *Let* $\mathfrak{g}$ *be the Lie algebra constructed in* Proposition II.21 *and* I *a complex structure on* M *which commutes with* $\varphi(\mathfrak{h}_1)$. *Set*

$$Q(x,y) := q(Ix,y) \quad \textit{for} \quad x,y \in M.$$

Then Q *is a symmetric bilinear form on* M *which satisfies*

1) $Q(Ix,x) = 0$ *and*
2) *the invariance condition* $Q(x,[h,y]) + Q(Iy,[h,Ix]) = 0$ *for all* $h \in \mathfrak{g}_1$. *Moreover,*

$$[(m,x,a),(m',x',a')] = \Big((\varphi(a)m' - \varphi(a')m), -Q(Im,m'), [a,a'] \Big)$$

and if $M = \bigoplus M_i$ *is a decomposition into isotypic* $\mathfrak{g}_1$*-modules we have that*

$$Q(M_i,M_j) = \{0\} \quad \textit{for} \quad i \neq j$$

whenever $Q(M_i) \neq \{0\}$.

Proof. The symmetry of Q follows from $Q(x,y) = q(Ix,y) = [Ix,y]$ and [HHL89, III.6.7]. It is clear that

$$Q(Ix,x) = [-x,x] = 0$$

and

$$Q(x,[h,y]) + Q(Iy,[h,Ix]) = \big[Ix,[h,y]\big] - \big[y,[h,Ix]\big] = \big[h,[Ix,y]\big] = 0$$

because $[Ix,y] \in Z(\mathfrak{g})$. Suppose that $Q(M_i,M_j) \neq \{0\}$. Then there exist simple $\mathfrak{g}_1$-modules $V_i \subseteq M_i$ and $V_j \subseteq M_j$ such that $[V_i,V_j] \neq \{0\}$. Let $\alpha \in \widehat{V}$ such that $\alpha\big([V_i,V_j]\big) \neq \{0\}$. Then the bilinear mapping

$$\beta : V_i \times V_j \to \mathbb{R}, (v,v') \mapsto \alpha([v,v'])$$

is a skew-symmetric $\mathfrak{g}_1$-invariant bilinear form. Let $x \in V_j$. Then $V_j^{\perp} = \{y \in V_i : \beta(y,V_j) = \{0\}\}$ is a $\mathfrak{g}_1$-submodule of V_i which has to be $\{0\}$. Therefore β is non-degenerate. This permits us to see that the $\mathfrak{g}_1$-modules V_i^* and V_j are isomorphic. If $Q(M_i) \neq \{0\}$ this implies that $V_i^* \cong V_i$. Therefore $Q(V_i,V_j) = \{0\}$ for $i \neq j$. ∎

Corollary II.24. *Let* $\mathfrak{h}_1$ *be a finite dimensional vector space and* $\Omega \subseteq \widehat{\mathfrak{h}}_1$ *a finite symmetric subset with* $\Omega^{\perp} = \{0\}$. *Let* $\Omega^+ \subseteq \Omega$ *be a positive system and* M_ω *be a vector space with a complex structure* I *for every* $\omega \in \Omega^+$. *We define an action of* $\mathfrak{h}_1$ *on* $M := \bigoplus_{\omega\in\Omega^+} M_\omega$ *by*

$$[h,x] = \omega(h)Ix \quad \textit{for all} \quad x \in M_\omega.$$

Suppose, in addition, that the mappings $Q_\omega : M_\omega \times M_\omega \to V$ *are symmetric and bilinear, where* V *is a fixed vector space, such that*

$$Q_\omega(Ix, x) = 0 \quad \textit{for all } x \in M_\omega.$$

Then the linear space $\mathfrak{g} := M \oplus V \oplus \mathfrak{h}_1$ *is a solvable Lie algebra with the bracket:*

$$[((m_\omega), x, h), ((m'_\omega), x', a')] = \Big((\omega(h)Im'_\omega - \omega'(h)Im_\omega), -\sum_{\omega \in \Omega^+} Q_\omega(Im_\omega, m'_\omega), 0\Big)$$

with

$$Z(\mathfrak{g}) = \{0\} \oplus V \oplus \{0\},$$
$$\mathfrak{n} = M \oplus V \oplus \{0\}, \quad \textit{and}$$
$$\mathfrak{h} = \{0\} \oplus V \oplus \mathfrak{h}_1$$

is a compactly embedded Cartan algebra of $\mathfrak{g}$.

Proof. This is a special case of Proposition II.21. The only thing we have to check is the invariance condition for the symmetric bilinear form Q (Lemma II.23). For $x, y \in M_\omega$ we have

$$Q_\omega(x, [h, y]) + Q_\omega(Iy, [h, Ix]) = \omega(h)\big(Q(x, Iy) + Q(Iy, -x)\big) = 0$$

since Q is symmetric. ■

Proposition II.25. *Let* $\mathfrak{g}$ *be a Lie algebra with cone potential. Take* $\mathfrak{g}_1 = \mathfrak{z}_1 + \mathfrak{s}$, *where* $\mathfrak{z}_1$ *is a vector space complement for the center in* $\mathfrak{h} \cap \mathfrak{r}$ *and define*

$$q : \mathfrak{r}_{\text{eff}} \times \mathfrak{r}_{\text{eff}} \to Z(\mathfrak{g}), \ (m, m') \mapsto [m, m']$$

and

$$\varphi(a)x := [a, x] \quad \textit{for} \quad a \in \mathfrak{g}_1, x \in \mathfrak{r}_{\text{eff}}.$$

Then

$$\mathfrak{g} \cong \operatorname{Lie}\big(\mathfrak{g}_1, \mathfrak{r}_{\text{eff}}, q, Z(\mathfrak{g})\big).$$

Proof. (cf. [Sp88, pp.79-82]) One checks easily that the assumptions of Proposition II.21 are satisfied for this data (Proposition II.11). For $m, m' \in \mathfrak{r}_{\text{eff}}, a, a' \in \mathfrak{g}_1$ and $x, x' \in Z(\mathfrak{g})$ this leads to

$$[m + x + a, m' + x' + a'] = [m + a, m' + a'] = \varphi(a)m' - \varphi(a')m + q(m, m') + [a, a']$$

and the assertion follows from the definition of the bracket in the Lie algebra $\operatorname{Lie}\big(\mathfrak{g}_1, \mathfrak{r}_{\text{eff}}, q, Z(\mathfrak{g})\big)$. ■

Remark II.26. On every finite dimensional complex vector space exists a non-degenerate skew symmetric bilinear form Q with $Q(x, ix) = 0$ and $Q(x, x) \neq 0$. One may take the imaginary part of any positive definite sesquilinear form. Therefore Proposition II.25 and Corollary II.24 imply that every symmetric subset $\Omega \subseteq \widehat{\mathfrak{h}}$ occurs as the system of real roots of a solvable Lie algebra with cone potential. The class of those Lie algebras may be parametrized by the data $\Omega \subseteq \widehat{\mathfrak{h}}$ and $\{Q_\omega : \omega \in \Omega\}$. ■

Remark II.27. Consider the Lie algebra constructed in Corollary II.24 and take $h \in \mathfrak{h}_1$ such that $\omega(h) \neq 0$ for all $\omega \in \Omega$. Then $h \in \operatorname{int} \operatorname{comp} \mathfrak{g}$ (Proposition II.18). This proves that $\mathfrak{g}$ is quasihermitean. The Lie algebra $\mathfrak{g}$ has cone potential if and only if $Q_\omega(x) \neq 0$ for $x \in \mathfrak{g}^\omega \setminus \{0\}$. ∎

Before we consider various examples we have to introduce another concept which will be essential to determine whether a Lie algebra contains pointed generating invariant cones or not.

Definition II.28. A Lie algebra with cone potential is said to have *strong cone potential* if there exists a positive system $\Omega^+ \subseteq \Omega$ and a linear functional $\alpha \in \widehat{Z}(\mathfrak{g})$ such that

$$\alpha\big(Q(x)\big) > 0 \quad \text{for every} \quad x \in \mathfrak{g}^\omega \setminus \{0\}, \omega \in \Omega_R^+.$$

∎

Example II.29. We consider a special case of Corollary II.24. Let $\mathfrak{h}_1 = \mathbb{R}$, $\Omega^+ = \{\omega\}$ with $\omega(1) = 1$, $Q_\omega(x) = \|x\|^2$, where $\|\cdot\|$ is an $\mathfrak{h}_1$-invariant euclidean norm on $\mathfrak{r}^\omega$, and $\dim \mathfrak{r}^\omega = 2m$. Then $A_{2m+2} = \mathfrak{g}$ is the $(2m+2)$-*dimensional oscillator algebra* (cf. Definition II.7). Since $Q_\omega(\mathfrak{r}^\omega) = \mathbb{R}^+$ it has strong cone potential. ∎

Example II.30. We give a construction of an interesting class of Lie algebras with cone potential. These algebras may serve as examples in various contexts.

Let $\mathfrak{s}$ be a semisimple Lie algebra with compactly embedded Cartan algebra $\mathfrak{h}_S$. We set $\mathfrak{g}_1 := \mathbb{R}i \oplus \mathfrak{s}$ and consider the $\mathfrak{g}_1$-module $M := \mathfrak{s}_{\mathbb{C}} = \mathbb{C} \otimes \mathfrak{s}$ with

$$\varphi(i,a)(z \otimes s) = (i,a).z \otimes s = (iz) \otimes [a,s] \quad \text{for } a, s \in \mathfrak{s}.$$

We denote the Cartan-Killing form on $\mathfrak{s}$ with B and define

$$q : M \times M \to \mathbb{R}, \ (z \otimes x, z' \otimes x') \mapsto \operatorname{Im}(z'\overline{z})B(x,x').$$

This is a skew-symmetric bilinear form on M which is invariant under $\mathfrak{g}_1$:

$$\begin{aligned} q\big((i,a).z \otimes s, z' \otimes s'\big) &= q\big(iz \otimes [a,s], z' \otimes s'\big) \\ &= \operatorname{Im}(z'\overline{iz})B([a,s],s') \\ &= -\operatorname{Im}(iz'\overline{z})(-1)B(s,[a,s']) \\ &= \operatorname{Im}(iz'\overline{z})B(s,[a,s']) \\ &= q\big(z \otimes s, (i,a)z' \otimes s'\big). \end{aligned}$$

Therefore $\operatorname{Lie}(\mathfrak{g}_1, M, q, \mathbb{R})$ is a Lie algebra with the bracket

$$[(m,t,a),(m',t',a')] = \big(a.m' - a'.m, q(m,m'), [a,a']\big)$$

and $\mathfrak{h} := \{0\} \oplus \mathbb{R} \oplus \mathfrak{h}_1$ with $\mathfrak{h}_1 := \mathbb{R}i \oplus \mathfrak{h}_S$ is a compactly embedded Cartan algebra of L.

Let $\mathfrak{s} = \mathfrak{h}_S \oplus \bigoplus_{\omega \in \Omega^+} \mathfrak{s}^\omega$ be a real root decomposition with respect to $\mathfrak{h}_S$. Then

$$\mathbb{C} \otimes \mathfrak{s}^\omega = \mathfrak{s}_{\mathbb{C}}^\lambda \oplus \mathfrak{s}_{\mathbb{C}}^{-\lambda},$$

where $\lambda|_{\mathfrak{h}} = i\omega$. We conclude that

$$M = \mathbb{C} \otimes \mathfrak{h}_S \oplus \bigoplus_{\lambda \in \Lambda} \mathfrak{s}_{\mathbb{C}}^\lambda,$$

where Λ is the root system of $\mathfrak{s}_{\mathbb{C}}$ with respect to $(\mathfrak{h}_S)_{\mathbb{C}}$, is the decomposition of M into isotypic $\mathfrak{h}_1$-modules. Setting $I(z \otimes a) := (iz) \otimes a$ we get

$$[h, z \otimes a] = -i\lambda(h) I(z \otimes a) \quad \text{for every} \quad z \otimes a \in \mathfrak{s}_{\mathbb{C}}^\lambda, h \in \mathfrak{h}_1,$$

where $-i\lambda(ai, h_0) = a - i\lambda(h_0) \in \mathbb{R}$ for $h = (ai, h_0) \in H_1$. For every root vector $z \otimes x \in M$ we have

$$Q(z \otimes x) := q\big(I(z \otimes x), z \otimes x\big) = q\big(iz \otimes x, z \otimes x\big) = -|z|^2 B(x, x).$$

For $x \in \mathfrak{h}_S$ or $x \in \mathfrak{s}^\omega$ with $\omega \in \Omega_K^+$ this leads to $Q(z \otimes x) > 0$ and to $Q(z \otimes x) < 0$ for $x \in \mathfrak{s}^\omega$ with $\omega \in \Omega_P^+$ because the Cartan Killing form B of $\mathfrak{s}$ is positive definite on the space $\mathfrak{p}$ and negative definite on the algebra $\mathfrak{k}$, where $\mathfrak{s} = \mathfrak{k} + \mathfrak{p}$ is a Cartan decomposition of $\mathfrak{s}$ (Proposition II.18, [Hel78, p.184]). This proves that $\mathfrak{g}$ has cone potential. It has strong cone potential if and only if there exists an element $h = (ai, h_0) \in \mathfrak{h}_1$ such that

$$a \pm i\lambda(h_0) \begin{cases} > 0, & \text{for } i\lambda \in \Omega_K^+ \\ < 0, & \text{for } i\lambda \in \Omega_P^+ \end{cases}.$$

We conclude that $a > 0$ and that $\Omega_P^+ = \emptyset$, i.e., $\mathfrak{s}$ is a compact Lie algebra. If, conversely, this is true, then one may set $h = (i, 0)$.
The real subspaces $z \otimes \mathfrak{s}$ with $z \in \mathbb{C}$ are abelian since

$$[z \otimes \mathfrak{s}, z \otimes \mathfrak{s}] = q(z \otimes \mathfrak{s}, z \otimes \mathfrak{s}) = 0 \cdot B(\mathfrak{s}, \mathfrak{s}) = \{0\}$$

and they are invariant under the action of $\mathfrak{s}$ on M. Therefore we have constructed Lie algebras with cone potential where the Levi algebra acts on the nilradical as on the adjoint module. To see which of those Lie algebras are quasihermitean we use Proposition II.20. A necessary condition is that $\mathfrak{s}$ is quasihermitean. We claim that it is sufficient. Suppose that $\mathfrak{s}$ is quasihermitean. We choose $z_S \in Z(\mathfrak{k}_{\mathfrak{h}}) \cap \mathfrak{s}$ such that $\omega(z) \neq 0$ for all $\omega \in \Omega_S^+ \cap \Omega_P^+$. Set $z := 0 \oplus \lambda i \oplus z_S \in \mathfrak{h}$. For $\omega \in \Omega_S^+ \cap \Omega_P^+$ we have $\omega(z) = \omega(z_S) \neq 0$ and for $\omega \in \Omega_R^+$ the relation

$$\omega(z) = \lambda + \omega(z_S)$$

holds. If we choose $\lambda > \max_{\omega \in \Omega_S^+} |\omega(z_S)|$ this expression is always positive. Thus $\mathfrak{g}$ is quasihermitean. ∎

There is another interesting source for quasihermitean Lie algebras which have cone potential (Lemma II.41).

Lemma II.31. *Let V be a real module of the compact group G and $q: V \times V \to \mathbb{R}$ a non-degenerate skew-symmetric invariant bilinear form. Then there exists a G-invariant complex structure J on V and a G-invariant hermitean product $\langle \cdot, \cdot \rangle$ such that V is a unitary G-module and*

$$q(x,y) = \operatorname{Im}\langle x, y\rangle \quad \textit{for all} \quad x, y \in V.$$

Proof. We choose a G-invariant scalar product $\langle \cdot, \cdot \rangle_0$ on V. Then we find a linear operator

$$J_1 : V \to V \quad \text{such that} \quad q(x,y) = \langle x, J_1 y\rangle_0 \quad \text{for all} \quad x, y \in V.$$

The skew symmetry of q implies that $J_1^\top = -J_1$ and the invariance that J_1 commutes with G. Therefore $\operatorname{Spec}(J) \subseteq i\mathbb{R}$. Let $V := \bigoplus_\alpha V_\alpha$ be the decomposition of the J_1-module V into isotypical components such that $\operatorname{Spec}(J_1|_{V^\alpha}) \subseteq \{\pm i\alpha\}$ and $\alpha \in \mathbb{R}^+$. Since J_1 commutes with the action of G on V the subspaces V_α are invariant under G and orthogonal because $x \in V_\alpha$ and $y \in V_\beta$ imply that

$$\langle x, y\rangle_0(\alpha^2 - \beta^2) = \langle -J_1^2 x, y\rangle_0 + \langle x, J_1^2 y\rangle_0 = -\langle x, J_1^2 y\rangle_0 + \langle x, J_1^2 y\rangle_0 = 0.$$

The non-degeneracy of q shows that $V_0 = 0$. We define a new G-invariant scalar product on V by

$$\langle x, y\rangle_{\mathbb{R}} := \sum_\alpha \alpha \langle x_\alpha, y_\alpha\rangle_0 \quad \text{for} \quad x = \sum_\alpha x_\alpha, y = \sum_\alpha y_\alpha, \quad x_\alpha, y_\alpha \in V_\alpha.$$

We set

$$Jx := \sum_\alpha \frac{1}{\alpha} J_1 x_\alpha.$$

Then $J^2 = -\mathrm{id}_V$ and

$$q(x,y) = \sum_\alpha \langle x_\alpha, J_1 y_\alpha\rangle_0 = \sum_\alpha \alpha \langle x_\alpha, J y_\alpha\rangle_0 = \langle x, Jy\rangle_{\mathbb{R}}.$$

Therefore J is a G-invariant complex structure on V. Setting

$$\langle x, y\rangle := \langle x, y\rangle_{\mathbb{R}} + i\langle x, Jy\rangle_{\mathbb{R}}$$

we get a G-invariant Hilbert space structure on V because

$$\begin{aligned}\langle Jx, y\rangle &= \langle Jx, y\rangle_{\mathbb{R}} + i\langle Jx, Jy\rangle_{\mathbb{R}} \\ &= i(\langle x, J^\top Jy\rangle_{\mathbb{R}} - i\langle x, J^\top y\rangle_{\mathbb{R}}) \\ &= i(-\langle x, J^2 y\rangle_{\mathbb{R}} + i\langle x, Jy\rangle_{\mathbb{R}}) \\ &= i\langle x, y\rangle\end{aligned}$$

and

$$\langle y, x\rangle = \langle y, x\rangle_{\mathbb{R}} + i\langle y, Jx\rangle_{\mathbb{R}} = \langle x, y\rangle_{\mathbb{R}} - i\langle x, Jy\rangle_{\mathbb{R}} = \overline{\langle x, y\rangle}.$$

■

Proposition II.32. *Let $\{A_i, i = 1, \ldots, k\}$ be a finite set of hermitean operators on $\mathbb{C}^n$. Suppose that*

$$d := \max\left\{\operatorname{rank} A : A \in \operatorname{span}_{\mathbb{R}}\{A_i : i = 1, \ldots, k\}\right\} < n.$$

Then there exists an element $z \in \mathbb{C}^n$ such that

$$\langle A_i z, z\rangle = 0 \quad \text{for} \quad i = 1, \ldots, k.$$

Proof. We choose an element $A \in \operatorname{span}_{\mathbb{R}}\{A_i : i = 1, \ldots, k\}$ with maximal rank d. Then we find an orthonormal basis $e_1, \ldots, e_n$ of $\mathbb{C}^n$ such that $Ae_i = \alpha_i e_i$ with $\alpha_i \in \mathbb{R}$, $\alpha_i = 0$ for $i > d$ and $\alpha_i \neq 0$ for $i \leq d$. We claim that

$$\langle A_i e_{d+1}, e_{d+1}\rangle = 0 \quad \text{for} \quad i = 1, \ldots, k.$$

We may consider the subspace $B := \operatorname{span}\{e_1, \ldots, e_{d+1}\} \subseteq \mathbb{C}^n$ and therefore assume that $n = d + 1$ since all assumptions carry over to this subspace. Let $i \in \{1, \ldots, k\}$. Then

$$\det(A_i + \lambda A) = \lambda^n \det A + \lambda^{n-1}\alpha_1 \cdot \ldots \cdot \alpha_d \cdot b_{d+1} + p(\lambda),$$

where $b_{d+1} = \langle A_i e_{d+1}, e_{d+1}\rangle$ and p is a polynomial of degree smaller than $n-2$. According to our assumption $\operatorname{rank}(A_i + \lambda A) \leq d$ for all real λ and therefore $\det(A_i + \lambda A) = 0$. We conclude that $b_{d+1} = 0$ since $\alpha_1 \cdot \ldots \cdot \alpha_d \neq 0$. ■

Corollary II.33. *Let $\mathfrak{g}$ be a Lie algebra with cone potential. Then there exists $\alpha \in \widehat{Z(\mathfrak{g})}$ such that*

$$(x, y) \mapsto \alpha([x, y]), \quad \mathfrak{r}_{\text{eff}} \times \mathfrak{r}_{\text{eff}} \to \mathbb{R}$$

is a non-degenerate skew-symmetric bilinear form. Moreover, there exists a $\mathfrak{k}_{\mathfrak{h}}$-invariant complex structure J on $\mathfrak{r}_{\text{eff}}$ and a positive definite hermitean form on $\mathfrak{r}_{\text{eff}}$ such that

$$\alpha([x, y]) = \operatorname{Im}\langle x, y\rangle.$$

This form satisfies

$$\alpha([Jx, x]) > 0 \quad \text{for all} \quad x \in \mathfrak{r}_{\text{eff}} \setminus \{0\}.$$

Proof. We choose an $\mathfrak{h}$-invariant complex structure I on $\mathfrak{r}_{\text{eff}} \cong \mathbb{C}^n$ and a positive definite I-hermitean form $\langle\cdot,\cdot\rangle$. We take a basis $\alpha_i \in \widehat{Z(\mathfrak{g})}$. Then the symmetric mapping $Q : R^+ \times R^+ \to Z(\mathfrak{g})$ defines k symmetric forms

$$q_i := \alpha_i \circ Q : \mathfrak{r}_{\text{eff}} \times \mathfrak{r}_{\text{eff}} \to \mathbb{R}$$

such that $q_i(Ix, x) = 0$ for all $x \in \mathfrak{r}_{\text{eff}}$. Now we find real linear mappings $A_i : \mathfrak{r}_{\text{eff}} \to \mathfrak{r}_{\text{eff}}$ such that

$$q_i(x, y) = \operatorname{Re}\langle x, Ay\rangle \quad \text{for all} \quad x, y \in \mathbb{C}^n.$$

If $x, y \in \mathfrak{r}^\omega$ and $h \in \mathfrak{h}$ with $\omega(h) = 1$ then

$$\begin{aligned} q_i(Ix, y) &= \alpha_i([I^2x, y]) = \alpha_i(\big[[h, Ix], y\big]) \\ &= -\alpha_i(\big[Ix, [h, y]\big]) = -\alpha_i([Ix, Iy]) = -q_i(x, Iy). \end{aligned}$$

Hence $\mathrm{Re}(\langle Ix, A_i y\rangle + \langle x, A_i Iy\rangle) = 0$ which implies that A_i commutes with I, and therefore is a complex linear mapping. Moreover

$$\mathrm{Re}(\langle x, A_i y\rangle) = \mathrm{Re}(\langle y, A_i x\rangle) = \mathrm{Re}(\langle x, A_i^* y\rangle \quad \text{for all } x, y \in \mathfrak{r}_{\mathrm{eff}},$$

i.e., $A^* = A$ is hermitean. For every $z \in \mathfrak{r}_{\mathrm{eff}}$ we know that $Q(z, z) \neq 0$ and now Proposition II.32 implies that there exist real numbers $\beta_1, \ldots, \beta_k$ such that

$$A := \sum_i \beta_i A_i$$

has maximal rank equal to n in $\mathrm{span}_{\mathbb{R}}\{A_i : i = 1, \ldots, k\}$. We set $\alpha := \sum_i \beta_i \alpha_i \in \widehat{Z(\mathfrak{g})}$ and see that

$$\begin{aligned} \alpha([x, y]) &= \sum_i \beta_i \alpha_i([x, y]) = -\sum_i \beta_i \alpha_i \circ Q(Ix, y) \\ &= -\sum_i \beta_i q_i(Ix, y) = -\sum_i \beta_i \,\mathrm{Re}\langle Ix, A_i y\rangle \\ &= \mathrm{Im}\langle x, Ay\rangle. \end{aligned}$$

This skew-hermitean form is non-degenrate since rank $A = n$. The last assertion follows from Lemma II.31 and the fact that the compact group $\mathrm{INN}_{\mathfrak{g}}\, \mathfrak{k}_{\mathfrak{h}}$ leaves the skew symmetric non-degenerate form $\alpha \circ [\cdot, \cdot]$ invariant. Then we have that

$$\alpha([Jx, x]) = \mathrm{Im}\langle Jx, x\rangle = \mathrm{Im}(i\|x\|^2) = \|x\|^2 > 0.$$

■

Corollary II.34. *Suppose that* $\mathfrak{g}$ *has cone potential. Then the following assertions are equivalent:*

(1) $\mathfrak{g}$ *has strong cone potential.*

(2) *The functional* $\alpha \in \widehat{Z(\mathfrak{g})}$ *and the complex structure* J *on* $\mathfrak{r}_{\mathrm{eff}}$ *from* Corollary II.33 *can be chosen such that*

$$\Omega(J) := \{\omega \in \widehat{\mathfrak{h}} : \{x \in \mathfrak{g} : (\forall h \in \mathfrak{h})[h, x] = \omega(h)Jx\} \neq \{0\}\}$$

is contained in a half space, i.e., $\Omega(J) \subseteq \Omega$ *is a positive system and* $I|_{\mathfrak{r}_{\mathrm{eff}}} = J$

Proof. (1) $\Rightarrow$ (2): If $\mathfrak{g}$ has strong cone potential we find $h_0 \in \mathfrak{h}$ and $\alpha \in \widehat{Z(\mathfrak{g})}$ such that

$$\alpha([Ix, x]) > 0 \quad \text{for every} \quad x \in \mathfrak{r}^\omega \setminus \{0\}.$$

Then the skew symmetric bilinear form $q : \mathfrak{r}_{\text{eff}} \times \mathfrak{r}_{\text{eff}} \to \mathbb{R}$ defined by $q(x,y) = \alpha([x,y])$ is $\mathfrak{k}_{\mathfrak{h}}$-invariant and non-degenerate because for $x = \sum_\omega x_\omega$ with $x_\omega \in \mathfrak{r}^\omega$ the relation

$$q(Ix,x]) = \alpha([Ix,x]) = \sum_\omega \alpha([Ix_\omega, x_\omega]) > 0 \quad \text{for} \quad x \in \mathfrak{r}_{\text{eff}} \setminus \{0\}$$

follows from Proposition II.10. Using Lemma II.31 we find a $\mathfrak{k}_{\mathfrak{h}}$-invariant complex structure J on $\mathfrak{r}_{\text{eff}}$ and a positive definite hermitean form $\langle\cdot,\cdot\rangle$ such that

$$\alpha([x,y]) = \operatorname{Im}\langle x,y\rangle = \operatorname{Re}\langle x, Jy\rangle.$$

It follows from the definition of J that $J\mathfrak{r}^\omega \subseteq \mathfrak{r}^\omega$ because J commutes with the action of H on $\mathfrak{r}^+$. Let $\omega \in \Omega_R^+$ and choose $h \in \mathfrak{h}$ with $\omega(h) = 1$. Then we find subspaces $\mathfrak{r}^\omega_\pm$ such that

$$Ix = [h,x] = \begin{cases} Jx, & \text{for } x \in \mathfrak{r}^\omega_+ \\ -Jx, & \text{for } x \in \mathfrak{r}^\omega_- \end{cases}.$$

Then we find for $x \in \mathfrak{r}^\omega_-$ that

$$0 \leq \alpha\big(Q(x)\big) = \alpha([Ix,x]) = \operatorname{Re}\langle Ix, Jx\rangle = -\langle Ix, Ix\rangle \leq 0.$$

This implies that $\mathfrak{r}^\omega_- = \{0\}$ and that $J = I\,|_{\mathfrak{r}_{\text{eff}}}$. Hence $\Omega(J) = \Omega^+ \subseteq \Omega$ is a positive system.
(2) $\Rightarrow$ (1): If $\Omega(J) \subseteq \Omega$ is a positive system, we set $\Omega^+ := \Omega(J)$. Then $I_{\mathfrak{r}^\omega} = J\,|_{\mathfrak{r}^\omega}$ and

$$\alpha\big(Q(x)\big) = \alpha([Jx,x]) > 0 \quad \text{for} \quad x \in \mathfrak{r}^\omega \setminus \{0\}.$$

Consequently $\mathfrak{g}$ has strong cone potential. ■

Remark II.35. To visualize the contents of Corollary II.33 we consider the Lie algebra

$$\mathfrak{g} := \operatorname{Lie}(\mathbb{R}, \mathbb{C}^2, q, \mathbb{R}^3),$$

where the action of $\mathbb{R}$ on $\mathbb{C}^2$ is defined by

$$\varphi(t)z = tiz \quad \text{and} \quad q(z,z') = \big(\operatorname{Re}\langle z, J_1 z'\rangle, \operatorname{Re}\langle z, J_2 z'\rangle, \operatorname{Re}\langle z, J_3 z'\rangle\big)$$

with

$$J_1 = \begin{pmatrix} i & 0 \\ 0 & -i \end{pmatrix}, \quad J_2 = \begin{pmatrix} 0 & 1 \\ -1 & 0 \end{pmatrix}, \quad \text{and} \quad J_3 = \begin{pmatrix} 0 & i \\ i & 0 \end{pmatrix}.$$

Then L has cone potential because $q(iz,z) = 0$ implies that

$$0 = q(iz,z) = \big(|z_1|^2 - |z_2|^2, -2\operatorname{Im}(z_1\overline{z}_2), 2\operatorname{Re}(z_1\overline{z}_2)\big)$$

and consequently

$$|z_1|^2 = |z_2|^2 = |z_1\overline{z}_2| = 0, \quad \text{i.e.,} \quad z = 0.$$

For every $\alpha \in \widehat{\mathbb{R}^3}$ the skew symmetric bilinear form $\alpha \circ q$ does not satisfy the condition for strong cone potential because

$$\alpha \circ q(z, z') = \operatorname{Re}\langle z, Az' \rangle \quad \text{with} \quad A \in \operatorname{span}_{\mathbb{R}}\{J_1, J_2, J_3\} = \operatorname{su}(2).$$

Thus $\operatorname{tr}(A) = 0$ and there exists $B \in \operatorname{SU}(2)$ with

$$B^{-1}AB = \begin{pmatrix} i\alpha & 0 \\ 0 & -i\alpha \end{pmatrix}, \quad \alpha \in \mathbb{R}.$$

Therefore

$$\alpha \circ q(iBz, Bz) = \operatorname{Re}\langle iz, B^{-1}ABz \rangle = \alpha \cdot (|z| - |z|) = 0.$$

But if we set $J := J_1$, then this is an $\mathbb{R}$-invariant complex structure on $\mathbb{C}^2$ with

$$q_1([J_1 z, z]) := \operatorname{Re}\langle J_1 z, J_1 z \rangle > 0 \quad \text{for} \quad z \neq 0.$$

■

Example II.36. We consider the Lie algebra

$$\mathfrak{g} := \operatorname{Lie}\left(i\mathbb{R} + \operatorname{sl}(2,\mathbb{R}), \operatorname{sl}(2,\mathbb{C}), q, \mathbb{R}\right)$$

from Example II.30. We have shown that $\mathfrak{g}$ does not have cone potential because $\operatorname{sl}(2,\mathbb{R})$ is not compact. We have seen that

$$\Omega_R^+ = \{\omega_0, \omega_0 + \omega_1, \omega_0 - \omega_1\},$$

where

$$\omega_0\left(ai + b\begin{pmatrix} 0 & 1 \\ -1 & 0 \end{pmatrix}\right) = a \quad \text{and} \quad \omega_1\left(ai + b\begin{pmatrix} 0 & 1 \\ -1 & 0 \end{pmatrix}\right) = 2b$$

and $Q(\mathfrak{g}^{\omega_0}) \subseteq \mathbb{R}^+$, $Q(\mathfrak{g}^{\omega_0 \pm \omega_1}) \subseteq \mathbb{R}^-$. Defining the complex structure J on $\operatorname{sl}(2,\mathbb{C})$ by changing the sign of I on the root spaces $\mathfrak{g}^{\omega_0 \pm \omega_1}$ we obtain that

$$[Jx, x] > 0 \quad \text{for} \quad x \in \operatorname{sl}(2,\mathbb{C}) \setminus \{0\}$$

but $\Omega(J) = \{\omega_0, -\omega_0 \pm \omega_1\}$ is not a positive system in Ω. ■

Example II.37. Let $\mathfrak{g} = \mathrm{sp}(n, \mathbb{R}) = \{X \in \mathfrak{gl}(2n, \mathbb{R}) : X^\top J + JX = 0\}$, where

$$J = \begin{pmatrix} 0 & -\mathbf{1}_n \\ \mathbf{1}_n & 0 \end{pmatrix}.$$

A compactly embedded Cartan algebra is

$$\mathfrak{h} = \left\{ \begin{pmatrix} 0 & D \\ -D & 0 \end{pmatrix} : D \text{ is diagonal} \right\}.$$

The center $\mathfrak{z}_{\mathfrak{k}}$ agrees with $\mathbb{R}J \subseteq \mathfrak{h}$.

Now we endow $\mathbb{R}^{2n}$ with the usual scalar product. Then the skew-symmetric bilinear form

$$q : \mathbb{R}^{2n} \times \mathbb{R}^{2n} \to \mathbb{R}, (x, y) \mapsto \langle x, Jy \rangle$$

is invariant under the action of $\mathfrak{g}$. All other assumptions of Proposition II.21 are satisfied and therefore the Lie algebra

$$\mathfrak{g}_J := \mathrm{Lie}\left(\mathrm{sp}(n, \mathbb{R}), \mathbb{R}^{2n}, q, \mathbb{R}\right)$$

is well defined. Note that $\mathfrak{g}_J \cong \mathfrak{h}_n \rtimes \mathrm{sp}(n, \mathbb{R})$ is a semidirect product of $\mathrm{sp}(n, \mathbb{R})$ and the $(2n+1)$-dimensional Heisenberg algebra. We choose Ω^+ such that $\omega(J) > 0$ for every non-compact positive root. That this is possible implies that $\mathfrak{g}_J$ is quasihermitean. Then we see that $\mathfrak{g}_J$ has strong cone potential since

$$[Ix, x] = q(Jx, x) = \langle Jx, Jx \rangle > 0 \quad \text{for} \quad x \in \mathbb{R}^{2n} \setminus \{0\}.$$

■

Theorem II.38. (Structure Theorem for Lie Algebra with Strong Cone Potential) *Let $\mathfrak{g}$ be a Lie algebra with strong cone potential and $\alpha \in \widehat{\mathfrak{z}}$ such that*

$$\alpha(Q(x)) > 0 \quad \textit{for} \quad x \in \mathfrak{r}^\omega \setminus \{0\}.$$

Suppose that the $\mathfrak{h}$-invariant Levi algebra acts effectively on the radical $\mathfrak{r}$. Then $\mathfrak{g}$ is a central extension of a subalgebra of the Lie algebra $\mathfrak{g}_J$ (Example II.37). *The homomorphism $\pi : \mathfrak{g} \to \mathfrak{g}_J$ can be constructed by choosing a $\mathfrak{k}_{\mathfrak{h}}$-invariant scalar product and a complex structure I on $\mathfrak{r}_{\mathrm{eff}}$ such that*

$$\alpha([x, y]) = \langle x, Iy \rangle \quad \textit{for} \quad x, y \in \mathfrak{r}_{\mathrm{eff}}.$$

Proof. Using Corollaries II.33 and II.34 we find a $\mathfrak{k}_{\mathfrak{h}}$-invariant complex structure I on $\mathfrak{r}_{\mathrm{eff}}$ and a positive definite hermitean form on $\mathfrak{r}_{\mathrm{eff}}$ such that

$$\alpha([x, y]) = \mathrm{Im}\langle x, y \rangle = \mathrm{Re}\langle x, Iy \rangle.$$

Therefore we find an isometry $\Psi : \mathfrak{r}_{\text{eff}} \to \mathbb{R}^{2n}$ such that $\Psi \circ I \circ \Psi^{-1}$ is represented by the matrix J in Example II.37. Choose a subspace $\mathfrak{h}_1 \subseteq \mathfrak{h} \cap \mathfrak{r}$ which is complementary to $Z(\mathfrak{g})$ and set $\mathfrak{g}_1 := \mathfrak{h}_1 \oplus \mathfrak{s}$. Then

$$\pi : X \mapsto \Psi \circ \operatorname{ad} X \,|_{\mathfrak{r}_{\text{eff}}} \circ \Psi^{-1}, \mathfrak{g}_1 \to \operatorname{sp}(n, \mathbb{R})$$

is a homomorphism of Lie algebras which permits a continuation to a homomorphism

$$\widetilde{\pi} : \mathfrak{g} \to \mathfrak{g}_J.$$

According to the assumption that $\mathfrak{s}$ acts effectively on $\mathfrak{r}_{\text{eff}}$ we know that $\ker \widetilde{\pi} \cap \mathfrak{s} = \{0\}$. Thus $\ker \pi \subseteq \mathfrak{h}$ and

$$\ker \pi = \ker \alpha \oplus (\ker \pi \cap \mathfrak{h}_1).$$

We show that $\ker \pi \cap \mathfrak{h}_1 = \{0\}$. Then $\ker \pi \subseteq Z(\mathfrak{g})$ and the theorem follows. If $h \in \mathfrak{h}_1 \subseteq \mathfrak{h} \cap \mathfrak{r}$ with $\pi(h) = 0$, all solvable roots vanish on h. But $[\mathfrak{h} \cap \mathfrak{r}, \mathfrak{s}] = \{0\}$ implies that all semisimple roots vanish on h, too. Thus $h \in Z(\mathfrak{g}) \cap \mathfrak{h}_1 = \{0\}$.■

Remark II.39. Note that the assumption that $\mathfrak{s}$ acts effectively on $\mathfrak{r}_{\text{eff}}$ is not very restrictive because in the general case $\mathfrak{s}$ decomposes into two semisimple ideals $\mathfrak{s} = \mathfrak{s}_0 \oplus \mathfrak{s}_1$ such that $\mathfrak{s}_0$ acts effectively on $\mathfrak{r}_{\text{eff}}$ and $\mathfrak{s}_1 = \{x \in \mathfrak{s} : [x, \mathfrak{r}] = \{0\}\}$. Then

$$\mathfrak{g} \cong (\mathfrak{r} \rtimes \mathfrak{s}_0) \oplus \mathfrak{s}_1$$

is a direct sum decomposition of $\mathfrak{g}$. ■

Remark II.40. As a consequence of Theorem II.38 we see that the Lie algebra $\mathfrak{g}_J = \operatorname{Lie}\big(\operatorname{sp}(n, \mathbb{R}), \mathbb{R}^{2n}, q, \mathbb{R}\big)$ is in some sense the prototype of a mixed Lie algebra with strong cone potential. ■

We have seen that in a Lie algebra with cone potential the space $\mathfrak{r}_{\text{eff}}$ carries the structure of a unitary $\mathfrak{k}_{\mathfrak{h}}$-module. Now we consider the converse problem. Which conditions on a unitary module V of a compact Lie algebra $\mathfrak{k}$ guarantee that the Lie algebra constructed in Proposition II.21 from this data has strong cone potential? One should expect a condition which is expressable in terms of the weights of V with respect to a Cartan algebra $\mathfrak{h}$ of $\mathfrak{k}$. So, let $\mathfrak{k}$ be a compact Lie algebra, $\mathfrak{h}$ a Cartan algebra of $\mathfrak{k}$ and a non-zero complex module V of $\mathfrak{k}$ where every element of $\mathfrak{k}$ acts as a skew-hermitean operator. Then V decomposes into weight spaces

$$V^\lambda := \{x \in V : h.x = \lambda(h)x \ \text{ for all } \ h \in \mathfrak{h}\}.$$

Setting $\Lambda := \{\lambda \in \operatorname{Hom}_{\mathbb{R}}(\mathfrak{h}, \mathbb{C}) : V^\lambda \neq \{0\}\}$ we have

$$V = \bigoplus_{\lambda \in \Lambda} V^\lambda.$$

We assume that $V^0 = \{0\}$ and that $\mathfrak{k}$ acts effectively on M. The V^λ are the isotypic $\mathfrak{h}$-modules in V. From the above assumption it is clear that $\lambda(\mathfrak{h}) \subseteq i\mathbb{R}$ for every $\lambda \in \Lambda$. Therefore $\omega := -i\lambda \in \widehat{\mathfrak{h}}$. We set

$$\Omega := -i\Lambda, \quad V^\omega := V^{-\omega} := V^\lambda + V^{-\lambda}$$

and choose a positive system $\Omega^+ \subseteq \Omega$. Then

$$V = V^0 \oplus \bigoplus_{\omega \in \Omega^+} V^\omega$$

is a decomposition of the real $\mathfrak{h}$-module V into isoptypical components.

Lemma II.41. *For the skew-symmetric bilinear form*

$$q : V \times V \to \mathbb{R}, (x,y) \mapsto \operatorname{Im}\langle x, y\rangle$$

the following conditions are equivalent:

(1) *$q(h.x, x) \neq 0$ for all $h \in \mathfrak{h}$ and $x \in V^\omega$ with $h.x \neq 0$.*

(2) *$V^\lambda = \{0\}$ or $V^{-\lambda} = \{0\}$ for $\lambda = i\omega$.*

(3) *The Lie algebra $\mathfrak{g} := \operatorname{Lie}(\mathfrak{k}, V, q, \mathbb{R})$ has cone potential.*

$\mathfrak{g}$ has strong cone potential iff Ω^+ may be chosen such that $V^{-\lambda} = 0$ for all $\lambda \in \Lambda$.

Proof. (1) $\Leftrightarrow$ (3): This follows from the definition of the bracket in L.
(1) $\Rightarrow$ (2): Suppose that $V^\lambda \neq \{0\} \neq V^{-\lambda}$ and choose $x_\lambda \in V^\lambda$, $x_{-\lambda} \in V^{-\lambda}$ such that $\|x_\lambda\| = \|x_{-\lambda}\| = 1$. We set $x := x_\lambda + x_{-\lambda} \in V^\omega$ and select $h \in \mathfrak{h}$ such that $\omega(h) = 1$. Then

$$\begin{aligned} q(h.x, x) &= \operatorname{Im}\langle h.x_\lambda + h.x_{-\lambda}, x_\lambda + x_{-\lambda}\rangle \\ &= \operatorname{Im}\langle ix_\lambda - ix_{-\lambda}, x_\lambda + x_{-\lambda}\rangle \\ &= \operatorname{Re}\langle x_\lambda - x_{-\lambda}, x_\lambda + x_{-\lambda}\rangle \\ &= \|x_\lambda\|^2 - \|x_{-\lambda}\|^2 = 0 \end{aligned}$$

for $\|x_\lambda\| = \|x_\lambda\|$ because $V^\lambda \perp V^{-\lambda}$.
(2) $\Rightarrow$ (1): We may assume that $V^{-\lambda} = \{0\}$. Then we have for $x \in V^\omega$ that

$$q(h.x, x) = \operatorname{Im}\langle h.x, x\rangle = \operatorname{Im}(i\lambda(h)\|x\|^2) = \omega(h)\|x\|^2 \neq 0$$

for $\omega(h) \neq 0$, i.e., $h.x \neq 0$. To prove the last assertion, let $\omega = -i\lambda \in \Omega$. Then

$$[Ix, x] \begin{cases} > 0, & \text{for } x \in V^\lambda \setminus \{0\} \\ < 0, & \text{for } x \in V^{-\lambda} \setminus \{0\} \end{cases}.$$

Now the assertion follows because $\mathfrak{g}$ has strong cone potential if one finds a positive system such that $[Ix, x] > 0$ for all $x \in V \setminus \{0\}$. ∎

Example II.42. Let $\mathfrak{k} = \mathrm{su}(2)$ and $V = \mathbb{C}^2$. Then $\mathfrak{h} = \mathbb{R}h_0$ with $h_0 = \begin{pmatrix} i & 0 \\ 0 & -i \end{pmatrix}$ is a Cartan algebra of $\mathfrak{k}$. Therefore

$$V^\lambda = \mathbb{C} \times \{0\} \quad \text{and} \quad V^{-\lambda} = \{0\} \times \mathbb{C}$$

for $\lambda(h_0) = i$ and the conditions of the previous Lemma II.41 are not satisfied.■

Example II.43. Let $\mathfrak{k} = \mathbb{R}i$ and $V = \mathbb{C}$. Then $V^\lambda = \mathbb{C}$ with $\lambda(i) = i$ and the condition of Lemma II.41 is satisfied. ■

Example II.44. Let $\mathfrak{k} = \mathrm{su}(3)$. Then $\dim H = 2$ and we may choose a basis $\{\alpha_1, \alpha_2\}$ of the root system such that

$$\{\pm\alpha_1, \pm\alpha_2, \pm(\alpha_1 + \alpha_2)\}$$

is the root system of K. Then

$$\lambda_1 = \frac{2}{3}\alpha_1 + \frac{1}{3}\alpha_2 \quad \text{and} \quad \lambda_2 = \frac{1}{3}\alpha_1 + \frac{2}{3}\alpha_2$$

is a basis of the system of dominant integral weights. Let V be the irreducible complex K-module with highest weight λ_1. Then

$$\Lambda := \{\frac{2}{3}\alpha_1 + \frac{1}{3}\alpha_2, -\frac{1}{3}\alpha_1 + \frac{1}{3}\alpha_2, -\frac{1}{3}\alpha_1 - \frac{2}{3}\alpha_2\}$$

is the weight system of V. Each weight has multiplicity 1. Therefore $\dim V = 6 = 2\dim_{\mathbb{C}} V$.

We construct the Lie algebra $\mathfrak{g}_0 = \mathrm{Lie}(\mathfrak{k}, V, q, \mathbb{R})$ as in Proposition II.21. Then $\mathfrak{g}_0$ has cone potential because the condition of Lemma II.32 is satisfied. The Lie algebra $\mathfrak{g}_0$ is not quasihermitean since $Z(\mathfrak{k}_\mathfrak{h}) = \{0\} \oplus \mathbb{R} \oplus \{0\} = Z(\mathfrak{g}_0)$. We will see later in Section III that this implies that $\mathfrak{g}_0$ does not have strong cone potential. If we set $\mathfrak{g}_1 := \mathbb{R}i \oplus \mathfrak{k}$, then V is an irreducible $\mathfrak{g}_1$-module and the Lie algebra $\mathfrak{g} = \mathrm{Lie}(\mathfrak{g}_1, V, q, \mathbb{R})$ is quasihermitean. ■

III. INVARIANT CONES IN LIE ALGEBRAS

In this section we investigate invariant cones in Lie algebras and their characterization by intersections with compactly embedded Cartan algebras. We generalize the reconstruction theorems in [HHL89] to the case of ideals, non-pointed generating and non-generating pointed invariant cones, and characterize those Lie algebras which contain invariant pointed generating cones. Throughout this section $\mathfrak{g}$ denotes a Lie algebra with cone potential and $\mathfrak{h}$ is a compactly embedded Cartan algebra of $\mathfrak{g}$.

For the next theorem we recall the definitions of $\mathrm{INN}_{\mathfrak{g}}\,\mathfrak{a}$ and $\mathrm{Inn}_{\mathfrak{g}}\,\mathfrak{a}$ from Definition II.2.

Proposition III.1. *If $N(\mathfrak{h}) := \{g \in \mathrm{INN}_{\mathfrak{g}}\,\mathfrak{g} : I_g(\mathrm{INN}_{\mathfrak{g}}\,\mathfrak{h}) \subseteq \mathrm{INN}_{\mathfrak{g}}\,\mathfrak{h}\}$ is the normalizer of the maximal torus $\mathrm{INN}_{\mathfrak{g}}\,\mathfrak{h}$ in $\mathrm{INN}_{\mathfrak{g}}\,\mathfrak{g}$, then $N(\mathfrak{h}) \subseteq \mathrm{INN}_{\mathfrak{g}}\,\mathfrak{k}_{\mathfrak{h}}$ and the quotient group*

$$\mathcal{W} := \mathcal{W}(\mathfrak{h},\mathfrak{g}) := N(\mathfrak{h})/\mathrm{INN}_{\mathfrak{g}}(\mathfrak{h})$$

is finite. It agrees with the classical Weyl group of the compact group $\mathrm{INN}_{\mathfrak{g}}\,\mathfrak{k}_{\mathfrak{h}}$. For $\nu \in \mathcal{W}$ with $\nu = n\,\mathrm{INN}_{\mathfrak{g}}\,\mathfrak{h}$, $n \in N(\mathfrak{h})$ and $h \in \mathfrak{h}$ we have

$$\nu.h = n(h) \quad \text{and} \quad e^{\operatorname{ad} \nu.h} = n \circ e^{\operatorname{ad} h} \circ n^{-1}.$$

The group $\mathcal{W}' := \{\nu|_{\mathfrak{h}} : \nu \in \mathcal{W}\}$ agrees with the group generated by reflections on the hyperplanes $\ker \omega$ with $\omega \in \Omega_K^+$.

Proof. In view of [HHL89, III.5.6, III.5.7] it only remains to prove the last statement. Let $\widetilde{\mathfrak{h}} := \mathbf{L}(\mathrm{INN}_{\mathfrak{g}}\,\mathfrak{h})$ and $\widetilde{\mathfrak{k}} := \mathbf{L}(\mathrm{INN}_{\mathfrak{g}}\,\mathfrak{k}_{\mathfrak{h}})$. Then $\widetilde{\mathfrak{h}} \subseteq \widetilde{\mathfrak{k}}$ is a Cartan algebra and $\mathrm{Inn}_{\mathfrak{g}}\,\mathfrak{k}_{\mathfrak{h}}$ is dense in $\mathrm{INN}_{\mathfrak{g}}\,\mathfrak{k}_{\mathfrak{h}}$. Therefore $\mathbf{L}(\mathrm{Inn}_{\mathfrak{g}}\,\mathfrak{k}_{\mathfrak{h}}) = \operatorname{ad}\mathfrak{k}_{\mathfrak{h}}$ contains the commutator algebra $\widetilde{\mathfrak{k}}' = [\widetilde{\mathfrak{k}},\widetilde{\mathfrak{k}}]$. We consider the homomorphism

$$\varphi : \mathfrak{k}_{\mathfrak{h}} \to \widetilde{\mathfrak{k}}, \quad x \mapsto \operatorname{ad} x.$$

Then $\varphi(\mathfrak{h}) \subseteq \widetilde{\mathfrak{h}}$ and $\varphi(\mathfrak{k}_{\mathfrak{h}}') = \widetilde{\mathfrak{k}}'$. If $\langle\cdot,\cdot\rangle$ is a positive definite invariant bilinear form on $\widetilde{\mathfrak{k}}$ we find that $\varphi(\mathfrak{k}_{\mathfrak{h}})^{\perp} \subseteq \widetilde{\mathfrak{k}}'^{\perp} \subseteq Z(\widetilde{\mathfrak{k}})$ and $\varphi' : \mathfrak{k}_{\mathfrak{h}}' \to \widetilde{\mathfrak{k}}' \subseteq \varphi(\mathfrak{k}_{\mathfrak{h}}), x \mapsto \operatorname{ad} x$ is an isomorphism of Lie algebras. It is clear that

$$\varphi'(\nu.h) = \operatorname{ad}(\nu.h) = n \circ \operatorname{ad} h \circ n^{-1} = \operatorname{Ad}(n)(\operatorname{ad} h)$$

for $\nu = n\,\mathrm{INN}_{\mathfrak{g}}\,\mathfrak{h}$ and $n \in N(\mathfrak{h})$. Therefore φ' is equivariant with respect to the action of $\mathcal{W}$ on $\mathfrak{h}$ and the action of the Weyl group of $\widetilde{\mathfrak{k}}$ on $\widetilde{\mathfrak{h}}$. Now we apply [Sp88, p.151] to see that $\mathcal{W}'$ is generated by the reflections on the hyperplanes $\ker\omega$ for $\omega \in \Omega_K^+$ since this holds for the compact Lie group $\mathrm{INN}_{\mathfrak{g}}\,\mathfrak{k}_{\mathfrak{h}}$. ■

Definition III.2. We call $\mathcal{W} = \mathcal{W}(\mathfrak{h}, \mathfrak{g})$ the *Weyl group* of $\mathfrak{g}$ with respect to the compactly embedded Cartan algebra $\mathfrak{h}$. ∎

Lemma III.3. *For $x \in \mathfrak{g}^\omega$ and $h \in \mathfrak{h}$ we have that*

$$(\operatorname{ad} x)^2(h) = \omega(h)Q(x).$$

Proof. This follows from

$$(\operatorname{ad} x)^2 h = \big[x, [x, h]\big] = -\big[x, [h, x]\big] = -\omega(h)[x, Ix] = \omega(h)Q(x).$$

∎

Definition III.4. For a subset A of a vector space V we set

$$\operatorname{cone}(A) := \bigcap \{W : A \subseteq W, W \text{ is a wedge in } V\}.$$

We write

$$C := C(\mathfrak{h}, \mathfrak{g}) := \operatorname{cone}\{(\operatorname{ad} x)^2 \mid_{\mathfrak{h}} : x \in \mathfrak{g}^\omega,\ \omega \in \Omega_P^+\} \subseteq \operatorname{End}(\mathfrak{h}).$$

According to Lemma III.3 this is a cone spanned by rank one operators on $\mathfrak{h}$. From now on we write p for the orthogonal projection $p : \mathfrak{g} \to \mathfrak{h}$ along $\mathfrak{g}_{\text{eff}}$. This is exactly the averaging operator of the action of the compact group $\operatorname{INN}_{\mathfrak{g}} \mathfrak{h}$ on $\mathfrak{g}$ (cf. Theorem I.10). ∎

Lemma III.5. *Let $W \subseteq \mathfrak{g}$ be an invariant wedge. Then*

$$p(W) = W \cap \mathfrak{h}.$$

Proof. (cf. [HHL89, III.5.4]) This is a consequence of Theorem I.10 because W is invariant under the action of the compact group $\operatorname{INN}_{\mathfrak{g}}(\mathfrak{h})$ and $\mathfrak{h} = \mathfrak{g}_{\text{fix}}$ with respect to this action. ∎

Proposition III.6. (Orbit projections) *Let $x \in \mathfrak{g}^\omega$ and $h \in \mathfrak{h}$. Then*

$$p(e^{\mathbb{R} \operatorname{ad} x} h) = h + \begin{cases} \frac{\omega(h)}{\omega(Q(x))}[-2, 0]Q(x), & \text{if } x \in \Omega_K^+ \\ \omega(h)\mathbb{R}^+ Q(x), & \text{if } x \in \Omega_P^+. \end{cases}$$

Proof. [HHL89, III.7.9] ∎

Proposition III.7. *Let $\mathfrak{g}$ be a Lie algebra with compactly embedded Cartan algebra $\mathfrak{h}$, $W \subseteq \mathfrak{g}$ an invariant wedge and $C := \mathfrak{h} \cap W$. Then*

$$(W) \qquad \mathcal{W}C \subseteq C$$

and

$$(C) \qquad \mathcal{C}C \subseteq C.$$

The condition (C) is equivalent to

$$\omega(C)Q(x) \subseteq C \quad \textit{for every} \quad x \in \mathfrak{g}^\omega, \omega \in \Omega_P^+.$$

Proof. Let $w \in C = W \cap \mathfrak{h}$ and $\nu = n\,\mathrm{INN}_{\mathfrak{g}}\,\mathfrak{h} \in \mathcal{W}$. Then we see with Proposition III.1 that

$$\nu.w = n(w) \in n(W) \cap n(\mathfrak{h}) = W \cap \mathfrak{h} = C.$$

Hence $\mathcal{W}C = C$. Let $x \in \mathfrak{g}^\omega \setminus \{0\}$ and $\omega \in \Omega_P^+$. Then, using Lemma III.5 and Proposition III.6, we get

$$p(e^{\mathbb{R}\,\mathrm{ad}\,x}w) = w + \omega(w)\mathbb{R}^+Q(x) \subseteq p(W) = C.$$

We conclude with Lemma III.3 that

$$(\mathrm{ad}\,x)^2 w = \omega(w)Q(x) = \lim_{t\to\infty} \frac{1}{t}w + \omega(w)Q(x) \in \overline{C} = C$$

since $C = W \cap \mathfrak{h}$ is closed. The last assertion follows from Lemma III.3. ∎

Our objective is to find necessary and sufficient conditions for a wedge $C \subseteq \mathfrak{h}$ to be the trace of an invariant wedge. This is already known if C is pointed and generating ([HHL89, III.9.18]). First we consider the case where C is a vector space.

Lemma III.8. *Let $\mathfrak{k}_{\mathfrak{h}}' = \bigoplus_{i=1}^l \mathfrak{k}_i$ be a decomposition into simple ideals and $\mathfrak{h}_{\mathfrak{k}_i} := \mathfrak{k}_i \cap \mathfrak{h}$. We consider the action of the finite group $\mathcal{W}$ on $\mathfrak{h}$. Then*

$$\mathfrak{h}_{\mathrm{fix}} = Z(\mathfrak{k}_{\mathfrak{h}}) = (\mathfrak{h} \cap \mathfrak{r}) \oplus \big(Z(\mathfrak{k}_{\mathfrak{h}}) \cap \mathfrak{s}\big) \quad \textit{and} \quad \mathfrak{h}_{\mathrm{eff}} = \bigoplus_{i=1}^l \mathfrak{h}_{\mathfrak{k}_i},$$

where the spaces $\mathfrak{h}_{\mathfrak{k}_i}$ are irreducible pairwise non-isomorphic $\mathcal{W}$-modules.

Proof. In view of Proposition III.1 we have to show the following. Let $\mathfrak{k}$ be a compact Lie algebra, $\mathfrak{h}$ a Cartan algebra of $\mathfrak{k}$, $\mathfrak{k}' = \bigoplus_{i=1}^l \mathfrak{k}_i$ a decomposition into simple ideals and $\mathcal{W}$ the Weyl group of $\mathfrak{k}$. Then

$$\mathfrak{h}_{\mathrm{fix}} = Z(\mathfrak{k}) \quad \text{and} \quad \mathfrak{h}_{\mathrm{eff}} = \bigoplus_{i=1}^l \mathfrak{h}_{\mathfrak{k}_i}$$

is a decomposition of $\mathfrak{h}_{\mathrm{eff}}$ into irreducible summands. It is clear that $\mathfrak{k}_i$ and therefore $\mathfrak{h}_i$ is invariant under $\mathcal{W}$. Hence we may assume that $\mathfrak{k}$ is simple. If $\mathfrak{h}$ decomposes into two non-trivial invariant subspaces $\mathfrak{h} = \mathfrak{h}_1 \oplus \mathfrak{h}_2$, then the root system of $\mathfrak{k}$ decomposes into two orthogonal non-empty subsets. This is impossible since $\mathfrak{k}$ is simple ([Hum72, p.73]). ∎

Proposition III.9. *The Weyl group $\mathcal{W}$ operates on Ω via*

$$[\alpha] * \omega := \omega \circ \alpha^{-1} = \widehat{\alpha}^{-1}(\omega),$$

where $[\alpha]$ is the class of $\alpha \in N(\mathfrak{h})$ in $\mathcal{W}$. The following assertions hold

(1) $\mathfrak{g}^{\omega \circ \alpha^{-1}} = \alpha(\mathfrak{g}^{\omega})$.

(2) *For all $[\alpha] \in \mathcal{W}$ the map $\alpha|_{\mathfrak{g}_{\text{eff}}}$ commutes with I.*

(3) *The sets Ω_K, Ω_P, Ω_R and Ω_S are invariant under the action of $\mathcal{W}$.*

Proof. (1), (2) [Sp89, p.13]
(3) Let $[\alpha] \in \mathcal{W}$ with $\alpha \in N(\mathfrak{h}) \subseteq \text{INN}_{\mathfrak{g}} \mathfrak{k}_{\mathfrak{h}}$. Then

$$\alpha(\mathfrak{r}) \subseteq \mathfrak{r}, \ \alpha(\mathfrak{s}) \subseteq \mathfrak{s}, \ \alpha(\mathfrak{k}_{\mathfrak{h}}) \subseteq \mathfrak{k}_{\mathfrak{h}} \quad \text{and} \quad \alpha(\mathfrak{p}_{\mathfrak{h}}) \subseteq \mathfrak{p}_{\mathfrak{h}}.$$

In view of (1), this proves the invariance of the sets $\Omega_K, \Omega_P, \Omega_R$ and Ω_S. ■

Example III.10. (cf. [Pa81, p.325], Example II.37) Let $\mathfrak{g} = \text{sp}(n, \mathbb{R}) = \{X \in \text{gl}(2n, \mathbb{R}) : X^{\top} J + JX = 0\}$, where

$$J = \begin{pmatrix} 0 & -\mathbf{1}_n \\ \mathbf{1}_n & 0 \end{pmatrix}.$$

Then JX is symmetric for every $X \in \mathfrak{g}$ and

$$W_J := \{X \in \mathfrak{g} : JX \text{ is positive semidefinite }\}$$

is a pointed generating cone in the simple Lie algebra $\mathfrak{g}$. A compactly embedded Cartan algebra is

$$\mathfrak{h} = \left\{ \begin{pmatrix} 0 & D \\ -D & 0 \end{pmatrix} : D \text{ is diagonal} \right\}.$$

The intersection $C_J = W_J \cap \mathfrak{h}$ consists of those matrices where all diagonal entries in D are non-negative. ■

Proposition III.11. *Under the assumptions of* Theorem II.38 *the set*

$$W = \{x \in \mathfrak{h}_1 + \mathfrak{s} : -I \operatorname{ad} x|_{\mathfrak{r}_{\text{eff}}} \text{ is positive semi-definite}\}$$

is a pointed generating invariant wedge in the reductive Lie algebra $\mathfrak{g}_1 = \mathfrak{h}_1 \oplus \mathfrak{s}$ with

$$W \cap \mathfrak{h} = \{h \in \mathfrak{h} \cap \mathfrak{g}_1 : \omega(h) \geq 0 \ \text{ for all } \ \omega \in \Omega_R^+\}.$$

Proof. The cone W is the preimage of the invariant cone $-W_J \subseteq \text{sp}(n, \mathbb{R})$ with respect to the homomorphism $\pi : \mathfrak{g}_1 \to \text{sp}(n, \mathbb{R})$ constructed in Theorem II.38. For $h \in \mathfrak{h}$ we find that

$$-I \operatorname{ad} h(x) = \omega(h)x \quad \text{for} \quad x \in \mathfrak{r}^{\omega}$$

Therefore every elememt $h_0 \in \mathfrak{h}$ with $\omega(h_0) > 0$ for all $\omega \in \Omega_R^+$ is contained in the interior of W which cannot be empty because such elements always exists according to the definition of a positive system. The edge $H(W)$ of W is contained in the kernel of π which is $\{0\}$ (Theorem II.38). Thus W is pointed and generating. ■

Definition III.12. Let $\mathfrak{g}$ be a Lie algebra with strong cone potential and $\Omega^+ \subseteq \Omega$ be a positive system. We define

$$C_{\min} := C_{\min}(\Omega^+) := \operatorname{cone}\{Q(x) : x \in \mathfrak{g}^\omega, \omega \in \Omega_P^+\},$$

$$C_{\min,z} := C_{\min,z}(\Omega^+) := C_{\min} \cap Z(\mathfrak{g}) = \operatorname{cone}\{Q(x) : x \in \mathfrak{g}^\omega, \omega \in \Omega_R^+\},$$

and

$$C_{\max} := C_{\max}(\Omega^+) := \{h \in \mathfrak{h} : \omega(h) \geq 0, \text{ for all } \omega \in \Omega_P^+\}.$$

■

The following proposition provides a geometric interpretation of the concept of strong cone potential.

Proposition III.13. *Let $\mathfrak{g}$ be a Lie algebra with cone potential. For a positive system $\Omega^+ \subseteq \Omega$ the following are equivalent:*

(1) *The cone $C_{\min,z} \subseteq Z(\mathfrak{g})$ is pointed.*

(2) *There exists $\alpha \in \widehat{Z}(\mathfrak{g})$ such that*

$$\alpha(Q(x)) > 0 \quad \textit{for all} \quad x \in \mathfrak{r}_{\text{eff}} \setminus \{0\},$$

i.e., L has strong cone potential with respect to Ω^+ and α.

Proof. We set $C := C_{\min,z}$.

(1) $\Rightarrow$ (2): Suppose that (1) holds and pick $\alpha \in \operatorname{algint} C^*$. For $x = \sum x_\omega \in \mathfrak{r}_{\text{eff}}$ this leads to

$$\sum \alpha(Q(x_\omega)) > 0$$

because $\alpha(Q(x_\omega)) > 0$ holds whenever $x_\omega \neq 0$. But $x_\omega \neq 0$ is equivalent to $Q(x_\omega) \neq 0$ since $\mathfrak{g}$ has cone potential.

(2) $\Rightarrow$ (1): Conversely, suppose that $\alpha(Q(x)) > 0$ for all $x \in \mathfrak{r}_{\text{eff}} \setminus \{0\}$, i.e., that $\mathfrak{r}_{\text{eff}}$ is a real Hilbert space with the scalar product defined by $\langle x, y \rangle := \alpha(Q(x,y))$. Let $x \in \mathfrak{g}^\omega \setminus \{0\}$ and $\omega \in \Omega_R^+$. Assume that $\mathfrak{r}_{\text{eff}} \neq \{0\}$. To prove that

$$C = C_{\min,z} = \overline{\sum_{\omega \in \Omega_R^+} \mathbb{R}^+ \operatorname{conv}(Q(\mathfrak{g}^\omega))}$$

is pointed we assume that $H(C) \neq \{0\}$. Then we find a sequence of convex combinations $\sum_{i=1}^n \lambda_i^m Q(x_i^m)$ with $x_i^m \in \mathfrak{g}^\omega$ such that

$$\lim_{m \to \infty} \alpha\Big(\sum_{i=1}^n \lambda_i^m Q(x_i^m)\Big) = 0$$

and

$$\lim_{m \to \infty} \sum_{i=1}^n \lambda_i^m Q(x_i^m) \in H(C) \setminus \{0\}.$$

In view of Caratheodory's theorem we only need n summands in convex combinations to get the convex hull if $\dim V = n-1$. Therefore it suffices to show that

$$\lim_{m\to\infty} \lambda_i^m Q(x_i^m) = 0 \qquad \text{for every } i \in \{1,\ldots,n\}.$$

This follows immediately from

$$0 \le \|\sqrt{\lambda_i^m} x_i^m\|^2 = \lambda_i^m \alpha\big(Q(x_i^m)\big) \le \alpha\big(\sum_{i=1}^n \lambda_i^m Q(x_i^m)\big) \to 0.$$

■

Definition III.14. Let $\mathfrak{g}$ be a Lie algebra with compactly embedded Cartan algebra. We say that a positive system $\Omega^+ \subseteq \Omega$ is *nice* if there exists an element $z \in \mathfrak{z}_{\mathfrak{k}}$ such that

$$\omega(z) > 0 \quad \text{for all} \quad \omega \in \Omega_P^+$$

and $C_{\min,z}$ is pointed. It follows from the preceding proposition that only Lie algebras which are quasihermitean (Proposition II.18) and which have strong cone potential admit nice positive systems. ■

The notion of strong cone potential is justified by the following observation and the Characterization Theorem of Lie Algebras with Invariant Cones (Theorem III.36 below).

Proposition III.15. *Let $\mathfrak{g}$ be a Lie algebra which contains a pointed generating invariant wedge W. Then there exists a compactly embedded Cartan algebra $\mathfrak{h} \subseteq \mathfrak{g}$ and a nice positive system Ω^+ of real roots with respect to $\mathfrak{h}$ such that*

$$C_{\min} \subseteq C := W \cap \mathfrak{h} \subseteq C_{\max}.$$

In particular, $\mathfrak{g}$ is quasihermitean and has strong cone potential.

Proof. We use [HHL89,III.2.14, III.6.18] to see that $\mathfrak{g}$ has cone potential. We choose $h \in \operatorname{int} C \setminus \bigcup_{\omega\in\Omega} \ker\omega$ and set

$$\Omega^+ := \{\omega \in \Omega : \omega(h) > 0\}.$$

Then we observe that $\omega(C) = \mathbb{R}^+$ because

$$(\operatorname{ad} x)^2(C) = \omega(C)Q(x) \subseteq C \quad \text{for} \quad x \in \mathfrak{g}^\omega, \omega \in \Omega_P^+$$

(Proposition III.7), $Q(x) \neq 0$ and C is pointed. Therefore $C \subseteq C_{\max}$. But $Q(x) \in \omega(C)Q(x) \subseteq C(C) \subseteq C$ for $\omega \in \Omega_P^+$ which implies that $C_{\min} \subseteq C$. Thus $C_{\min,z} \subseteq C_{\min}$ is pointed. Set

$$z := \frac{1}{\operatorname{card}\mathcal{W}} \sum_{w\in\mathcal{W}} w.h.$$

Then the invariance of C under $\mathcal{W}$ (Proposition III.7) implies that $z \in \operatorname{int} C$ and the invariance of z under $\mathcal{W}$ shows that $z \in \mathfrak{z}_{\mathfrak{k}}$ (Lemma III.8). For every $\omega \in \Omega_P^+$ this leads to $\omega(z) > 0$ since $z \in \operatorname{algint} C_{\max}$. ■

We give a criterion which makes it very easy to check whether a Lie algebra with strong cone potential is quasihermitean or not.

Proposition III.16. *A Lie algebra $\mathfrak{g}$ with strong cone potential is quasihermitean if and only if the ideal*

$$Z_{\mathfrak{s}}(\mathfrak{r}) = \{x \in \mathfrak{s} : [x, \mathfrak{r}] = \{0\}\}$$

is quasihermitean. In this case $\mathfrak{g}$ posseses a nice positive system and an ideal $\mathfrak{g}_0$ such that $\mathfrak{g} \cong \mathfrak{g}_0 \oplus Z_{\mathfrak{s}}(\mathfrak{r})$. The minimal and maximal cones are adjusted to this decomposition, i.e.,

$$C_{\min} = C_{\min} \cap \mathfrak{g}_0 + C_{\min} \cap Z_{\mathfrak{s}}(\mathfrak{r}) \quad and \quad C_{\max} = C_{\max} \cap \mathfrak{g}_0 + C_{\max} \cap Z_{\mathfrak{s}}(\mathfrak{r}).$$

Proof. Using Remark II.39 and Proposition III.11 we find that $\mathfrak{g}$ has a direct sum decomposition

$$\mathfrak{g} \cong \mathfrak{g}_0 \oplus Z_{\mathfrak{s}}(\mathfrak{r})$$

and $\mathfrak{g}_0 = \mathfrak{r}_{\text{eff}} \oplus Z(\mathfrak{g}) \oplus \mathfrak{g}_1$, where $\mathfrak{g}_1$ is a reductive Lie algebra which contains a pointed generating invariant cone W. If $\mathfrak{g}$ is quasihermitean, then, according to Proposition II.20, $Z_{\mathfrak{s}}(\mathfrak{r})$ is quasihermitean. Suppose that this is true. We choose $z_1 \in \mathfrak{z}_{\mathfrak{k}} \cap Z_{\mathfrak{s}}(\mathfrak{r})$ such that $\omega(z_1) \neq 0$ for every non-compact root with $\mathfrak{g}^{\omega} \subseteq Z_{\mathfrak{s}}(\mathfrak{r})$. In view of Propositions III.11 and III.15 we find a nice positive system in $\Omega_1 := \{\omega \in \Omega : \mathfrak{g}^{\omega} \subseteq \mathfrak{g}_1\}$ such that $C := W \cap \mathfrak{h} \subseteq C_{\max, \mathfrak{g}_1}$, where

$$W = \{x \in \mathfrak{g}_1 : -I \operatorname{ad} x\,|_{\mathfrak{r}_{\text{eff}}} \text{ is positive semidefinite}\}.$$

Let $z_0 \in \operatorname{int}_{\mathfrak{g}_1} W$ such that $\omega(z) > 0$ for all non-compact roots with $\mathfrak{g}^{\omega} \subseteq \mathfrak{g}_1$. Setting $z := z_0 + z_1$ we claim that no non-compact root ν of $\mathfrak{g}$ vanishes on z. We have to consider three cases. If $\mathfrak{g}^{\nu} \subseteq Z_{\mathfrak{s}}(\mathfrak{r})$, then $\nu(z) = \nu(z_1) \neq 0$, if $\mathfrak{g}^{\nu} \subseteq \mathfrak{g}_1$, then $\nu(z) = \nu(z_0) > 0$, and if $\mathfrak{g}^{\nu} \subseteq \mathfrak{r}$, we know from the definition of $W \subseteq \mathfrak{g}_1$ that $\nu(C) \subseteq \mathbb{R}^+$. But $z_0 \in \operatorname{int}_{\mathfrak{h} \cap \mathfrak{g}_1} C$ and therefore $\nu(z) = \nu(z_0) > 0$. This proves that $\mathfrak{g}$ is quasihermitean (Proposition II.18). Moreover, we find a regular element $h \in \mathfrak{h}$ near z such that $\omega(h)$ has the same sign as $\omega(z)$ for all non-compact roots $\omega \in \Omega_P^+$. Setting

$$\Omega^+ := \{\omega \in \Omega : \omega(h) > 0\}$$

it follows from the definition of W that this defines a nice positive system for $\mathfrak{g}$. That the cone $C_{\min}$ is adjusted to the direct decomposition of $\mathfrak{g}$ is clear, and that this also holds for $C_{\max}$ follows from the fact that every non-compact root vanishes either on $\mathfrak{h} \cap Z_{\mathfrak{s}}(\mathfrak{r})$ or on $\mathfrak{h} \cap \mathfrak{g}_0$. ∎

Lemma III.17. *Let $\mathfrak{g}$ be a simple hermitean Lie algebra, $\mathfrak{h} \subseteq \mathfrak{g}$ a compactly embedded Cartan algebra, and $\{0\} \neq \widetilde{\mathfrak{h}} \subseteq \mathfrak{h}$ a subspace with*

$$\mathcal{W}\widetilde{\mathfrak{h}} \subseteq \widetilde{\mathfrak{h}} \quad and \quad \mathcal{C}\widetilde{\mathfrak{h}} \subseteq \widetilde{\mathfrak{h}}.$$

Then $\widetilde{\mathfrak{h}} = \mathfrak{h}$.

Proof. Let $\mathfrak{k}_\mathfrak{h} = \mathfrak{z}_\mathfrak{k} \oplus \mathfrak{k}'_\mathfrak{h}$ and $\mathfrak{k}'_\mathfrak{h} = \bigoplus_{i=1}^k \mathfrak{k}_i$ be a decomposition into simple ideals. Then, according to Lemma III.8, we know that

$$\widetilde{\mathfrak{h}} = (\widetilde{\mathfrak{h}} \cap \mathfrak{z}_\mathfrak{k}) \oplus \bigoplus_{i=1}^{k} (\widetilde{\mathfrak{h}} \cap \mathfrak{h}_{\mathfrak{k}_i}). \tag{3.1}$$

Assume that $\widetilde{\mathfrak{h}} \neq \{0\}$. We claim that $\widetilde{\mathfrak{h}} \cap \mathfrak{z}_\mathfrak{k} = \mathfrak{z}_\mathfrak{k}$. Suppose that this is false. Then $\widetilde{\mathfrak{h}} \cap \mathfrak{z}_\mathfrak{k} = \{0\}$ because $\dim \mathfrak{z}_\mathfrak{k} = 1$ (Proposition II.19). Therefore $\widetilde{\mathfrak{h}} \subseteq \mathfrak{k}'_\mathfrak{h}$. We choose $z \in \mathfrak{z}_\mathfrak{k}$ such that $\omega(z) = 1$ holds for all $\omega \in \Omega_P^+$ ([Ne90, 2.4]) and denote the Cartan Killing form of $\mathfrak{g}$ with B. For every $\omega \in \Omega_P^+$ and $x \in \mathfrak{g}^\omega \setminus \{0\}$ we conclude with [Ne90, 2.5] that

$$B\big(Q(x_\omega), z\big) = -\omega(z)B(x_\omega) = -B(x_\omega) < 0.$$

Therefore $Q(x_\omega) \not\in \mathfrak{z}_\mathfrak{k}^\perp = \mathfrak{k}'_\mathfrak{h} \oplus \mathfrak{p}_\mathfrak{h}$. It follows in particular, that $Q(x_\omega) \not\in \widetilde{\mathfrak{h}}$. Now Lemma III.3 implies that $\omega(\widetilde{\mathfrak{h}}) = \{0\}$ for all $\omega \in \Omega_P^+$. Choose $i \in \{1, \ldots, k\}$ with $\mathfrak{h}_{\mathfrak{k}_i} \subseteq \widetilde{\mathfrak{h}}$. Then $\omega(\mathfrak{h}_{\mathfrak{k}_i}) = \{0\}$ for all $\omega \in \Omega_P^+$. Consequently

$$\Omega^+ = \big(\Omega_P^+ \cup \{\omega \in \Omega_K^+ : \mathfrak{g}^\omega \not\subseteq \mathfrak{k}_i\}\big) \cup \{\omega \in \Omega_K^+ : \mathfrak{g}^\omega \subseteq \mathfrak{k}_i\}$$

is a decomposition of Ω^+ into two orthogonal subsets of roots. This contradicts the simplicity of $\mathfrak{g}$ ([Hum72, p.73]). This contradiction proves our claim that $\mathfrak{z}_\mathfrak{k} \subseteq \widetilde{\mathfrak{h}}$. Then $\omega(\widetilde{\mathfrak{h}}) \supseteq \omega(\mathfrak{z}_\mathfrak{k}) = \mathbb{R}$ for every $\omega \in \Omega_P^+$ since $\omega(z) = 1$. The invariance under C, together with Lemma III.3, leads to $Q(x) \in \widetilde{\mathfrak{h}}$ for all $\omega \in \Omega_P^+$. The above argument applies again and shows that

$$\mathfrak{h}_{\mathfrak{k}_i} \not\subseteq \{Q(x) : x \in \mathfrak{g}^\omega, \omega \in \Omega_P^+\}^\perp \quad \text{for all } i = 1, \ldots, k.$$

Thus $\mathfrak{h}_{\mathfrak{k}_i} \subseteq \widetilde{\mathfrak{h}}$ holds for all i since the decomposition (3.1) is orthogonal with respect to B. Now we have that

$$\mathfrak{z}_\mathfrak{k} \oplus \bigoplus_{i=1}^{k} \mathfrak{h}_{\mathfrak{k}_i} = \mathfrak{h} \subseteq \widetilde{\mathfrak{h}}.$$

■

Lemma III.18. *Let $\mathfrak{g}$ be a quasihermitean Lie algebra with cone potential and $\widetilde{\mathfrak{h}} \subseteq \mathfrak{h}$ a subspace with*

$$\mathcal{W}\widetilde{\mathfrak{h}} \subseteq \widetilde{\mathfrak{h}} \quad \textit{and} \quad \mathcal{C}\widetilde{\mathfrak{h}} \subseteq \widetilde{\mathfrak{h}}.$$

Further, let $\mathfrak{s} = \bigoplus_{i=1}^n \mathfrak{s}_i$ be a decomposition of the $\mathfrak{h}$-invariant Levi algebra $\mathfrak{s}$ into simple ideals and $\mathfrak{h}_i := \mathfrak{h} \cap \mathfrak{s}_i$. Then there exists a subset $P \subseteq \{1, \ldots, n\}$ such that

$$\widetilde{\mathfrak{h}} = (\widetilde{\mathfrak{h}} \cap \mathfrak{r}) \oplus \bigoplus_{i \in P} \mathfrak{h}_i.$$

Proof. Using Lemma III.8 we see that

$$\widetilde{\mathfrak{h}} = (\widetilde{\mathfrak{h}} \cap \mathfrak{z}_{\mathfrak{k}}) \oplus \bigoplus_{j=1}^{l} (\mathfrak{h}_{\mathfrak{k}_j} \cap \widetilde{\mathfrak{h}}),$$

where $\mathfrak{k}'_{\mathfrak{h}} = \bigoplus_{j=1}^{l} \mathfrak{k}_j$ is a decomposition into simple ideals and $\mathfrak{h}_{\mathfrak{k}_j} \cap \widetilde{\mathfrak{h}} = \mathfrak{h}_{\mathfrak{k}_j}$ or $\{0\}$. If $\mathfrak{k}_j = \mathfrak{s}_i \subseteq \mathfrak{s}$ is a compact ideal of $\mathfrak{s}$, then $\mathfrak{h}_{\mathfrak{k}_j} = \mathfrak{h}_i \subseteq \widetilde{\mathfrak{h}}$ whenever $\widetilde{\mathfrak{h}} \cap \mathfrak{h}_{\mathfrak{k}_j} \neq \{0\}$. If $\mathfrak{k}_j \subseteq \mathfrak{s}_i$ and $\mathfrak{s}_i$ is a simple hermitean ideal, then it follows with Lemma III.17 that $\mathfrak{h}_i \subseteq \widetilde{\mathfrak{h}}$ if $\widetilde{\mathfrak{h}} \cap \mathfrak{h}_i \neq \{0\}$. Now we set $P := \{i \in \{1, \ldots, n\} : \mathfrak{h}_i \subseteq \widetilde{\mathfrak{h}}\}$ and choose $\mathfrak{h}_0 \subseteq \widetilde{\mathfrak{h}} \cap \mathfrak{z}_{\mathfrak{k}}$ such that

$$\widetilde{\mathfrak{h}} = \mathfrak{h}_0 \oplus \bigoplus_{i \in P} \mathfrak{h}_i$$

and $\mathfrak{h}_0 \subseteq (\mathfrak{h} \cap \mathfrak{r}) \oplus \bigoplus_{i \notin P} \mathfrak{h}_i$. We have to show that $\mathfrak{h}_0 \subseteq \mathfrak{h} \cap \mathfrak{r} = \bigcap_{\omega \in \Omega_S^+} \ker \omega$ (Corollary II.15). If this is false, we find $\omega \in \Omega_S^+$ with $\omega(\mathfrak{h}_0) \neq \{0\}$. Then $\omega \in \Omega_P^+$ since $\mathfrak{h}_0 \subseteq \mathfrak{z}_{\mathfrak{k}}$. Now we pick $x \in \mathfrak{g}^\omega \setminus \{0\}$. Then

$$Q(x) \in \omega(\mathfrak{h}_0) Q(x) \subseteq C\mathfrak{h}_0 \cap \mathfrak{s}_i \subseteq \widetilde{\mathfrak{h}} \cap \mathfrak{s}_i = \mathfrak{h}_i,$$

where $\mathfrak{g}^\omega \subseteq \mathfrak{s}_i$. Then $\mathfrak{h}_i \subseteq \widetilde{\mathfrak{h}}$ and consequently $i \in P$. Thus

$$[\mathfrak{h}_0, \mathfrak{g}^\omega] \subseteq [\mathfrak{h} \cap \mathfrak{r}, \mathfrak{g}^\omega] + [\sum_{j \notin P} \mathfrak{s}_j, \mathfrak{s}_i] = \{0\}$$

contradicts $\omega(\mathfrak{h}_0) \neq \{0\}$ and therefore implies that $\mathfrak{h}_0 \subseteq \mathfrak{h} \cap \mathfrak{r}$. ∎

Lemma III.19. *Suppose that $\Omega^+ \subseteq \Omega$ is a positive system such that there exists an element $z \in \mathfrak{z}_{\mathfrak{k}}$ with*

$$\omega(z) > 0 \quad \textit{for all} \quad \omega \in \Omega_P^+.$$

Then the set Ω_P^+ of non-compact positive roots is invariant under the Weyl group $\mathcal{W}$.

Proof. Let $w = [\alpha] \in \mathcal{W}$ with $\alpha \in N(\mathfrak{h}) \subseteq \mathrm{INN}_{\mathfrak{g}}\, \mathfrak{k}_{\mathfrak{h}}$ and $\omega \in \Omega_P^+$. Then

$$(w * \omega)(z) = \omega \circ \alpha^{-1}(z) = \omega(z)$$

since $\mathrm{INN}_{\mathfrak{g}}\, \mathfrak{k}_{\mathfrak{h}}$ fixes all points of $\mathfrak{z}_{\mathfrak{k}}$. Hence $w * \omega \in \Omega_P^+$. ∎

Now we have the tools to derive further information about the minimal and maximal cones $C_{\min}$ and $C_{\max}$ in a Lie algebra with strong cone potential.

Proposition III.20. *Suppose that $\mathfrak{g}$ is quasihermitean and has strong cone potential. Then there exists a nice positive system $\Omega^+ \subseteq \Omega$ and the following assertions hold:*

(1) $C_{\min}$ *is pointed.*

(2) $C_{\max}$ *is generating.*

(3) $C_{\min} \subseteq C_{\max}$.

(4) $C_{\min}$ *and* $C_{\max}$ *are invariant under the Weyl group* $\mathcal{W}$.

(5) $C_{\min}$ *and* $C_{\max}$ *are invariant under* $\mathcal{C}$.

(6) $H(C_{\max}) = Z(\mathfrak{g}) \oplus \bigoplus_{i=1}^m \mathfrak{h}_i$, *where* $\mathfrak{s}_1, \ldots, \mathfrak{s}_m$ *are the compact simple ideals of* $\mathfrak{s}$ *with* $[\mathfrak{s}_i, \mathfrak{r}] = \{0\}$ *and* $\mathfrak{h}_i = \mathfrak{h} \cap \mathfrak{s}_i$.

(7) $C_{\min} \cap \mathfrak{r} = C_{\min,z} \subseteq Z(\mathfrak{g})$.

(8) *If* $\mathfrak{g}$ *is a simple hermitean Lie algebra, the cones* $C_{\min}$ *and* $C_{\max}$ *are pointed and generating. Moreover,*

$$C_{\min} = \sum_{w \in \mathcal{W}} \mathbb{R}^+ w\big(Q(x)\big) \quad \textit{for all} \quad x \in \mathfrak{g}^\omega,$$

where $\omega \in \Omega_p^+$ *is a non-compact root which is not the arithmetical mean of two other non-compact roots.*

Proof. Proposition III.16 implies the existence of a nice positive system for $\mathfrak{g}$ and an element $z \in \mathfrak{z}_\mathfrak{k}$ with $\omega(z) > 0$ for all $\omega \in \Omega_p^+$.+

(1) First we note that

$$C_{\min} = C_{\min,z} \oplus \bigoplus_{i=1}^n (C_{\min} \cap \mathfrak{s}_i),$$

where $\mathfrak{s} = \bigoplus_{i=1}^n \mathfrak{s}_i$ is the decomposition into simple ideals. According to the definition of a nice positive system and Proposition III.13 the cone $C_{\min,z}$ is pointed. If $\mathfrak{s}_i$ is compact, then the cone $C_{\min}$ intersects $\mathfrak{s}_i$ only in $\{0\}$. Therefore we may assume that $\mathfrak{s}_i$ is hermitean simple and Ω^+ is chosen such that $\omega(z) > 0$ for a non-zero element of the one-dimensional subspace $\mathfrak{z}_\mathfrak{k}$. Now we use [Ne90, 2.5] to see that $-B\big(Q(x), z\big) = \omega(z)B(x,x) > 0$ for all $x \in \mathfrak{g}^\omega \subseteq \mathfrak{p}_\mathfrak{h}$, where B is the Cartan Killing form of $\mathfrak{s}_i$. So the functional $h \mapsto -B(h,z)$ is strictly positive on

$$C_{\min} = \sum_{\omega \in \Omega_p^+} Q(\mathfrak{g}^\omega)$$

and consequently $C_{\min}$ is pointed.

(2) That $C_{\max}$ is generating follows from the fact that it contains every element $h \in \mathfrak{h}$ defining the positive system Ω^+ by

$$\Omega^+ = \{\omega \in \Omega : \omega(h) > 0\}$$

in its interior.

(3) According to Proposition III.16 we have to prove that $C_{\min} \subseteq C_{\max}$ in the case where $Z_{\mathfrak{s}}(\mathfrak{r}) = \{0\}$ and in the case where $\mathfrak{g}$ is semisimple. If $\mathfrak{g}$ is semisimple, this follows from Corollary II.9 in [Ne90]. So we assume that $Z_{\mathfrak{s}}(\mathfrak{r}) = \{0\}$. It is clear that $C_{\min,\mathfrak{z}} \subseteq Z(\mathfrak{g}) \subseteq C_{\max}$. We find an element $z \in \mathfrak{z}_{\mathfrak{k}}$ such that $\omega(z) > 0$ for all $\omega \in \Omega_P^+$. Let $\mathfrak{h}_0 \subseteq \mathfrak{h} \cap \mathfrak{r}$ be a vector subspace complementary to $Z(\mathfrak{g})$ and $W \subseteq \mathfrak{h}_0 + \mathfrak{s}$ the pointed generating invariant cone defined in Proposition III.11. Then there exists $z_1 \in \big(z + Z(\mathfrak{g})\big) \cap \big(\mathfrak{h}_0 + (\mathfrak{h} \cap \mathfrak{s})\big)$. This element has the property that $\omega(z_1) > 0$ for all $\omega \in \Omega_P^+$. Therefore $z_1 \in W \cap \mathfrak{h}$. Let $\omega \in \Omega_P^+ \cap \Omega_S^+$. Then $\mathfrak{g}^\omega \subseteq \mathfrak{h}_0 + \mathfrak{s}$ and consequently

$$Q(\mathfrak{g}^\omega) \subseteq \omega(\mathbb{R}^+ z_1) Q(\mathfrak{g}^\omega) \subseteq \omega(W \cap \mathfrak{h}) Q(\mathfrak{g}^\omega) \subseteq W \cap \mathfrak{h}$$

(Proposition III.7). This proves that $\nu\big(Q(\mathfrak{g}^\omega)\big) \subseteq \mathbb{R}^+$ for every $\nu \in \Omega_R^+$. For $\nu \in \Omega_P^+ \cap \Omega_S^+$ the fact that $W \cap \mathfrak{h}$ is pointed (Proposition III.11) and the relations

$$\nu(W \cap \mathfrak{h}) Q(\mathfrak{g}^\nu) \subseteq W \cap \mathfrak{h}, \qquad Q(\mathfrak{g}^\nu) \neq \{0\}$$

show that $\nu(W \cap \mathfrak{h}) = \mathbb{R}^+$ because $z_1 \in W \cap \mathfrak{h}$. Thus

$$\nu\big(Q(\mathfrak{g}^\omega)\big) \subseteq \nu(W \cap \mathfrak{h}) = \mathbb{R}^+.$$

Putting these facts together we have proved that

$$C_{\min} \subseteq C_{\max}.$$

(4) Let $\alpha \in N(\mathfrak{h}) \subseteq \operatorname{INN}_{\mathfrak{g}} \mathfrak{k}_{\mathfrak{h}}$ and $x \in \mathfrak{g}^\omega$. Then, according to Proposition III.9,

$$\alpha\big(Q(x)\big) = \alpha([Ix, x]) = [\alpha(Ix), \alpha(x)] = [I\alpha(x), \alpha(x)] = Q\big(\alpha(x)\big)$$

with $\alpha(x) \in \mathfrak{g}^{\omega \circ \alpha^{-1}}$ and $\omega \circ \alpha^{-1} \in \Omega_P^+$ (Lemma III.19). This proves the invariance of $C_{\min}$ under $\mathcal{W}$. Let $h \in C_{\max}$, $\omega \in \Omega_P^+$ and $\alpha \in N(\mathfrak{h})$. Then

$$\langle \omega, \alpha(h) \rangle = \langle \widehat{\alpha}(\omega), h \rangle \geq 0$$

because $\widehat{\alpha}(\omega) \in \Omega_P^+$ (Lemma III.19). Hence $\mathcal{W} C_{\max} \subseteq C_{\max}$.
(5) Let $\omega \in \Omega_P^+$ and $x \in \mathfrak{g}^\omega \setminus \{0\}$. Then, in view of Lemma III.3,

$$(\operatorname{ad} x)^2 C_{\max} = \omega(C_{\max}) Q(x) \subseteq \mathbb{R}^+ Q(x) \subseteq C_{\min}.$$

This implies, in view of (3) above, that $\mathcal{C} C_{\max} \subseteq C_{\min} \subseteq C_{\max}$ and therefore the invariance of both cones under $\mathcal{C}$.
(6) According to Lemma III.18 we find $P \subseteq \{1, \ldots, n\}$ such that

$$H(C_{\max}) = \big(H(C_{\max}) \cap \mathfrak{r}\big) \oplus \bigoplus_{i \in P} \mathfrak{h}_i$$

because $C_{\max}$ and therefore $H(C_{\max})$ is invariant under $\mathcal{W}$ and $\mathcal{C}$ (cf. (4),(5)). From $[\mathfrak{h} \cap \mathfrak{r}, \mathfrak{s}] = \{0\}$ we conclude further that

$$H(C_{\max}) \cap \mathfrak{r} = \{h \in \mathfrak{h} \cap \mathfrak{r} : \omega(h) = 0 \text{ for all } \omega \in \Omega_P^+\} = Z(\mathfrak{g}).$$

If $\mathfrak{s}_i \subseteq \mathfrak{s}$ is a hermitean ideal, then $\mathfrak{s}_i \cap H(C_{\max}) = \{0\}$ and if $\mathfrak{s}_i \subseteq \mathfrak{s}$ is a compact simple ideal, there are two cases:
Case 1: $[\mathfrak{s}_i, \mathfrak{r}] = \{0\}$. Then $[\mathfrak{s}_i, \mathfrak{p}_{\mathfrak{h}}] = \{0\}$ and $\omega(\mathfrak{h} \cap \mathfrak{s}_i) = \{0\}$ for all $\omega \in \Omega_P^+$, hence $\mathfrak{h}_i \subseteq C_{\max}$ and $i \in P$.
Case 2: $[\mathfrak{s}_i, \mathfrak{r}] \neq \{0\}$. Then there exists $h \in \mathfrak{h} \cap \mathfrak{s}_i$ and $\omega \in \Omega_R^+ \subseteq \Omega_P^+$ with $\omega(h) \neq 0$. Whence $i \notin P$.
(7) This follows from the definition of $C_{\min,\mathfrak{z}}$ and the fact that $Q(\mathfrak{g}^\omega) \subseteq Z(\mathfrak{g})$ for $\omega \in \Omega_R^+$ and $Q(\mathfrak{g}^\omega) \subseteq \mathfrak{s}$ for $\omega \in \Omega_S^+$.
(8) The representation of $C_{\min}$ is a consequence of [Ne90, 2.15, 2.16]. Point (6) above implies that $C_{\max}$ and therefore $C_{\min} \subseteq C_{\max}$ is pointed. The fact that

$$C_{\max} = C_{\min}^\star = \{h \in \mathfrak{h} : -B(h, C_{\min}) \subseteq \mathbb{R}^+\}$$

([Ne90, 2.13]) shows that $C_{\min}$ is generating as a dual of a pointed cone. ∎

Lemma III.21. *Let $\mathfrak{g}$ be a Lie algebra with cone potential. Then*

(1) $\mathrm{Inn}_{\mathfrak{g}}\, \mathfrak{g} = e^{\mathrm{ad}\, \mathfrak{r}_{\mathrm{eff}}} \mathrm{Inn}_{\mathfrak{g}}\, \mathfrak{g}_1$, *where*

$$\mathfrak{g}_1 = \mathfrak{h}_1 \oplus \mathfrak{s}, \quad \mathfrak{h} \cap \mathfrak{r} = Z(\mathfrak{g}) \oplus \mathfrak{h}_1$$

and $\mathrm{ad}\, \mathfrak{r}_{\mathrm{eff}}$ *is an abelian ideal in* $\mathbf{L}(\mathrm{Inn}_{\mathfrak{g}}\, \mathfrak{g}) = \mathrm{ad}\, \mathfrak{g}$.

(2) *For* $m \in \mathfrak{r}_{\mathrm{eff}}$ *and* $h \in \mathfrak{h}$ *we have*

$$e^{\mathrm{ad}\, m} h = h + [m,h] + \frac{1}{2}[m,[m,h]] \quad \text{and} \quad [m,[m,h]] \in \mathcal{C}(h).$$

(3) $p(g.h) \in p(g_1.h) + \mathcal{C}(h)$ *for* $g = e^{\mathrm{ad}\, m} g_1 \in \mathrm{Inn}_{\mathfrak{g}}\, \mathfrak{g}$, $g_1 \in \mathrm{Inn}_{\mathfrak{g}}\, \mathfrak{g}_1$, *and* $h \in \mathfrak{h}$.

Proof. (cf. [HHL89, III.6.40]) (1) Clearly $\mathfrak{g} = \mathfrak{r}_{\mathrm{eff}} + \mathfrak{g}_1$, $[\mathfrak{g}_1, \mathfrak{r}_{\mathrm{eff}}] \subseteq \mathfrak{r}_{\mathrm{eff}}$ and $[\mathfrak{r}_{\mathrm{eff}}, \mathfrak{r}_{\mathrm{eff}}] \subseteq Z(\mathfrak{g})$ (Propositions II.10, II.11). Thus

$$[\mathrm{ad}\, \mathfrak{r}_{\mathrm{eff}}, \mathrm{ad}\, \mathfrak{g}] = \mathrm{ad}[\mathfrak{r}_{\mathrm{eff}}, \mathfrak{g}] \subseteq \mathrm{ad}\, \mathfrak{r}_{\mathrm{eff}} \quad \text{and} \quad [\mathrm{ad}\, \mathfrak{r}_{\mathrm{eff}}, \mathrm{ad}\, \mathfrak{r}_{\mathrm{eff}}] = \mathrm{ad}[\mathfrak{r}_{\mathrm{eff}}, \mathfrak{r}_{\mathrm{eff}}] = \{0\}.$$

(2) Firstly $[m,[m,h]] \in [\mathfrak{r}_{\mathrm{eff}}, \mathfrak{r}_{\mathrm{eff}}] \subseteq Z(\mathfrak{g})$ implies that $(\mathrm{ad}\, m)^3(h) = 0$ and therefore

$$e^{\mathrm{ad}\, m} h = h + [m,h] + \frac{1}{2}[m,[m,h]].$$

We write $m = \sum_{\omega \in \Omega_R^+} m_\omega$. Then

$$\begin{aligned} [m,[m,h]] &= -\sum_{\omega \in \Omega_R^+} \omega(h)[m, I m_\omega] \\ &= -\sum_{\omega \in \Omega_R^+} \omega(h)[m_\omega, I m_\omega] = \sum_{\omega \in \Omega_R^+} (\mathrm{ad}\, m_\omega)^2(h) \in \mathcal{C}(h). \end{aligned}$$

(3) Let $m' := g_1^{-1}m$. Then $[m, g_1.h] = g_1.[m', h] \in g_1.\mathfrak{r}_{\text{eff}} \subseteq \mathfrak{r}_{\text{eff}}$. Thus $p([m, g_1.h]) = 0$ for all $m \in \mathfrak{r}_{\text{eff}}$, $g_1 \in \text{Inn}_{\mathfrak{g}}\, \mathfrak{g}_1$ and $h \in \mathfrak{g}$. Further

$$\big[m, [m, g_1.h]\big] = g_1.\big[m', [m', h]\big] \quad \text{and} \quad \big[m', [m', h]\big] \in Z(\mathfrak{g}).$$

Therefore $\big[m, [m, g_1.h]\big] = \big[m', [m', h]\big] \in \mathcal{C}(h)$ and

$$p(g.h) = p(e^{\text{ad}\, m} g_1.h) = p(g_1.h + [m, g_1.h] + \frac{1}{2}\big[m, [m, g_1.h]\big]\big) \in p(g_1.h) + \mathcal{C}(h).$$

■

Lemma III.22. *With the notation of* Lemma III.21 *the following are equivalent for a $\mathcal{C}$-invariant wedge $C \subseteq \mathfrak{h}$:*

(1) $p\big(\text{Inn}_{\mathfrak{g}}\, \mathfrak{g}(C)\big) \subseteq C$, *and*

(2) $p\big(\text{Inn}_{\mathfrak{g}}\, \mathfrak{g}_1(C)\big) \subseteq C$.

Proof. It is clear that (1) implies (2). If (2) holds, then (1) follows from $\mathcal{C}(C) \subseteq C$ and Lemma III.21(3). ■

Theorem III.23. (Reconstruction of Ideals) *Let $\mathfrak{g}$ be a quasihermitean Lie algebra with cone potential and $\widetilde{\mathfrak{h}} \subseteq \mathfrak{h}$ be a subspace. Then there exists an ideal $J \subseteq \mathfrak{g}$ with $J \cap \mathfrak{h} = \widetilde{\mathfrak{h}}$ iff*

$$\mathcal{W}\widetilde{\mathfrak{h}} \subseteq \widetilde{\mathfrak{h}} \quad \textit{and} \quad \mathcal{C}\widetilde{\mathfrak{h}} \subseteq \widetilde{\mathfrak{h}}.$$

Conversely, let $J \subseteq \mathfrak{g}$ be an ideal and set $\widetilde{\mathfrak{h}} := J \cap \mathfrak{h}$ and $\Omega_1 := \big\{\omega \in \Omega^+ : \omega(\widetilde{\mathfrak{h}}) \neq \{0\}\big\}$. Then the subalgebra

$$\mathfrak{h}_J := \widetilde{\mathfrak{h}} \oplus \bigoplus_{\omega \notin \Omega_1} \mathfrak{g}^{\omega} \cap J$$

is a Cartan algebra of J and $\widetilde{\mathfrak{h}} \neq \{0\}$ if $J \neq \{0\}$.

Proof. In view of Proposition III.7, we only have to show that the above conditions are sufficient. According to Lemma III.18 the subspace $\widetilde{\mathfrak{h}}$ decomposes as

$$\widetilde{\mathfrak{h}} = (\widetilde{\mathfrak{h}} \cap \mathfrak{r}) \oplus \bigoplus_{i \in P} \mathfrak{h}_i,$$

where $P = \big\{i \in \{1, \ldots, n\} : \mathfrak{s}_i \cap \widetilde{\mathfrak{h}} \neq \{0\}\big\}$. We set

$$J_1 := (\widetilde{\mathfrak{h}} \cap \mathfrak{r}) \oplus \bigoplus_{i \in P} \mathfrak{s}_i.$$

Then $J_1 \subseteq \mathfrak{g}_1$ is an ideal with $J_1 \cap \mathfrak{h} = p(J_1) = \widetilde{\mathfrak{h}}$. Now the preceding lemma implies that

$$p\big(\text{Inn}_{\mathfrak{g}}\, \mathfrak{g}(\widetilde{\mathfrak{h}})\big) \subseteq \widetilde{\mathfrak{h}}$$

and hence that $p(J) = \widetilde{\mathfrak{h}}$, where $J := \operatorname{span} \operatorname{Inn}_{\mathfrak{g}} \mathfrak{g}(\widetilde{\mathfrak{h}})$ is the smallest ideal of $\mathfrak{g}$ containing $\widetilde{\mathfrak{h}}$.

Now we prove the last statement. First we note that, in view of the first part of the proof, $\mathfrak{g}^\omega \cap J \subseteq \mathfrak{r}_{\text{eff}}$ holds for $\omega \not\in \Omega_1$. Hence

$$[\mathfrak{h}_J, \mathfrak{h}_J] \subseteq [J \cap \mathfrak{r}_{\text{eff}}, J \cap \mathfrak{r}_{\text{eff}}] \subseteq J \cap Z(\mathfrak{g}) \subseteq \widetilde{\mathfrak{h}} \cap Z(\mathfrak{g})$$

and therefore $\mathfrak{h}_J$ is a nilpotent subalgebra of J which is self-normalizing since

$$N(\mathfrak{h}_J) \cap \bigoplus_{\omega \in \Omega_1} \mathfrak{g}^\omega \subseteq N(\widetilde{\mathfrak{h}}) \cap \bigoplus_{\omega \in \Omega_1} \mathfrak{g}^\omega = \{0\}.$$

Thus $\mathfrak{h}_J \subseteq J$ is a Cartan algebra. To see that $\widetilde{\mathfrak{h}} \neq \{0\}$, we notice that

$$J = (\mathfrak{h} \cap J) \oplus \bigoplus_{\omega \in \Omega+} (\mathfrak{g}^\omega \cap J)$$

and that $\widetilde{\mathfrak{h}} = \{0\}$ would imply that there exists an $\omega \in \Omega^+$ and $0 \neq x \in J \cap \mathfrak{g}^\omega$. Then $0 \neq Q(x) \in J \cap \mathfrak{h}$, a contradiction. ■

Remark III.24. It follows from the Proof of the preceding proposition that

$$J = J_1 + [J_1, \mathfrak{r}_{\text{eff}}]$$

because

$$g.h = e^{\operatorname{ad} m} g_1.h \in g_1.h + [m, g_1.h] + C(h) \subseteq J_1 + [\mathfrak{r}_{\text{eff}}, J_1]$$

for $g = e^{\operatorname{ad} m} g_1 \in \operatorname{Inn}_{\mathfrak{g}} \mathfrak{g}$ and $h \in \widetilde{\mathfrak{h}}$. If $\mathfrak{g}$ is solvable, this implies that

$$J = \widetilde{\mathfrak{h}} + [\widetilde{\mathfrak{h}}, \mathfrak{g}_{\text{eff}}] \quad \text{and} \quad H_J = \widetilde{H}.$$

■

Having the above characterization of the intersections of ideals with compactly embedded Cartan algebras we need an additional characterization of the intersections with $\mathfrak{h}$ of those ideals which have cone potential and which are quasihermitean, to handle non-generating pointed cones.

Lemma III.25. *Let $\omega \in \Omega_R^+$. Then $Q|_{\mathfrak{r}^\omega \times \mathfrak{r}^\omega}$ is a symmetric bilinear form and*

$$[\mathfrak{r}^\omega, \mathfrak{r}^\omega] = \operatorname{span} Q(\mathfrak{r}^\omega).$$

Proof. The symmetry of Q follows from [HHL89, III.6.7]. Now we have that

$$Q(x + y) = Q(x) + Q(y) + 2Q(x, y)$$

and therefore

$$[x, y] = -Q(Ix, y) = -\frac{1}{2}\big(Q(Ix + y) - Q(Ix) - Q(y)\big) \in \operatorname{span} Q(\mathfrak{r}^\omega).$$

■

Proposition III.26. *Let $\mathfrak{g}$ be a Lie algebra with cone potential, $J \subseteq \mathfrak{g}$ an ideal, $\widetilde{\mathfrak{h}} = J \cap \mathfrak{h}$ and suppose that J is minimal with this property. Set $\Omega_1 := \{\omega \in \Omega : \omega|_{\widetilde{\mathfrak{h}}} \neq 0\}$. Then the following are equivalent:*

(1) *J has a compactly embedded Cartan algebra*

(2)

$$(CA) \qquad \omega|_{\widetilde{\mathfrak{h}}} \neq \pm\nu|_{\widetilde{\mathfrak{h}}} \quad \text{for} \quad \omega \in \Omega_1 \cap \Omega_{\mathfrak{r}_{\text{eff}}} \quad \text{and} \quad \nu \in \Omega_1 \cap \Omega_S^+.$$

If this condition is satisfied, then

$$J = \widetilde{\mathfrak{h}} \oplus \bigoplus_{\omega \in \Omega_1} \mathfrak{g}^\omega, \qquad \mathfrak{k}_{\mathfrak{h},J} = \mathfrak{k}_{\mathfrak{h}} \cap J, \qquad \mathfrak{z}_{\mathfrak{k},J} = \mathfrak{z}_{\mathfrak{k}} \cap \widetilde{\mathfrak{h}}$$

and J is quasihermitean iff

$$(QH) \qquad \omega(\mathfrak{z}_{\mathfrak{k}} \cap \widetilde{\mathfrak{h}}) \neq \{0\} \quad \text{for} \quad \omega \in \Omega_1 \cap \Omega_P^+.$$

Proof. "$\Rightarrow$": Suppose that J has a compactly embedded Cartan algebra $\mathfrak{h}_0$ which has to be abelian. Then the subalgebra $\mathfrak{h}_0 \otimes \mathbb{C} \subseteq J \otimes \mathbb{C}$ is an abelian Cartan algebra ([Bou75, p.19]) and all Cartan algebras in the complex Lie algebra $J \otimes \mathbb{C}$ are conjugate ([Bou75, p.29]). Thus $\mathfrak{h}_J$ (cf. Theorem III.23) must be an abelian Cartan algebra of J. This implies that

$$J = \widetilde{\mathfrak{h}} \oplus \bigoplus_{\omega \in \Omega_1} \mathfrak{g}^\omega$$

and that $\mathfrak{h}_J = \widetilde{\mathfrak{h}}$ is a compactly embedded Cartan algebra of J. Assume that (CA) does not hold and that there exist $\omega \in \Omega_1 \cap \Omega_R^+$ and $\nu \in \Omega_1 \cap \Omega_S^+$ such that $\omega|_{\widetilde{\mathfrak{h}}} = \nu|_{\widetilde{\mathfrak{h}}}$. Choose $x_\nu \in \mathfrak{g}^\nu \subseteq \mathfrak{s}$ with $|\nu(Q(x_\nu))| = 2$. Now Corollary II.15 implies that $\omega \pm \nu \neq 0$ and therefore that

$$[\mathfrak{g}^\nu, \mathfrak{r}^\omega] \subseteq (\mathfrak{r}^{\omega+\nu} + \mathfrak{r}^{\omega-\nu}) \cap J = \mathfrak{r}^{\omega+\nu} \subseteq \mathfrak{r}_{\text{eff}}$$

because $\omega + \nu \in \Omega_1$ and $\omega - \nu \notin \Omega_1$. Set

$$M := \bigoplus_{k \in \mathbb{N}_0} \mathfrak{r}^{\omega + k\nu} \subseteq J.$$

Then $[\mathfrak{g}^\nu, M] \subseteq M$, since

$$[\mathfrak{g}^\nu, \mathfrak{r}^{\omega+k\nu}] \subseteq (\mathfrak{r}^{\omega+(k+1)\nu} + \mathfrak{r}^{\omega+(k-1)\nu}) \cap J \subseteq M$$

and $k = 0$ implies that $\mathfrak{r}^{\omega+(k-1)\nu} = \mathfrak{r}^{\omega-\nu} \not\subseteq J$. Consequently M is a module of the simple three dimensional Lie algebra $\langle x_\nu \rangle = \mathbb{R}Q(x_\nu) + \mathfrak{g}^\nu \subseteq \mathfrak{s}$ and

$$(\omega + k\nu)(Q(x_\nu)) \in 2\mathbb{Z} \setminus \{0\}.$$

Thus $Z_M(Q(x_\nu)) = \{0\}$ and $\operatorname{Spec}(\operatorname{ad} Q(x_\nu)|_M) \subseteq 2i\mathbb{Z} \setminus \{0\}$. This contradicts Lemma II.12. In a similar way we prove that $\omega|_{\widetilde{\mathfrak{h}}} = -\nu|_{\widetilde{\mathfrak{h}}}$ is impossible. In this case on has to set

$$M := \bigoplus_{k \in \mathbb{N}_0} \mathfrak{r}^{-\omega+k\nu} \subseteq J.$$

"$\Leftarrow$": Assume that (CA) holds, i.e., $(\omega \pm \nu)(\widetilde{\mathfrak{h}}) \neq \{0\}$ for $\omega \in \Omega_1 \cap \Omega_R^+, \nu \in \Omega_1 \cap \Omega_S^+$. We set

$$\widetilde{J} := \widetilde{\mathfrak{h}} \oplus \bigoplus_{\omega \in \Omega_1} \mathfrak{g}^\omega$$

and claim that this is an ideal of $\mathfrak{g}$ which has cone potential, hence $J = \widetilde{J}$ has cone potential since J was minimal with the property that $J \cap \mathfrak{h} = \widetilde{\mathfrak{h}}$. Firstly $[\mathfrak{h}, \widetilde{J}] \subseteq \widetilde{J}$ is clear. Let $\nu \in \Omega^+ \setminus \Omega_1$. Then

$$[\mathfrak{g}^\nu, \widetilde{J}] \subseteq [\widetilde{\mathfrak{h}}, \mathfrak{g}^\nu] + \sum_{\omega \in \Omega_1} [\mathfrak{g}^\omega, \mathfrak{g}^\nu] \subseteq \widetilde{J}$$

since

$$[\widetilde{\mathfrak{h}}, \mathfrak{g}^\nu] = \nu(\widetilde{\mathfrak{h}})\mathfrak{g}^\nu = \{0\} \quad \text{and} \quad [\mathfrak{g}^\omega, \mathfrak{g}^\nu] \subseteq \mathfrak{g}^{\omega+\nu} + \mathfrak{g}^{\omega-\nu}$$

with $(\omega \pm \nu)(\widetilde{\mathfrak{h}}) = \omega(\widetilde{\mathfrak{h}}) \neq \{0\}$. The crucial fact is that $\widetilde{J}$ is a subalgebra. To see this, we pick $\nu \in \Omega_1$. Then

$$[\mathfrak{g}^\nu, \widetilde{J}] \subseteq \mathfrak{g}^\nu + \sum_{\omega \in \Omega_1} [\mathfrak{g}^\omega, \mathfrak{g}^\nu]$$

and it remains to show that $[\mathfrak{g}^\omega, \mathfrak{g}^\nu] \subseteq \widetilde{J}$ for $\omega \in \Omega_1$. If $\nu \in \Omega_S^+$ and $\omega \in \Omega_R^+$ or $\nu \in \Omega_R^+$ and $\omega \in \Omega_S^+$, this follows from our assumption $(\omega \pm \nu)(\widetilde{\mathfrak{h}}) \neq \{0\}$ and

$$[\mathfrak{g}^\omega, \mathfrak{g}^\nu] \subseteq \mathfrak{g}^{\omega+\nu} + \mathfrak{g}^{\omega-\nu}.$$

If $\nu, \omega \in \Omega_S^+$, then

$$[\mathfrak{g}^\omega, \mathfrak{g}^\nu] \subseteq [\mathfrak{s} \cap J, \mathfrak{s} \cap J] \subseteq \mathfrak{s} \cap J = \sum_{i \in P} \mathfrak{s}_i.$$

For $\omega, \nu \in \Omega_R^+$ there are two cases. If $\omega \neq \nu$, then $[\mathfrak{g}^\omega, \mathfrak{g}^\nu] = \{0\}$ (Proposition II.10) and if $\omega = \nu$ then

$$[\mathfrak{g}^\omega, \mathfrak{g}^\nu] = [\mathfrak{g}^\omega, \mathfrak{g}^\omega] = \operatorname{span} Q(\mathfrak{g}^\omega) \subseteq \widetilde{\mathfrak{h}}$$

since $C(\widetilde{\mathfrak{h}}) = \operatorname{span}\{Q(\mathfrak{g}^\omega) : \omega \in \Omega_1\}$ (Lemma III.22).

Now we assume that the above assumptions are satisfied. Then the set of real roots is

$$\Omega_J := \{\omega|_{\widetilde{\mathfrak{h}}} : \omega \in \Omega\}.$$

Moreover, it is clear that $\mathfrak{s}_J = \bigoplus_{i\in P} \mathfrak{s}_i$ is an $\widetilde{\mathfrak{h}}$-invariant Levi algebra. Therefore

$$\Omega_{J,S} = \{\omega|_{\widetilde{\mathfrak{h}}} : \omega \in \Omega_S\} \quad \text{and} \quad \Omega_{J,R} = \{\omega|_{\widetilde{\mathfrak{h}}} : \omega \in \Omega_R\}.$$

Hence

$$\mathfrak{k}_{\mathfrak{h},J} = \widetilde{\mathfrak{h}} \oplus \bigoplus_{\omega\in\Omega_1\cap\Omega_K^+} \mathfrak{g}^\omega = \mathfrak{k}_\mathfrak{h} \cap J$$

and

$$\mathfrak{z}_{\mathfrak{k},J} := Z(\mathfrak{k}_{\mathfrak{h},J}) = (\widetilde{\mathfrak{h}} \cap \mathfrak{r}) \oplus \bigoplus_{i\in P} Z(\mathfrak{k}_\mathfrak{h} \cap \mathfrak{s}_i) = \mathfrak{z}_\mathfrak{k} \cap \widetilde{\mathfrak{h}}.$$

It follows from Proposition II.20 that J is quasihermitean if and only if no root $\omega \in \Omega_P^+ \cap \Omega_1$ vanishes on $\mathfrak{z}_{\mathfrak{k},J} = \mathfrak{z}_\mathfrak{k} \cap \widetilde{\mathfrak{h}}$. ∎

Example III.27. This example shows that in general it is not possible to conclude, with the notation of Proposition III.26, that the ideal J has cone potential if it satisfies the condition (CA). We consider the Lie algebra

$$\mathfrak{g} := \text{Lie}\left(i\mathbb{R} + \text{su}(2), \text{sl}(2,\mathbb{C}), q, \mathbb{R}\right)$$

from Example II.30. Then $\mathfrak{g}$ has strong cone potential because su(2) is a compact simple Lie algebra. We have seen that

$$\Omega_R^+ = \{\omega_0, \omega_0 + \omega_1, \omega_0 - \omega_1\} \quad \text{and} \quad \Omega_S^+ = \{\omega_1\},$$

where

$$\omega_0\left(ai \oplus b\begin{pmatrix} i & 0\\ 0 & -i\end{pmatrix}\right) = a \quad \text{and} \quad \omega_1\left(ai \oplus b\begin{pmatrix} i & 0\\ 0 & -i\end{pmatrix}\right) = 2b$$

and $Q(\mathfrak{g}^\nu) \subseteq \mathbb{R}^+$ for $\nu \in \Omega_R^+$. Set $U := \begin{pmatrix} i & 0\\ 0 & -i\end{pmatrix}$ and $\widetilde{\mathfrak{h}} := Z(\mathfrak{g}) + \mathbb{R}U$. Then $\mathcal{W}\widetilde{\mathfrak{h}} \subseteq \widetilde{\mathfrak{h}}$ and $\mathcal{C}\widetilde{\mathfrak{h}} \subseteq \widetilde{\mathfrak{h}}$ because $\mathcal{C}(\widetilde{\mathfrak{h}}) \subseteq Z(\mathfrak{g})$ and $\mathcal{W}$ is a group with two elements which acts by reflections on the plane $Z(\mathfrak{g}) \oplus \mathbb{R}i \oplus \{0\} \subseteq \mathfrak{h}$. Now

$$\Omega_J := \{\omega|_{\widetilde{\mathfrak{h}}} : \omega \in \Omega\} = \{\pm\omega_1|_{\widetilde{\mathfrak{h}}}\}.$$

Thus $(\omega_0 + \omega_1)|_{\widetilde{\mathfrak{h}}} = \omega_1|_{\widetilde{\mathfrak{h}}}$ and Proposition III.26 implies that no ideal J of $\mathfrak{g}$ with $J \cap \mathfrak{h} = \widetilde{\mathfrak{h}}$ has cone potential. Now we set

$$\mathfrak{g}_0 := \text{Lie}(\mathbb{R}i \oplus \mathbb{R}U, \text{sl}(2,\mathbb{C}), q, \mathbb{R}) \subseteq \mathfrak{g}.$$

This is a solvable subalgebra of $\mathfrak{g}$ which has strong cone potential, too. The subspace $\widetilde{\mathfrak{h}} \subseteq \mathfrak{g}_0$ satisfies (CA) and

$$J := \widetilde{\mathfrak{h}} \oplus \mathfrak{g}_0^{\omega_0+\omega_1} \oplus \mathfrak{g}_0^{\omega_0-\omega_1} \subseteq \mathfrak{g}_0$$

is an ideal with $J \cap \mathfrak{h} = \widetilde{\mathfrak{h}}$. Moreover,

$$J^\nu = \mathfrak{g}_0^{\omega_0+\omega_1} \oplus \mathfrak{g}_0^{\omega_0-\omega_1}$$

is the only real root space, where $\nu(U) = 2$ holds. If I is a $\widetilde{\mathfrak{h}}$-invariant complex structure on J^ν with

$$[h, x] = \nu(h)Ix$$

for $h \in \widetilde{\mathfrak{h}}$ and $x \in J^\nu$, we find that

$$Ix = \begin{cases} ix, & \text{for } x \in \mathfrak{g}^{\omega_0+\omega_1} \\ -ix, & \text{for } x \in \mathfrak{g}^{\omega_0-\omega_1} \end{cases}.$$

Hence

$$[Ix, x] \begin{cases} > 0, & \text{for } x \in \mathfrak{g}^{\omega_0+\omega_1} \\ < 0, & \text{for } x \in \mathfrak{g}^{\omega_0-\omega_1} \end{cases}$$

and J has no cone potential because there exists an element $y \in J^\nu$ with $Q(y) = 0$ since the function $Q : J^\nu \setminus \{0\} \to \mathbb{R}$ is continuous and $J^\nu \setminus \{0\}$ is connected. ∎

Lemma III.28. *Let $\widetilde{\mathfrak{h}} \subseteq \mathfrak{h}$ be a subspace, $\Omega_1 := \{\omega \in \Omega^+ : \widetilde{\mathfrak{h}} \not\subseteq \ker \omega\}$ and*

$$\mathfrak{g}_0 := \mathfrak{h} \oplus \bigoplus_{\omega \notin \Omega_1} \mathfrak{g}^\omega.$$

Then $\mathfrak{g}_0 \subseteq \mathfrak{g}$ is a subalgebra with $[\widetilde{\mathfrak{h}}, \mathfrak{g}_0] = \{0\}$. If $J \subseteq \mathfrak{g}$ is an ideal with $p(J) = \widetilde{\mathfrak{h}}$, then $\mathfrak{g} = \mathfrak{g}_0 + J$ and

$$\operatorname{Inn}_\mathfrak{g} \mathfrak{g} = \operatorname{Inn}_\mathfrak{g} \mathfrak{g}_0 \operatorname{Inn}_\mathfrak{g} J.$$

Suppose that $W_J \subseteq J$ is a J-invariant wedge. Then

$$W := \{x \in \mathfrak{g} : p\big(\operatorname{Inn}_\mathfrak{g} \mathfrak{g}(x)\big) \subseteq W_J\}$$

is an invariant wedge in $\mathfrak{g}$ with

$$p(W) = W \cap \mathfrak{h} = W_J \cap \mathfrak{h}.$$

Proof. That $\mathfrak{g}_0$ is a subalgebra follows from

$$[\mathfrak{g}^\omega, \mathfrak{g}^{\omega'}] \subseteq \mathfrak{g}^{\omega+\omega'} + \mathfrak{g}^{\omega-\omega'}$$

and $(\omega \pm \omega')(\widetilde{\mathfrak{h}}) = \{0\}$ for $\omega, \omega' \notin \Omega_1$. The fact that $J \subseteq \mathfrak{g}$ is an ideal implies that $\operatorname{Inn}_\mathfrak{g} J \subseteq \operatorname{Inn}_\mathfrak{g} \mathfrak{g}$ is a normal analytic subgroup and therefore

$$\operatorname{Inn}_\mathfrak{g} \mathfrak{g}_0 \operatorname{Inn}_\mathfrak{g} J \subseteq \operatorname{Inn}_\mathfrak{g} \mathfrak{g}$$

is a subgroup whose Lie algebra contains

$$\operatorname{ad} \mathfrak{g}_0 + \operatorname{ad} J = \operatorname{ad}(\mathfrak{g}_0 + J) = \operatorname{ad} \mathfrak{g} = \mathbf{L}(\operatorname{Inn}_\mathfrak{g} \mathfrak{g}).$$

Thus $\operatorname{Inn}_\mathfrak{g} \mathfrak{g} = \operatorname{Inn}_\mathfrak{g} \mathfrak{g}_0 \operatorname{Inn}_\mathfrak{g} J$.

If $W_J \subseteq J$ is an invariant wedge, then it is clear from the definitions that $W \subseteq \mathfrak{g}$ is an invariant wedge. For $c \in W_J \cap \mathfrak{h}$ we have $\operatorname{Inn}_\mathfrak{g} \mathfrak{g}_0(c) = \{c\}$ and therefore

$$p\big(\operatorname{Inn}_\mathfrak{g} \mathfrak{g}(c)\big) = p\big(\operatorname{Inn}_\mathfrak{g} J(c)\big) \subseteq p(W_J) = W_J \cap \mathfrak{h}.$$

Hence $W_J \cap \mathfrak{h} \subseteq W \cap \mathfrak{h}$. Now let $h \in W \cap \mathfrak{h}$. Then $p(h) = h \in W_J$ and therefore $W \cap \mathfrak{h} \subseteq W_J \cap \mathfrak{h}$. This proves that $p(W) = p(W_J)$. ∎

Corollary III.29. *Let $C \subseteq \mathfrak{h}$ be a wedge satisfying* (C) *and* (W), $\widetilde{\mathfrak{h}} := C - C$ *and $J \subseteq \mathfrak{g}$ an ideal with $J \cap \mathfrak{h} = \widetilde{\mathfrak{h}}$. Suppose that there exists a J-invariant wedge $W_J \subseteq J$ with $p(W_J) = C$. Then there exists a $\mathfrak{g}$-invariant wedge $W \subseteq J$ such that $p(W) = C$.*

Proof. It is clear that $\mathcal{W}(C) \subseteq C$ and $\mathcal{C}(C) \subseteq C$ imply that $\mathcal{W}\widetilde{\mathfrak{h}} \subseteq \widetilde{\mathfrak{h}}$ and $\mathcal{C}\widetilde{\mathfrak{h}} \subseteq \widetilde{\mathfrak{h}}$. Now the assertion follows from Lemma III.27 and Proposition III.23. ∎

Lemma III.30. *Let $\mathfrak{g}$ be a Lie algebra with cone potential. Suppose that there exists a pointed wedge $C \subseteq \mathfrak{h}$ with $CC \subseteq C$ and that $J \subseteq \mathfrak{g}$ is an ideal containing $J \cap \mathfrak{h} = (C - C)$ as a Cartan algebra. Then J has strong cone potential.*

Proof. Let $h_0 \in \operatorname{algint} C$ and $\Omega_J^+ \subseteq \Omega_J$ a positive system such that $\omega(h_0) > 0$ for all $\omega \in \Omega_J^+$. According to Proposition III.26 we know that

$$J = \widetilde{\mathfrak{h}} \oplus \bigoplus_{\omega \in \Omega_1} \mathfrak{g}^\omega,$$

where $\Omega_1 = \{\omega \in \Omega : \omega(C) \neq \{0\}\}$. Let $\nu \in \Omega_{J,R}^+$. Then

$$Q(J^\nu) \subseteq \nu(C) Q(J^\nu) \subseteq CC \subseteq C.$$

Pick $\alpha \in \operatorname{algint} C^\star \subseteq \widehat{\mathfrak{h}}$. Then every element $x \in J^\nu \setminus \{0\}$ has a decomposition as $x = \sum_\omega x_\omega$ with $\omega \in \Omega$, $\omega|_{\widetilde{\mathfrak{h}}} = \pm\nu$ and $x_\omega \in \mathfrak{r}^\omega$. Therefore

$$\alpha(Q(x)) = \sum_\omega \alpha(Q(x_\omega)) > 0$$

because there exists at least one ω with $Q(x_\omega) \in C \setminus \{0\}$. ∎

Theorem III.31. (Reconstruction Theorem for Pointed Wedges) *Let $\mathfrak{g}$ be a quasihermitean Lie algebra with cone potential and $C \subseteq \mathfrak{h}$ a pointed wedge. Then $C = W \cap \mathfrak{h}$ for an invariant wedge $W \subseteq \mathfrak{g}$ iff C satisfies* (W) *and* (C) *and $C - C$ satisfies* (CA). *Every invariant wedge W with $W \cap \mathfrak{h} = C$ is pointed.*

Proof. "$\Rightarrow$": Suppose that $W \subseteq \mathfrak{g}$ is an invariant wedge with $p(W) = C$. Then $H(W)$ is an ideal of $\mathfrak{g}$ and $p(J) = H(W) \cap \mathfrak{h} = \{0\}$ because $H(C) = \{0\}$. Now Theorem III.23 implies that $H(W) = \{0\}$. Therefore $J := W - W$ is an ideal of $\mathfrak{g}$ which contains a pointed generating invariant wedge and satisfies

$$p(J) = p(W) - p(W) = C - C.$$

According to Proposition III.15 the ideal J has cone potential and is quasihermitean. Thus $C - C$ satisfies (CA) (Proposition III.26). That C satisfies (C) and (W) is a consequence of Lemma III.18.

"$\Leftarrow$": Assume that C satisfies (W) and (C). Then $\widetilde{\mathfrak{h}} := C - C$ satisfies (W) and (C), too. Now the additional assumption (CA) guarantees, together with Theorem III.23 and Proposition III.26, that there exists an ideal $J \subseteq \mathfrak{g}$

which contains $p(J) = \widetilde{\mathfrak{h}}$ as a compactly embedded Cartan algebra. According to Lemma III.30 the ideal J has strong cone potential. It is quasihermitean since $J \cap \mathfrak{s} \subseteq \mathfrak{s}$ contains only quasihermitean simple ideals (Proposition III.16). Now we apply the Classification Theorem for Invariant Cones ([HHL89, III.9.18]) to see that there exists a pointed generating invariant wedge $W_J \subseteq J$ with $W_J \cap \widetilde{\mathfrak{h}} = C$. Corollary III.29 implies the existence of a $\mathfrak{g}$-invariant wedge $W \subseteq J$ with $W \cap \widetilde{\mathfrak{h}} = W_J \cap \widetilde{\mathfrak{h}}$. ■

We note that the previous theorem shows that $C_{\min}$ is reconstructable if $\mathfrak{g}$ is reductive. The Reconstruction Theorem for Pointed Wedges represents the solution of the reconstruction problem for pointed cones. It allows us to decide for a given pointed cone $C \subseteq \mathfrak{h}$ whether it is the trace of an invariant wedge $W \subseteq \mathfrak{g}$ or not only by considering the data directly connected with the triple $(C, \mathfrak{h}, \Omega^+)$. Our next step is to give similar conditions for non-pointed generating cones $C \subseteq \mathfrak{h}$ to be traces of invariant wedges. To solve this problem we firstly have to find conditions for ideals $J \subseteq \mathfrak{g}$ to guarantee that $\mathfrak{g}/J$ has cone potential and is quasihermitean.

Lemma III.32. *Let $\mathfrak{g}$ be a quasihermitean Lie algebra with cone potential and $J \subseteq \mathfrak{g}$ be an ideal, $\widetilde{\mathfrak{h}} := J \cap \mathfrak{h}$, $\mathfrak{g}_1 := \mathfrak{g}/J$, $\pi : \mathfrak{g} \to \mathfrak{g}_1$ the canonical projection, and $\mathfrak{h}_1 := \pi(\mathfrak{h})$. Then $\mathfrak{h}_1$ is a compactly embedded Cartan algebra of $\mathfrak{g}_1$ and the set of real roots of $\mathfrak{g}_1$ with respect to $\mathfrak{h}_1$ is given by*

$$\Omega_1^+ := \{\omega' \in \widehat{\mathfrak{h}}_1 : (\exists \omega \in \Omega^+)\omega' \circ \pi|_{\mathfrak{h}} = \omega\}.$$

The compact and semisimple roots are

$$\Omega_{1,K}^+ := \{\omega' \in \widehat{\mathfrak{h}}_1 : (\exists \omega \in \Omega_K^+)\omega' \circ \pi|_{\mathfrak{h}} = \omega\}$$

and

$$\Omega_{1,S}^+ := \{\omega' \in \widehat{\mathfrak{h}}_1 : (\exists \omega \in \Omega_S^+)\omega' \circ \pi|_{\mathfrak{h}} = \omega\}.$$

The Lie algebra $\mathfrak{g}_1$ has cone potential iff

$$(CP) \qquad Q(x) \not\in \widetilde{\mathfrak{h}} \quad \text{for all} \quad x \in \mathfrak{r}^\omega \setminus J.$$

The maximal compactly embedded subalgebra containing $\mathfrak{h}_1$ is

$$\mathfrak{k}_{\mathfrak{h}_1} = \pi(\mathfrak{k}_{\mathfrak{h}}) \quad \text{and} \quad \mathfrak{z}_{\mathfrak{k}_1} := Z(\mathfrak{k}_{\mathfrak{h}_1}) = \bigcap_{\omega \in \Omega_{K,1}^+} \ker \omega = \pi(\mathfrak{z}_{\mathfrak{k}}).$$

The Lie algebra $\mathfrak{g}_1$ is quasihermitean and for every element w_1 of the Weyl group $\mathcal{W}(\mathfrak{h}_1, \mathfrak{g}_1)$ there exists an element $w \in \mathcal{W}(\mathfrak{h}, \mathfrak{g})$ such that

$$\pi(w.h) = w_1.\pi(h) \quad \text{for all} \quad h \in \mathfrak{h}.$$

Moreover, $\pi(\mathfrak{g}^\omega) = \mathfrak{g}^{\omega'}$ if $\omega = \omega' \circ \pi$ and

$$\pi\big((\operatorname{ad} x)^2 h\big) = \big(\operatorname{ad} \pi(x)\big)^2 \pi(h) \quad \text{for all } h \in \mathfrak{h}, x \in \mathfrak{g}^\omega.$$

Proof. It follows from [Bou75, p.20] that $\mathfrak{h}_1$ is a Cartan algebra of $\mathfrak{g}_1$. The $\mathfrak{h}_1$-module $\mathfrak{g}_1$ is isomorphic to a complementary $\mathfrak{h}_1$-module for J in $\mathfrak{g}$. Therefore $\mathfrak{h}_1$ is compactly embedded. The roots are the functionals $\omega' \in \widehat{\mathfrak{h}_1}$ for which there exists a root $\omega \in \Omega^+$ such that $\omega' \circ \pi|_{\mathfrak{h}} = \omega$ and $\pi(\mathfrak{g}^\omega) = \mathfrak{g}^{\omega'}$ for $\omega' \circ \pi|_{\mathfrak{h}} = \omega$. According to Definition II.4 the Lie algebra $\mathfrak{g}_1$ has cone potential if and only if $Q(\pi(x)) \neq \{0\}$ for $x \in \mathfrak{g}^\omega \setminus J$ and $\omega \in \Omega_P^+$, i.e., if $Q(y) \neq \{0\}$ holds for $y \in \mathfrak{g}^{\omega'} \setminus \{0\}$. But this is equivalent to $Q(x) \not\subseteq \widetilde{\mathfrak{h}}$ for $x \in \mathfrak{g}^\omega \setminus J$ and $\omega \in \Omega_R^+$ since $x \in \mathfrak{g}^\nu$, $\nu(\widetilde{\mathfrak{h}}) = \{0\}$ and $\nu \in \Omega_P^+ \cap \Omega_S^+$ implies that $\langle x \rangle \cap J = \{0\}$ and therefore $Q(x) \not\subseteq J$. It is clear that

$$\mathfrak{k}_{\mathfrak{h},1} = \mathfrak{h}_1 \oplus \bigoplus_{\omega \in \Omega_{1,K}} \mathfrak{g}_1^\omega = \pi\Big(\mathfrak{h} \oplus \bigoplus_{\substack{\omega \in \Omega_K^+ \\ \omega(\widetilde{\mathfrak{h}}) = \{0\}}} \mathfrak{g}^\omega\Big)$$

$$= \pi\Big(\mathfrak{h} \oplus \bigoplus_{\omega \in \Omega_K^+} \mathfrak{g}^\omega\Big) = \pi(\mathfrak{k}_{\mathfrak{h}}).$$

Thus $\pi(\mathfrak{z}_{\mathfrak{k}}) \subseteq \mathfrak{z}_{\mathfrak{k},1}$. The restriction of π to $\mathfrak{k}_{\mathfrak{h}}$ is a surjective homomorphism of compact Lie algebras. Since a surjective homomorphism maps the radical onto the radical we find that $\pi(\mathfrak{z}_{\mathfrak{k}}) = \mathfrak{z}_{\mathfrak{k}_1}$ because the radical of a compact Lie algebra agrees with its center. Now let $\omega' \in \Omega_{1,P}^+$ and $\omega' \circ \pi \in \Omega_P^+$. Then

$$\omega'(\mathfrak{z}_{\mathfrak{k},1}) = \omega'(\pi(\mathfrak{z}_{\mathfrak{k}})) = (\omega' \circ \pi)(\mathfrak{z}_{\mathfrak{k}}) \neq \{0\}$$

because $\mathfrak{g}$ is quasihermitean and hence $\mathfrak{g}_1$ is quasihermitean. ∎

Theorem III.33. (Reconstruction Theorem for Generating Wedges) *Let $\mathfrak{g}$ be a quasihermitean Lie algebra with cone potential and $C \subseteq \mathfrak{h}$ a generating wedge. Then $C = W \cap \mathfrak{h}$ for an invariant wedge $W \subseteq \mathfrak{g}$ iff*

$$\mathcal{W}C = C \quad \textit{and} \quad \mathcal{C}C \subseteq C$$

and if there exists an ideal J with $J \cap \mathfrak{h} = H(C)$ such that

$$\text{(Rec)} \qquad Q(x) \not\subseteq H(C) \quad \textit{for all} \quad x \in \mathfrak{r}^\omega \setminus J.$$

Proof. "$\Rightarrow$": Suppose that there exists an invariant wedge $W \subseteq \mathfrak{g}$ with $C = W \cap \mathfrak{h}$. We set $J := H(W)$. Then J is an ideal of $\mathfrak{g}$ with $p(J) = J \cap \mathfrak{h} = H(C)$. Moreover, $\mathfrak{g}/J$ contains the pointed generating invariant wedge $W/H(W)$ and therefore $\mathfrak{g}/J$ has cone potential. Now Lemma III.32 implies that

$$Q(x) \not\subseteq H(C) \quad \text{for all} \quad x \in \mathfrak{r}^\omega \setminus J.$$

The invariance of C under $\mathcal{W}$ and $\mathcal{C}$ follows from Lemma III.7.
"$\Leftarrow$": We consider the Lie algebra $\mathfrak{g}/J$ with the canonical projection $\pi : \mathfrak{g} \to \mathfrak{g}/J$. Then $\mathfrak{g}/J$ has cone potential and is quasihermitean with $\pi(\mathfrak{h})$ as a compactly embedded Cartan algebra (Lemma III.32). The wedge $\pi(C) \subseteq \pi(\mathfrak{h})$

is pointed and generating since $C \subseteq \mathfrak{h}$ is generating and $H(C) = \ker \pi \cap \mathfrak{h}$. Now the Characterization Theorem for Invariant Cones [HHL89, III.9.18] implies the existence of a pointed generating invariant wedge $W_1 \subseteq \mathfrak{g}/J$ with $W_1 \cap \pi(\mathfrak{h}) = \pi(C)$. We set $W := \pi^{-1}(W_1)$. Then W is an invariant wedge in $\mathfrak{g}$ with $H(W) = J$ and

$$p(W) = W \cap \mathfrak{h} = (\pi|_{\mathfrak{h}})^{-1}(\pi(C)) = C.$$

■

We note that, if we take $J := \mathfrak{n} + \mathfrak{s}_k$, where $\mathfrak{s}_k$ is the sum of all compact simple ideals of $\mathfrak{g}$ and $\mathfrak{n}$ its nilradical, Theorem III.33 and Propositions II.10 and III.20 imply that there exists a generating invariant wedge $W_{\max} \subseteq \mathfrak{g}$ such that

$$W_{\max} \cap \mathfrak{h} = C_{\max}.$$

Proposition III.34. (Uniqueness of Reconstruction) *Let $\mathfrak{g}$ be a Lie algebra with a compactly embedded Cartan algebra, $C \subseteq \mathfrak{h}$ a wedge and $W, W' \subseteq \mathfrak{g}$ two invariant wedges with*

$$C = W \cap \mathfrak{h} = W' \cap \mathfrak{h} \quad \textit{and} \quad W - W = W' - W' = \operatorname{span}\big(\mathrm{INN}_{\mathfrak{g}}(\mathfrak{g})C\big).$$

Then $W = W'$. The assertion holds in particular if C is generating in $\mathfrak{h}$ and $C = W \cap \mathfrak{h} = W' \cap \mathfrak{h}$.

Proof. Set $C := W \cap \mathfrak{h}$ and write $p : \mathfrak{g} \to \mathfrak{h}$ for the projection along the real root spaces. We set $W'' := \operatorname{cone}\big(\mathrm{INN}_{\mathfrak{g}}(\mathfrak{g})C\big)$. Then $W'' \subseteq W \cap W'$ and therefore we may assume that $W \subseteq W'$ because $W - W = W'' - W''$. Let $\pi : \mathfrak{g} \to \mathfrak{g}_1 := \mathfrak{g}/H(W)$ be the quotient homomorphism. Then $\mathfrak{g}/H(W)$ contains the pointed invariant wedge $\pi(W)$ which is contained in the invariant wedge $p(W')$. Moreover $\pi(\mathfrak{h})$ is a compactly embedded Cartan algebra of $\mathfrak{g}_1$ ([Bou75, p.20]) and $p_1 \circ \pi = \pi \circ p$, where $p_1 : \mathfrak{g}_1 \to \pi(\mathfrak{h})$ is the projection onto $\pi(\mathfrak{h})$. Hence

$$p_1\big(\pi(W)\big) = \pi \circ p(W) = \pi(C) = p_1\big(\pi(W')\big).$$

Thus we may assume that W is pointed. The ideal $J := W - W = W' - W' \subseteq \mathfrak{g}$ contains a pointed generating invariant wedge. Therefore it has cone potential(Proposition III.15) and $p\big(H(W')\big) = H(C) = \{0\}$. We conclude that W' is pointed (Proposition III.23). Then [HHL89, III.2.15] shows that $W = W'$ bacause $\operatorname{int} W' \subseteq (\mathrm{INN}_{\mathfrak{g}}\, \mathfrak{g})(C) \subseteq W$. ■

In Section XI we only need the following special Reconstruction Theorem which we formulate now because the assumptions are much easier to formulate in this case.

Theorem III.35. (The Special Reconstruction Theorem) *Let $\mathfrak{g}$ be a finite dimensional Lie algebra and suppose that $W \subseteq \mathfrak{g}$ is a pointed generating invariant wedge, $\mathfrak{h} \subseteq \mathfrak{g}$ is a compactly embedded Cartan algebra, $C = W \cap \mathfrak{h}$ and $C^\star \subseteq \widehat{\mathfrak{h}}$ the dual wedge. We identify $\widehat{\mathfrak{h}} \subseteq \widehat{\mathfrak{g}}$ with the subspace of functionals fixed under the contragredient action of the compact group* $\mathrm{INN}_{\mathfrak{g}}(\mathfrak{h})$ (cf. Theorem I.10).

(1) *For a wedge* $F \subseteq C$ *there exists a pointed wedge* $V \subseteq \mathfrak{g}$ *with* $F = \mathfrak{h} \cap V$ *iff*

$$\mathcal{W}F \subseteq F \quad \text{and} \quad \mathcal{C}F \subseteq F.$$

(2) *For a face* $F \in \mathcal{F}(C^\star)$ *there exists an invariant wedge* $V \subseteq \mathfrak{g}$ *with* $V \cap \mathfrak{h} = F^\star$ *iff*

$$\mathcal{W}F^\star \subseteq F^\star \quad \text{and} \quad \mathcal{C}F^\star \subseteq F^\star.$$

For every invariant wedge V *with this property we have that* $V^* \in \mathcal{F}(W^*)$ *and* $V^* \cap \widehat{\mathfrak{h}} = F$.

Proof. (1) The invariance of F under $\mathcal{W}$ and $\mathcal{C}$ are clearly necessary (Proposition III.7). Let $\widetilde{\mathfrak{h}} := F - F$. Then $\mathcal{W}\widetilde{\mathfrak{h}} \subseteq \widetilde{\mathfrak{h}}$ and $\mathcal{C}\widetilde{\mathfrak{h}} \subseteq \widetilde{\mathfrak{h}}$. Using Theorem III.23 we find an ideal $I \subseteq \mathfrak{g}$ with $I \cap \mathfrak{h} = \widetilde{\mathfrak{h}}$. We assume that I is minimal with this property. Let $W' := W \cap I$. This is a pointed invariant wedge in I with $\widetilde{\mathfrak{h}} = F - F \subseteq W' - W'$. Hence the minimality of I implies that $I = W' - W'$, i.e., W' is generating in I. Thus I has a cone potential (Proposition III.15) and contains a compactly embedded Cartan algebra. Now Proposition III.26 together with the Reconstruction Theorem for Pointed Wedges (Theorem III.31) proves that there exists a pointed wedge $V \subseteq \mathfrak{g}$ with $V \cap \mathfrak{h} - F$.
(2) Again the necessity of the invariance of $F^\star$ under the Weyl group $\mathcal{W}$ and the cone $\mathcal{C}$ follows from Proposition III.7. Suppose that this condition is satisfied. Then the wedge $H(F^\star)$ is also invariant under $\mathcal{W}$ and $\mathcal{C}$. Consequently we find an ideal $I \subseteq \mathfrak{g}$ such that $I \cap \mathfrak{h} = H(F^\star)$ (Theorem III.23). Let

$$V := L_I(W) = \overline{W - I}.$$

Then

$$L_{H(F^\star)}(C) = \overline{C - H(F^\star)} \subseteq p(V) \subseteq \overline{p(W) - p(I)} = \overline{C - H(F^\star)}.$$

Thus $p(V) = L_{H(F^\star)}(C) = F^\star$ (Proposition I.5.5). It remains to prove that $V^* \in \mathcal{F}(W^*)$. We set $W_0 := W$, $F_0 := H(V) \cap W_0$ and inductively $W_{i+1} := L_{F_i}(W_i)$, $F_{i+1} := H(V) \cap W_{i+1}$. We prove several properties of the sequence of wedges W_i by induction on i:
(a) $W_i \subseteq V$: This is clear for $i = 0$. If $W_i \subseteq V$, then $W_{i+1} \subseteq V$ because $F_i \subseteq H(V)$.
(b) $F_i \in \mathcal{F}_e(W_i)$: This follows from (a) and the definition of exposed faces.
(c) $W_i^* \in \mathcal{F}(W^*)$: This is trivial for $i = 0$. For $i > 0$ we assume that $W_{i-1}^* \in \mathcal{F}(W^*)$ and use Proposition I.5 to see that

$$W_i^* = L_{F_{i-1}}(W_{i-1})^* \in \mathcal{F}_e(W_{i-1}^*) \subseteq \mathcal{F}(W^*)$$

since W_{i-1}^* is a face of W^*.
(d) $W_i^* \cap \widehat{\mathfrak{h}} \in \mathcal{F}(C^\star)$ and $F \subseteq W_i^* \cap \widehat{\mathfrak{h}}$: This follows from (a) and (c) above and Proposition I.13.

Suppose that $F \neq W_i^* \cap \widehat{\mathfrak{h}}$. Then $F \in \mathcal{F}(W_i^* \cap \widehat{\mathfrak{h}})$ (Proposition I.5) and consequently

$$\big(H(V) \cap W_i\big) \cap \mathfrak{h} = H(F^\star) \cap p(W_i) \not\subseteq H\big(p(W_i)\big) = H(W_i) \cap \mathfrak{h}.$$

This leads to $\dim H(W_{i+1}) > \dim H(W_i)$. But $\mathfrak{g}$ is finite dimensional and thus we find $k \in \mathbb{N}$ such that $F = W_k^* \cap \widehat{\mathfrak{h}}$. Then

$$V \cap \mathfrak{h} = F^\star = (W_k^* \cap \widehat{\mathfrak{h}})^\star = W_k \cap \mathfrak{h}$$

and the uniqueness of reconstruction (Proposition III.34) proves that $W_k = V$. Hence $V^* = W_k^* \in \mathcal{F}(W^*)$ by (c). ∎

We conclude this section with a characterization of those Lie algebras which contain pointed generating invariant cones:

Theorem III.36. (Characterization Theorem for Lie Algebras with Invariant Cones) *A finite dimensional Lie algebra $\mathfrak{g}$ contains a pointed generating invariant cone W iff it is quasihermitean, has strong cone potential, and is not compact semisimple.*

Proof. The necessity of these conditions follows from Proposition III.15 and the fact that a compact semisimple Lie algebra cannot contain a pointed generating invariant cone since $\mathfrak{g}_{\text{fix}} = \{0\}$ with respect to the adjoint action of the compact group $\text{INN}_{\mathfrak{g}}(\mathfrak{g})$ on $\mathfrak{g}$ (Theorem I.10). We prove that the above conditions are also sufficient. The cones

$$C_{\min,z} \subseteq Z(\mathfrak{g}) \quad \text{and} \quad C_{\min} \cap \mathfrak{s} = \sum_{i=1}^{n} C_{\min} \cap \mathfrak{s}_i = \sum_{i=1}^{m} C_{\min} \cap \mathfrak{s}_i,$$

where $\mathfrak{s}_1, \ldots, \mathfrak{s}_m \subseteq \mathfrak{s} = \bigoplus_{i=1}^n \mathfrak{s}_i$ are the hermitean simple ideals. But $C_{\min} \cap \mathfrak{s}_i$ is pointed (Proposition III.23). Therefore

$$C_{\min} = C_{\min,z} + \sum_{i=1}^{m} C_{\min} \cap \mathfrak{s}_i$$

is a pointed cone in $\mathfrak{h}$. According to the definition of a positive system $\Omega^+ \subseteq \Omega$ we find $h \in \mathfrak{h}$ with $\omega(h) > 0$ for all $\omega \in \Omega^+$.
We consider two cases:
Case 1: $C_{\max} \neq \mathfrak{h}$. Then $h \in \operatorname{int} C_{\max}$ and for every compact neighborhood B of h with $B \subseteq \operatorname{int} C_{\max}$ the set

$$K := \operatorname{conv}(\mathcal{W}B) \subseteq \operatorname{int} C_{\max}$$

is compact, convex and $\mathcal{W}$-invariant since $\mathcal{W}C_{\max} = C_{\max}$ (Proposition III.20). We set

$$C := \mathbb{R}^+ K + C_{\min}.$$

Now $\mathcal{W}C = C$ and $CC \subseteq C$ because for every $\omega \in \Omega_P^+$ and $x \in \mathfrak{g}^\omega$ we have

$$(\operatorname{ad} x)^2(C) = \omega(C)Q(x) \subseteq \mathbb{R}^+ Q(x) \subseteq C_{\min} \subseteq C.$$

The set C is closed since $\mathbb{R}^+ K$ is a pointed wedge with

$$\mathbb{R}^+ K \cap (-C_{\min}) \subseteq \{0\} \cup \operatorname{int} C_{\max} \cap (-C_{\min}) = \{0\}$$

because $C_{\min} \subseteq C_{\max}$ (Proposition III.20, Proposition I.7). Moreover C is generating because $\operatorname{int}_{\mathfrak{h}} K \neq \emptyset$ and C is pointed because

$$H(C) = H(\mathbb{R}^+ K) + H(C_{\min}) = \{0\}$$

(Propositon I.7). Now Theorem III.31 proves the existence of a pointed generating wedge $W \subseteq \mathfrak{g}$ with $W \cap \mathfrak{h} = C$.
Case 2: $C_{\max} = \mathfrak{h}$. Then $\Omega_P = \emptyset$ and $\mathfrak{g}$ is a compact Lie algebra. According to our assumption $Z(\mathfrak{g}) \neq \{0\}$ and we find $h \in Z(\mathfrak{g}) \setminus \{0\}$. We choose a compact convex neighborhood K of h which is invariant under $\mathrm{INN}_{\mathfrak{g}}\, \mathfrak{g}$ and which does not contain 0. Then $W = \mathbb{R}^+ K$ is a pointed generating invariant wedge in $\mathfrak{g}$. ■

It is an immediate consequence of Theorem III.36 that the property of a Lie algebra to contain pointed generating invariant cones, (IC) for short, does not depend heavily on the existence of compact semisimple ideals. More precisely, we have the following proposition.

Proposition III.37. *The following assertions hold:*

(1) *If* $\mathfrak{g}_1$ *satisfies* (IC) *and* $\mathfrak{k}$ *is compact semisimple, then* $\mathfrak{g} := \mathfrak{g}_1 \oplus \mathfrak{k}$ *satisfies* (IC).

(2) *If* $\mathfrak{g} = \mathfrak{g}_1 \oplus \mathfrak{k}$ *satisfies* (IC), *where* $\mathfrak{k}$ *is compact semisimple, then* $\mathfrak{g}_1$ *satisfies* (IC).

Proof. (1) Let $\mathfrak{h}_1 \subseteq \mathfrak{g}_1$ and $\mathfrak{h}_2 \subseteq \mathfrak{k}$ be compactly embedded Cartan algebras (Proposition III.15). Then $\mathfrak{h} := \mathfrak{h}_1 \oplus \mathfrak{h}_2$ is a compactly embedded Cartan algebra of $\mathfrak{g}$. It is clear from Definition II.28 that $\mathfrak{g}_1 \oplus \mathfrak{k}$ has strong cone potential because we find for every positive system Ω_1^+ of real roots of $\mathfrak{g}_1$ with respect to $\mathfrak{h}_1$, or equivalently of roots vanishing on $\mathfrak{h}_2$, a positive system Ω^+ of real roots with respect to $\mathfrak{h}$ such that

$$\Omega_1^+ = \{\omega \in \Omega^+ : \mathfrak{h}_2 \subseteq \ker \omega\}.$$

Thus $\mathfrak{g}$ is quasihermitean (Proposition III.16) and the assertion follows from Theorem III.36.
(2) Let $W \subseteq \mathfrak{g}$ be a pointed generating invariant wedge. Then

$$\mathfrak{g}_{\text{eff}} = \mathfrak{k} \quad \text{and} \quad \mathfrak{g}_{\text{fix}} = \mathfrak{g}_1$$

with respect to the action of the compact group $\mathrm{INN}_{\mathfrak{g}}\, \mathfrak{k}$ on $\mathfrak{g}$. Thus $W \cap \mathfrak{g}_1$ is generating in $\mathfrak{g}_1$ (Theorem I.10) and therefore a pointed generating invariant wedge in $\mathfrak{g}_1$. ■

We even have the following stronger result.

Corollary III.38. *Under the condition of* Proposition III.37(1) *we find for every pointed generating invariant wedge* $W \subseteq \mathfrak{g}_1$ *a pointed generating invariant wedge* $V \subseteq \mathfrak{g}$ *such that*

$$V \cap \mathfrak{g}_1 = W.$$

Proof. We consider a compactly embedded Cartan algebra $\mathfrak{h} = \mathfrak{h}_1 \oplus \mathfrak{h}_2$ with $\mathfrak{h}_1 := \mathfrak{h} \cap \mathfrak{g}_1$ and $\mathfrak{h}_2 := \mathfrak{h} \cap \mathfrak{k}$. Let $\mathcal{W}_1$ and $\mathcal{W}_2$ denote the corresponding subgroups of the Weyl group. Then $\mathcal{W} \cong \mathcal{W}_1 \times \mathcal{W}_2$, where $\mathcal{W}_1$ acts trivially on $\mathfrak{h}_2$ and $\mathcal{W}_2$ acts trivially on $\mathfrak{h}_1$. We choose a positive system Ω_1^+ of real roots of $\mathfrak{g}_1$ with respect to $\mathfrak{h}_1$ such that

$$C_{\min}(\Omega_1^+) \subseteq C_1 := V \cap \mathfrak{h}_1 \subseteq C_{\max}(\Omega_1^+)$$

(Proposition III.15). For a positive system Ω_2^+ of real roots with respect to $\mathfrak{h}_2$ we then have $\Omega^+ := \Omega_1^+ \cup \Omega_2^+$ as a positive system with respect to $\mathfrak{h}$. The facts that

$$C_{\min}(\Omega^+) = C_{\min}(\Omega_1^+) \quad \text{and} \quad C_{\max}(\Omega^+) = C_{\max}(\Omega_1^+) \oplus \mathfrak{h}_2$$

(Definition III.12) imply that $\mathfrak{g}$ has strong cone potential with respect to Ω^+ (Proposition III.13). Since $CC_1 \subseteq C_{\min}(\Omega_1^+)$ for every wedge $C_1 \subseteq C_{\max}(\Omega^+)$ we have to find a pointed generating invariant wedge $C \subseteq \mathfrak{h}$ such that C is $\mathcal{W}$-invariant,

$$C_{\min}(\Omega^+) \subseteq C \subseteq C_{\max}(\Omega^+), \quad \text{and} \quad C \cap \mathfrak{h}_1 = C_1.$$

First we note that

$$\mathfrak{h}_{\text{eff}} = \mathfrak{h}_2 \quad \text{and} \quad \mathfrak{h}_{\text{fix}} = \mathfrak{h}_1$$

with respect to the action of $\mathcal{W}_2$ on $\mathfrak{h}$. We have proved in Corollary I.11 that, if B is a compact base of the cone C_1 and B_1 is a compact convex $\mathcal{W}_2$-invariant neighborhood of 0 in $\mathfrak{h}_2$, the set

$$C := \mathbb{R}^+(B + B_1) \cap C_{\max}(\Omega^+) = \mathbb{R}^+(B + B_1)$$

is a pointed generating $\mathcal{W}_2$-invariant wedge in $\mathfrak{h}$ with $C \cap \mathfrak{h}_1 = C_1$. If we take a $\mathcal{W}_1$-invariant compact base of C_1, then this cone is also invariant under the action of $\mathcal{W}_1$ and therefore

$$\mathcal{W}C \subseteq C \quad \text{and} \quad C_{\min}(\Omega^+) \subseteq C \subseteq C_{\max}(\Omega^+).$$

Thus Theorems III.31 and III.33 guarantee the existence of a pointed generating invariant cone $V \subseteq \mathfrak{g}$ with $V \cap \mathfrak{h} = C$.

If $\pi : \mathfrak{g} \to \mathfrak{g}_1$ is the projection along $\mathfrak{k}$ and $p : \mathfrak{g} \to \mathfrak{h}$ is the orthogonal projection along $[\mathfrak{h}, \mathfrak{g}] = \mathfrak{g}_{\text{eff}}$ onto $\mathfrak{h}$, then

$$\pi|_{\mathfrak{h}} \circ p = p|_{\mathfrak{g}_1} \circ \pi$$

and therefore

$$\begin{aligned} p|_{\mathfrak{g}_1}(V \cap \mathfrak{g}_1) &= p|_{\mathfrak{g}_1} \circ \pi(V) \\ &= \pi|_{\mathfrak{h}} \circ p(V) = \pi|_{\mathfrak{h}}(C) = C \cap \mathfrak{h}_1 = C_1. \end{aligned}$$

Whence $V \cap \mathfrak{g}_1 = W$ (Proposition III.34). ∎

We give another formulation of the Characterization theorem which is easier to check in concrete cases.

Theorem III.39. *Let* $\mathfrak{g}$ *be a finite dimensional Lie algebra which is not compact semisimple. Then* $\mathfrak{g}$ *contains pointed generating invariant cones iff*

(1) $\mathfrak{g}$ *has strong cone potential, and*

(2) *all simple ideals contained in* $\mathfrak{g}$ *are either compact or hermitean.*

Proof. It follows from Proposition II.20 and Proposition III.16 that a Lie algebra $\mathfrak{g}$ with strong cone potential is quasihermitean if and only if all simple ideals are compact or hermitean. Hence the theorem follows from Theorem III.36. ■

Corollary III.40. *A solvable Lie algebra* $\mathfrak{g}$ *contains pointed generating invariant cones iff* $\mathfrak{g}$ *has strong cone potential.* ■

Remark III.41. We note that Example II.36 provides an example of a Lie algebra $\mathfrak{g}$ without strong cone potential, where $\mathrm{sl}(2,\mathbb{R})$ is a Levi algebra, $\mathfrak{g}$ contains a compactly embedded Cartan algebra $\mathfrak{h}$, and the radical $\mathfrak{r}$ has strong cone potential. According to Corollary III.40 this means that the radical contains pointed generating invariant cones. That the bigger Lie algebra $\mathfrak{g}$ does not have strong cone potential comes from the fact that the rank of $\mathfrak{g}$ is strictly smaller than the rank of $\mathfrak{r}$. In this example this entails that $\mathfrak{r}$ contains three different real root spaces with respect to $\mathfrak{h}$ but only one real root space with respect to the Cartan algebra $\mathfrak{h} \cap \mathfrak{r}$ of $\mathfrak{r}$. ■

IV. FACES OF LIE SEMIGROUPS

In the first three sections we have described an infinitesimal theory of invariant subsemigroups of Lie groups, the theory of invariant wedges in Lie algebras. Now we turn to the global theory. Some of the results we shall need later in the special case of invariant subsemigroups are true for general Lie semigroups. These results are contained in this section. Our main objective is to define faces of Lie semigroups S, to relate these objects to faces of the tangent wedge $\mathbf{L}(S)$ of S, to discuss some functorial aspects, and to give some characterizations of the different kinds of faces which will show how faces of Lie semigroups generalize faces of wedges in vector spaces. In the abelian case this was already done in [Rup87].

In the following G denotes a connected finite dimensional Lie group, $\mathbf{L}(G)$ its Lie algebra and $\exp : \mathbf{L}(G) \to G$ its exponential function. We write $\lambda_g : G \to G, x \mapsto gx$ for left multiplication with g and $\rho_g : G \to G, x \mapsto xg$ for right multiplication with g. For a subset $M \subseteq G$ we denote the semigroup generated by M with $\langle M \rangle = \bigcup_{n \in \mathbb{N}} M^n$.

Definition IV.1. A *Lie wedge* is a pair $(W, \mathfrak{g})$, where W is a wedge in a finite dimensional Lie algebra $\mathfrak{g}$ satisfying the condition

$$e^{\operatorname{ad} x} W = W \quad \text{for all} \quad x \in H(W).$$

We also write simply W for the Lie wedge $(W, \mathfrak{g})$. A Lie wedge $(W, \mathfrak{g})$ is said to be *Lie generating* if $\mathfrak{g}$ is the smallest subalgebra of $\mathfrak{g}$ containing W. A morphism $\varphi : (W, \mathfrak{g}) \to (W', \mathfrak{g}')$ of Lie wedges is a morphism $\varphi : \mathfrak{g} \to \mathfrak{g}'$ of Lie algebras satisfying $\varphi(W) \subseteq W'$. We write <u>*LWed*</u> for the so defined category of Lie wedges. The full subcategory of invariant Lie wedges is denoted with <u>*IWed*</u>.∎

Definition IV.2. Let $S \subseteq G$ be a closed subsemigroup. We define the *tangent wedge* of G by

$$\mathbf{L}(S) := \{x \in \mathbf{L}(G) : \exp(\mathbb{R}^+ x) \subseteq S\}.$$

∎

Definition IV.3. A *Lie semigroup* (*LSg*) is a pair (S, G), where G is a connected Lie group and $S \subseteq G$ a closed subsemigroup with

$$S = \overline{\langle \exp \mathbf{L}(S) \rangle}.$$

We also use the notion Lie semigroup for the semigroup S itself if it is clear in which group it sits. Note that our definition is consistent with the one given in [HoLa83]. For such a subsemigroup of G we define a left invariant order on G by setting

$$g \leq_S g' \quad \text{if} \quad g' \in gS.$$

We usually omit the subscript and write $\leq$ instead of $\leq_S$. Then $g \leq g' \leq g$ is equivalent to $gH(S) = g'H(S)$ and we get a partial order $\preceq$ on the homogeneous space $G/H(S)$ by defining

$$gH \preceq g'H \quad \text{if} \quad g \leq_S g'.$$

A LSg is said to be a *generating LSg* (GLSg) if $\mathbf{L}(S)$ is Lie generating in $\mathbf{L}(G)$. A morphism $\varphi : (S,G) \to (S',G')$ of *LSgs* is a homomorphism $\varphi : G \to G'$ of Lie groups such that $\varphi(S) \subseteq S'$. We denote the category of *LSgs* with $\underline{LSg}$. The full subcategory of $\underline{LSg}$ whose objects are the invariant Lie semigroups, *ILSgs*, is called $\underline{ILSg}$. We say that an *ILSg* (S,G) is *a GILSg* if it is a *GLSg*.

Proposition IV.4. *The prescription*

$$\mathbf{L} : \underline{LSg} \to \underline{LWed}, \quad (S,G) \to \big(\mathbf{L}(S), \mathbf{L}(G)\big), \quad \varphi \mapsto \mathbf{L}(\varphi) := d\varphi(\mathbf{1})$$

defines a covariant functor from the category $\underline{LSg}$ to the category $\underline{LWed}$. Moreover, the following assertions hold:

(1) *Suppose that $\alpha : \mathfrak{g} \to \mathfrak{g}_1$ is a surjective morphism of Lie algebras which induces a morphism $\alpha : (W,\mathfrak{g}) \to (V,\mathfrak{g}_1)$ of Lie wedges and that W is Lie generating. Then V is Lie generating.*

(2) *Let (S,G) be a LSg and $\varphi : G \to G_1$ a morphism of Lie groups. Then $\big(\overline{\varphi(S)}, G_1\big)$ is a Lie semigroup and $\varphi : (S,G) \to \big(\overline{\varphi(S)}, G_1\big)$ defines a morphism of LSgs. If (S,G) is a GLSg and φ is surjective, then $\big(\overline{\varphi(S)}, G_1\big)$ is a GLSg.*

Proof. The functoriality of $\mathbf{L}$ is a trivial consequence of the fact that

$$\mathbf{L}(\varphi_1 \circ \varphi_2) = \mathbf{L}(\varphi_1) \circ \mathbf{L}(\varphi_2),$$

for $\varphi_1 : (S_1,G_1) \to (S_2,G_2)$, $\varphi_2 : (S_2,G_2) \to (S_3,G_3)$. This is the chain rule for the derivatives of homomorphisms of Lie groups. It is clear that $\varphi(S_1) \subseteq S_2$ implies that $\mathbf{L}(\varphi)\mathbf{L}(S_1) \subseteq \mathbf{L}(S_2)$.

(1) Let $\mathfrak{a} \subseteq \mathfrak{g}_1$ be a subalgebra containing V. Then $\varphi^{-1}(\mathfrak{a})$ is a subalgebra of $\mathfrak{g}$ containing W. Thus $\varphi^{-1}(\mathfrak{a}) = \mathfrak{g}$ and therefore $\mathfrak{a} = \varphi\big(\varphi^{-1}(\mathfrak{a})\big) = \mathfrak{g}_1$.

(2) Set $S_1 := \overline{\varphi(S)}$. Then $\mathbf{L}(\varphi)\mathbf{L}(S) \subseteq \mathbf{L}(S_1)$. Therefore

$$\overline{\langle \exp \mathbf{L}(S_1)\rangle} \supseteq \overline{\langle \exp \mathbf{L}(\varphi)\mathbf{L}(S)\rangle} = \overline{\varphi(\langle \exp \mathbf{L}(S)\rangle)} = S_1.$$

Hence (S_1, G_1) is a LSg. The second assertion follows from (1) because the surjectivity of φ implies the surjectivity of $\mathbf{L}(\varphi) : \mathbf{L}(G) \to \mathbf{L}(G_1)$. ∎

Definition IV.5. For a Lie semigroup $S \subseteq G$ we define the set of *reachable elements* reach(S) to be the set of those elements $s \in S$ for which there exists an absolutely continuous S-monotone curve $\gamma : [0,T] \to G$ such that $\gamma(0) = \mathbf{1}$, $\gamma(T) = s$ and the derivative of γ, the function

$$u : [0,T] \to \mathbf{L}(G), \quad t \mapsto d\lambda_{\gamma(t)^{-1}}(\gamma(t))\gamma'(t)$$

is essentially bounded. We also write $\widetilde{\operatorname{comp}}(S)$ for the set of all elements $s \in S$ for which the order interval

$$[\pi(\mathbf{1}), \pi(s)] = \{x \in G/H(S) : \pi(\mathbf{1}) \preceq x \preceq \pi(s)\}$$

is compact and comp(S) for the set of all elements $s \in S$ for which the order interval

$$[\mathbf{1}, s] = S \cap sS^{-1} = \{g \in G : \mathbf{1} \leq g \leq s\}$$

is compact. Note that $\operatorname{comp}(S) = \emptyset$ whenever $H(S)$ is not compact, and that

$$\operatorname{comp}(S) = \widetilde{\operatorname{comp}}(S)$$

if $H(S)$ is compact. ■

Proposition IV.6. *Let (S,G) be a Lie semigroup and $G/H(S)$ the associated homogeneous space. Then the following assertions hold:*

(1) *The interior of S is a an ideal which is dense iff $\mathbf{L}(S)$ is Lie generating.*

(2) *The unit group $H(S) := S \cap S^{-1}$ is connected.*

(3) *There exists a $\mathbf{1}$-neighborhood U in G such that*

$$S \cap U \subseteq \operatorname{reach}(S).$$

(4) *For every element $x \in S \setminus \widetilde{\operatorname{comp}}(S)$ there exists a continuous monotone curve $\gamma : (\mathbb{R}^+, \leq) \to (G, \leq_S)$ such that $\gamma(0) = \mathbf{1}$, $\gamma(t) \leq_S x$ for all $t \in \mathbb{R}^+$ and $\gamma(\mathbb{R}^+)/H(S)$ is not relatively compact in $G/H(S)$.*

(5) $\widetilde{\operatorname{comp}}(S) \subseteq \operatorname{reach}(S)$.

(6) *If S is generating, then* $\operatorname{int} S \subseteq \langle \exp \mathbf{L}(S) \rangle$.

Proof. (1) [Ne91d, III.15].
(2) [Ne91d, III.2].
(3) [Ne91d, III.4].
(4) [Ne91b, 1.24].
(5) [Ne91d, IV.4].
(6) [HHL89, V.1.10]. ■

Proposition IV.7. *For a LSg (S, G) the following are equivalent:*

(1) *S is invariant under all inner automorphisms of G.*

(2) *$\mathbf{L}(S)$ is an invariant wedge.*

Suppose that (1) *holds, then the following are equivalent:*

(3) *(S, G) is a GLSg.*

(4) *$\mathbf{L}(S) - \mathbf{L}(S) = \mathbf{L}(G)$.*

Proof. (1) $\Rightarrow$ (2): For $x \in \mathbf{L}(G)$ we have that

$$\mathbf{L}(S) = \mathbf{L}\left(I_{\exp x}(S)\right) = e^{\operatorname{ad} x}\,\mathbf{L}(S).$$

Now the connectedness of G shows that $\mathbf{L}(S)$ is invariant under all inner automorphisms of $\mathbf{L}(G)$.

(2) $\Rightarrow$ (1): Suppose that $e^{\operatorname{ad} x}\,\mathbf{L}(S) = \mathbf{L}(S)$ for all $x \in \mathbf{L}(G)$. Then

$$I_{\exp x}(S) = \overline{\langle I_{\exp x}(\exp \mathbf{L}(S))\rangle} = \overline{\langle \exp\left(e^{\operatorname{ad} x}\,\mathbf{L}(S)\right)\rangle} = \overline{\langle \exp \mathbf{L}(S)\rangle} = S$$

and therefore $I_{\exp x}(S) = S$. This proves the invariance of S since G is connected.

(3) $\Leftrightarrow$ (4): If $\mathbf{L}(S)$ is an invariant wedge, then $\mathbf{L}(S) - \mathbf{L}(S)$ is an ideal, hence a subalgebra. Therefore $\mathbf{L}(S)$ is Lie generating in $\mathbf{L}(G)$ if and only if $\mathbf{L}(S) - \mathbf{L}(S) = \mathbf{L}(G)$. ∎

Corollary IV.8. *The functor $\mathbf{L} : \underline{LSg} \to \underline{LWed}$ from* Proposition IV.4 *maps the subcategory $\underline{ILSg}$ into the subcategory $\underline{IWed}$ of $\underline{LWed}$.* ∎

Now we turn to the geometric structure of Lie semigroups (S, G). We want to use the functor $\mathbf{L}$ to relate it to the geometric structure of the Lie wedge $\left(\mathbf{L}(S), \mathbf{L}(G)\right)$. Such objects have been studied in Section I. This functor reduces to the identity if G is a vector group. Then $\exp_G = \operatorname{id}_{\mathbf{L}(G)}$ and the Lie semigroups (S, G) are exactly the wedges in $\mathbf{L}(G)$.

Definition IV.9. Let (S, G) be a Lie semigroup and $F \subseteq S$ a subsemigroup. We set

$$L_F(S) := \overline{\langle SF^{-1}\rangle} \quad \text{and} \quad T_F(S) := H\left(L_F(S)\right).$$

Suppose, in addition, that F is closed. Then we say that F is

(1) *a face of S* if $S \setminus F$ is an ideal in S.

(2) *an exposed face of S* if $F = S \cap T_F(S)$.

(3) *a compact exposed face of S* if F is an exposed face and if there exists a continuous homomorphism $\varphi : S \to K$, where K is a compact monoid, such that $F = \varphi^{-1}\left(H(K)\right)$.

(4) *a normal exposed face of S* if $F = S \cap H(S_1)$, where $S_1 \subseteq G$ is a closed subsemigroup with $S \subseteq S_1$ and which has a normal group of units.

We denote the set of faces (exposed faces, compact exposed faces, normal exposed faces) of S with $\mathcal{F}(S)\left(\mathcal{F}_e(S),\ \mathcal{F}_c(S),\ \mathcal{F}_n(S)\right)$. ∎

To study the relations between these notions we need some preparation. The following result will be proved in the next section in a more general context.

Proposition IV.10. *Let (S,G) be a Lie semigroup such that $H(S)$ is compact. Then there exists a compact monoid K and a homomorphism $\varphi : S \to K$ such that $\varphi^{-1}(H(K)) = H(S)$.*

Proof. Using Corollary I.13 in [Ne91d] we find a compact $\mathbf{1}$-neighborhhod U in G and a closed right ideal $J \neq S$ in S such that

$$S \subseteq UH(S) \cup J.$$

We claim that the closed two sided ideal $I := \overline{SJ}$ is different from S. To see this, suppose that this is false. Then there exist sequences $s_n \in S$, $j_n \in J$ such that $s_n j_n \to \mathbf{1}$. Let $n_0 \in \mathbb{N}$ such that $s_n j_n \notin J$ whenever $n > n_0$. Then, for $n > n_0$, we have that $s_n \notin J$. So we find $u_n \in U$ and $h_n \in H(S)$ such that $s_n = u_n h_n$. Since both sets, U and $H(S)$, are compact, we may assume that $u_n \to u$ and $h_n \to h$. It follows that $s_n \to uh$, so that j_n converges to $j := s^{-1}$. We conclude that $j \in H(S)$, a contradiction to $J \neq S$. This proves our claim that $I \neq S$.

The set $I \times I \cup \Delta_S \subseteq S \times S$, where Δ_S is the diagonal, defines a closed congruence R on S ([CHK83, p.39]) and since S is locally compact and σ-compact (G is σ-compact), the semigroup $K := S/R$ is a topological monoid ([CHK83, p.48]). We write $\varphi : S \to S/R, s \mapsto [s]$, where $[s]$ denotes the congruence class of the element $s \in S$. Then $K = \varphi(S) = \varphi(UH(S)) \cup \varphi(I)$ is compact. Moreover $H(K) = \varphi(H(S))$ and $\varphi^{-1}(H(K)) = H(S)$ because

$$[\mathbf{1}] = [s][s'] = [ss']$$

implies that $ss' = \mathbf{1}$ and $s, s' \in H(S)$. ∎

Example IV.11. The assumption that $H(S)$ is compact is crucial because there are Lie semigroups with the property that every morphism of topological semigroups $\varphi : S \to K$, where K is compact, is constant. An easy example is the semigroup $S = \mathrm{sl}(2,\mathbb{R})\exp(iC) \subseteq \mathrm{Sl}(2,\mathbb{C})$, where $C \subseteq \mathrm{sl}(2,\mathbb{R})$ is a pointed generating invariant cone (cf. [HHL89, V.4]). To see why this is true, let $\varphi : S \to K$ be a morphism into a compact monoid K. Then $\varphi(H(S)) \subseteq H(K)$ and therefore $\varphi(H(S)) = \{\mathbf{1}\}$ because $\mathrm{Sl}(2,\mathbb{R})^\flat = \{\mathbf{1}\}$, i.e., every morphism of $\mathrm{Sl}(2,\mathbb{R})$ into a compact group is constant (Proposition X.1). Thus $\varphi(S) = \varphi(\exp iC)$. For $h \in H(S)$ this shows that

$$\varphi\Big(\exp\big(i\,\mathrm{Ad}(h)c\big)\Big) = \varphi\big(h\exp(ic)h^{-1}\big) = \varphi(h)\varphi\big(\exp(ic)\big)\varphi(h)^{-1} = \varphi\big(\exp(ic)\big).$$

The set $\partial C \setminus \{0\}$ consists of exactly one $\mathrm{Ad}\,(\mathrm{Sl}(2,\mathbb{R}))$-orbit and therefore we conclude that $\varphi(\exp(i\partial C)) = \{\mathbf{1}\}$ because $0 \in \overline{\partial C \setminus \{0\}}$ and $\varphi(\exp 0) = \mathbf{1}$. Let $T := \overline{\langle H(S)\exp(i\partial C)\rangle} \subseteq S$. We know already that $\varphi(T) = \{\mathbf{1}\}$. But T is a Lie semigroup with $\mathrm{sl}(2,\mathbb{R}) + i\partial C \subseteq \mathbf{L}(T)$. Hence $\mathbf{L}(T) = \mathbf{L}(S)$ and $T = S$. This shows that $\varphi(S) = \{\mathbf{1}\}$. ∎

Proposition IV.12. *Let (S,G) be a LSg. Then we have the following characterization of exposed and normal exposed faces.*

(1) *A closed subsemigroup $F \subseteq S$ is an exposed face iff there exists a continuous monotone function $f:(G,\leq_S) \to (\mathbb{R},\leq)$ which is also monotone as a function $(G,\leq_{F^{-1}}) \to (\mathbb{R},\leq)$ and satisfies $f(\mathbf{1})=0$ and*

$$F = f^{-1}(0) \cap S.$$

(2) *A closed subsemigroup $F \subseteq S$ is a normal exposed face iff there exists a morphism $\varphi:(S,G) \to (S_1,G_1)$ of Lie semigroups such that $H(S_1)=\{\mathbf{1}\}$ and*

$$F = \varphi^{-1}(\mathbf{1}) \cap S.$$

Proof. (1) Let $f: G \to \mathbb{R}$ be such a function. Then we set

$$T := \{t \in G : f(gt) \geq f(g) \ \text{ for all } \ g \in G\}.$$

This is a closed submonoid of G which contains S because $s, s' \in T$ and $g \in G$ implies that

$$f(gss') \geq f(gs) \geq f(g).$$

We also have that $H(T) \subseteq \{t \in G : f(t) = f(\mathbf{1})\} = f^{-1}(0)$ and

$$F \subseteq H(T) \cap S \subseteq f^{-1}(0) \cap S = F.$$

Thus $F \in \mathcal{F}_e(S)$. If, conversely, $F \in \mathcal{F}_e(S)$, then $F = T_F(S) \cap S$ and it suffices to find a continuous $L_F(S)$-monotone function $f: G \to \mathbb{R}$ such that $f(\mathbf{1}) = 0$ and $f^{-1}(0) \cap L_F(S) = T_F(S)$. We set $f(g) := d(gL_F(S), \mathbf{1})$, where d is a left invariant metric on G. Then $g \mapsto f(g) = d(L_F(S), g^{-1})$ is continuous, $L_F(S)$-monotone and

$$f^{-1}(0) \cap S = \{s \in S : s^{-1} \in L_F(S)\} = T_F(S) \cap S = F.$$

(2) Suppose that $F \in \mathcal{F}_n(S)$ and $F = S \cap H(S_2)$, where S_2 is a closed subsemigroup of G such that $H(S_2)$ is normal. Then the quotient morphism $\varphi: (S,G) \to \big(\overline{\varphi(S)}, G/H(S_1)\big)$ has the desired properties because

$$\big(\overline{\varphi(S)}, G/H(S_1)\big)$$

is a Lie semigroup (Proposition IV.4) and

$$H\big(\overline{\varphi(S)}\big) \subseteq H\big(\varphi(S_2)\big) = \{\mathbf{1}\}.$$

To see that the converse also holds, we assume that $\varphi:(S,G) \to (S_1,G_1)$ is a Lie semigroup morphism with $\ker\varphi \cap S = F$ and $H(S_1) = \{\mathbf{1}\}$. We set $S_2 := \varphi^{-1}(S_1)$. This is a closed subsemigroup of G with the normal group of units $H(S_2) = \ker\varphi$ and $H(S_2) \cap S = F$. Thus $F \in \mathcal{F}_n(S)$. ∎

Proposition IV.13. (The Hierarchy of Faces) *For a Lie semigroup* (S, G) *we have the following hierarchy of faces:*

$$\textit{normal exposed face} \Rightarrow \textit{compact exposed face} \Rightarrow \textit{exposed face} \Rightarrow \textit{face}\,.$$

Proof. (1) Let $F \subseteq S$ be a normal exposed face and $\varphi : (S, G) \to (S_1, G_1)$ a morphism of Lie semigroups with $H(S_1) = \{\mathbf{1}\}$ and $\ker\varphi \cap S = F$ (Proposition IV.12). Then we use Lemma IV.10 to find a morphism $\psi : S_1 \to K$ of S_1 into a compact monoid such that $\psi^{-1}\big(H(K)\big) = \{\mathbf{1}\}$. Now $\psi \circ \varphi\,|_S : S \to K$ is a morphism of S into a compact monoid with

$$(\psi \circ \varphi\,|_S)^{-1}\big(H(K)\big) = (\varphi\,|_S)^{-1}\psi^{-1}\big(H(K)\big) = (\varphi\,|_S)^{-1}(\mathbf{1}) = \varphi^{-1}(\mathbf{1}) \cap S = F.$$

To see that F is an exposed face we note that $T_F(S) \subseteq H(S_1)$. Therefore

$$F \subseteq T_F(S) \cap S \subseteq H(S_1) \cap S = F.$$

Thus $F \in \mathcal{F}_c(S)$.
(2) $\mathcal{F}_c(S) \subseteq \mathcal{F}_e(S)$ is contained in the definition of $\mathcal{F}_c(S)$.
(3) Let $F = T_F(S) \cap S$ be an exposed face. Then $S \setminus F = S \setminus T_F(S)$ and the fact that $L_F(S) \setminus T_F(S)$ is an ideal in $L_F(S)$ shows that $S \setminus F$ is an ideal in S, i.e., F is a face of S. ∎

Remark IV.14. We will see in Section IX that $\mathcal{F}_n(S) = \mathcal{F}_c(S)$ if S is a generating invariant Lie semigroup. ∎

Remark IV.15. Suppose that G is abelian. Then every exposed face is normal because every subgroup of G is normal. Hence

$$\mathcal{F}_n(S) = \mathcal{F}_c(S) = \mathcal{F}_e(S).$$

Let $G \cong \mathbb{R}^n$ and (S, G) be a LSg. Then $W := S$ is a wedge since $\exp_G = \mathrm{id}_G$ and $\exp_G\big(\mathbf{L}(S)\big)$ is a closed subsemigroup of G. Let $F \in \mathcal{F}(S)$ and $f \in F$. For every $t \in [0,1]$ we find that $tf + (1-t)f \in F$ and $tf, (1-t)f \in S$. Hence $tf \in F$ and therefore $\mathbb{R}^+ F = F$. Consequently F is a wedge. This shows that the definition of a face, in this case, is consistent with the definition of a face of a wedge in Section I. A closed subsemigroup $F \subseteq S$ is an exposed face iff

$$F = S \cap T_F(S) = S \cap H(\overline{S - F})$$

which is also consistent with the definition from Section I. We notice also that an exposed face F of a wedge W may be defined by the existence of a linear functional $\omega \in W^*$ such that

$$F = \omega^{-1}(0) \cap W.$$

It is interesting to compare this with the characterization in Proposition IV.12 because, in the abelian case, a functional $\omega \in W^*$ is a W-monotone function

$(\mathbb{R}^n, \leq_W) \to (\mathbb{R}, \leq)$ and also a homomorphism of Lie semigroups $(W, \mathbb{R}^n) \to (\mathbb{R}^+, \mathbb{R})$. So we may identify W^* with the set

$$\operatorname{Hom}\big((W, \mathbb{R}^n), (\mathbb{R}^+, \mathbb{R})\big).$$

This illuminates the splitting of the notion of an exposed face into three different ones in the non-abelian case. The set

$$\operatorname{Hom}\big((S, G), (\mathbb{R}^+, \mathbb{R})\big)$$

is in general very small for a Lie semigroup (S, G). Indeed, all such morphisms vanish on the commutator group G' which implies that they are trivial whenever $G = G'$. ∎

Proposition IV.16. *The sets $\mathcal{F}(S), \mathcal{F}_e(S), \mathcal{F}_c(S), \mathcal{F}_n(S)$ are stable under formation of arbitrary intersections.*

Proof. (1) $\mathcal{F}(S)$: Let $F_i \in \mathcal{F}(S)$ for $i \in I$ and set $F := \bigcap_{i \in I} F_i$. Then F is a closed subsemigroup of S and its complement is the union of the ideals $S \setminus F_i$, hence is an ideal and therefore $F \in \mathcal{F}(S)$.

(2) $\mathcal{F}_e(S)$: Let $F_i \in \mathcal{F}_e(S)$ for $i \in I$ and set $F := \bigcap_{i \in I} F_i$. Then $F_i = S \cap T_{F_i}(S)$ and (2) follows from

$$F \subseteq S \cap T_F(S) \subseteq \bigcap_{i \in I} S \cap T_{F_i}(S) = \bigcap_{i \in I} F_i = F.$$

(3) $\mathcal{F}_c(S)$: Let $F_i \in \mathcal{F}_c(S)$ for $i \in I$ and set $F := \bigcap_{i \in I} F_i$. We find compact monoid K_i and homomorphisms $\varphi : S \to K_i$ such that $\varphi_i^{-1}\big(H(K_i)\big) = F_i$. Then the homomorphism

$$\varphi : S \to \prod_{i \in I} K_i, \quad s \mapsto \big(\varphi_i(s)\big)$$

is a continuous homomorphism into the compact monoid $\prod_{i \in I} K_i$ and

$$\varphi^{-1}\big(H(\prod_{i \in I} K_i)\big) = \varphi^{-1}\big(\prod_{i \in I} H(K_i)\big) = \bigcap_{i \in I} \varphi_i^{-1}\big(H(K_i)\big) = F.$$

(4) $\mathcal{F}_n(S)$: Let $F_i \in \mathcal{F}_n(S)$ for $i \in I$ and set $F := \bigcap_{i \in I} F_i$. We find closed subsemigroups $S_i \subseteq G$ such that the groups $H(S_i)$ are normal and $F_i = H(S_i) \cap S$. Set $T := \bigcap_{i \in I} S_i$. Then T contains S and

$$H(T) \subseteq \bigcap_{i \in I} H(S_i) \subseteq H(T).$$

Consequently $H(T) = \bigcap_{i \in I} H(S_i)$ is a normal subgroup of G. Moreover $H(T) \cap S = \bigcap_{i \in I} \big(H(S_i) \cap S\big) = F$. ∎

Corollary IV.17. *The sets $\mathcal{F}(S), \mathcal{F}_e(S), \mathcal{F}_c(S), \mathcal{F}_n(S)$ are complete lattices. $H(S)$ is minimal in $\mathcal{F}(S)$ and $\mathcal{F}_e(S)$, and also in $\mathcal{F}_c(S)$ and $\mathcal{F}_n(S)$ whenever it is normal. The whole semigroup S is the maximal element in all these lattices.*■

Proposition IV.18. *Let $\varphi : (S,G) \to (S',G')$ an LSg-morphism. Then the mapping*

$$\mathcal{F}_e(\varphi) : \mathcal{F}_e(S') \to \mathcal{F}_e(S), \quad F' \mapsto \varphi^{-1}(F') \cap S$$

preserves infs and maps $\mathcal{F}_n(S')$ into $\mathcal{F}_n(S)$. If $\varphi : (S',G') \to (S'',G'')$ is another LSg-morphism, we have

$$\mathcal{F}_e(\varphi' \circ \varphi) = \mathcal{F}_e(\varphi) \circ \mathcal{F}_e(\varphi'),$$

i.e., $\mathcal{F}_e : \underline{LSg} \to \underline{CISL}$ is a contravariant functor into the category $\underline{CISL}$ of complete inf-semilattices, whose objects are the partially ordered sets in which arbitrary infs exist and the morphisms are the inf-preserving mappings. Moreover,

$$\varphi(L_F(S)) \subseteq L_{\varphi(F)}(\varphi(S)), \quad \varphi(T_F(S)) \subseteq T_{\varphi(F)}(\varphi(S))$$

and $\mathcal{F}_e(\varphi)$ is an isomorphism of complete lattices if $\ker\varphi \subseteq S$ and $S' = \varphi(S)$. This isomorphism induces also an isomorphism of $\mathcal{F}_n(S')$ onto $\mathcal{F}_n(S)$ if $\varphi(G) = G'$.

Proof. We firstly notice that

$$\varphi(L_F(S)) = \varphi(\overline{\langle SF^{-1}\rangle}) \subseteq \overline{\langle \varphi(S)\varphi(F)^{-1}\rangle} \subseteq L_{\varphi(F)}(\varphi(S))$$

and

$$\begin{aligned}\varphi(T_F(S)) &= \varphi(H(L_F(S)) \subseteq H\Big(\varphi(L_F(S))\Big) \\ &\subseteq H\Big(L_{\varphi(F)}(\varphi(S))\Big) = T_{\varphi(F)}(\varphi(S)).\end{aligned}$$

Let $F' \in \mathcal{F}_e(S')$ and $F := \varphi^{-1}(F') \cap S$. We claim that $F \in \mathcal{F}_e(S)$. We have that $F \subseteq S \cap T_F(S)$ and

$$\varphi(S \cap T_F(S)) \subseteq S' \cap T_{\varphi(F)}(\varphi(S)) \subseteq S' \cap T_{F'}(S') = F'.$$

We conclude that $S \cap T_F(S) \subseteq F$, hence that $F \in \mathcal{F}_e(S)$. If, in addition, $F' \in \mathcal{F}_n(S')$ and $S_1' \subseteq G'$ a closed subsemigroup which contains S' such that $H(S_1')$ is normal and $H(S_1') \cap S' = F'$, then $S_1 := \varphi^{-1}(S_1')$ is a closed subsemigroup of G which contains S, which has the normal group of units $H(S_1) = \varphi^{-1}(H(S_1'))$, and which satisfies

$$H(S_1) \cap S = (\varphi|_S)^{-1} H(S_1') = (\varphi|_S)^{-1}(F') = F.$$

Thus $F \in \mathcal{F}_n(S)$. Moreover, we have

$$\begin{aligned}\mathcal{F}_e(\varphi)\Big(\bigcap_{i\in I} F_i'\Big) &= \varphi^{-1}\Big(\bigcap_{i\in I} F_i'\Big) \cap S = \bigcap_{i\in I} \varphi^{-1}(F_i) \cap S \\ &= \bigcap_{i\in I} (\varphi^{-1}(F_i) \cap S) = \bigcap_{i\in I} \mathcal{F}_e(\varphi)(F_i').\end{aligned}$$

If $\varphi' : (S', G') \to (S'', G'')$ is another LSg-morphism, we have

$$\begin{aligned}\mathcal{F}_e(\varphi' \circ \varphi) &= (\varphi' \circ \varphi)^{-1}(F'') \cap S = \varphi^{-1}\big(\varphi'^{-1}(F'')\big) \cap S \\ &= \varphi^{-1}\big(\varphi'^{-1}(F'') \cap S'\big) \cap S = \mathcal{F}_e(\varphi) \circ \mathcal{F}_e(\varphi')F''.\end{aligned}$$

To prove the last assertion, we assume that $\ker \varphi \subseteq H(S)$ and $\varphi(S) = S'$. $\mathcal{F}_e(\varphi)$ is injective: Suppose that $F', F'' \in \mathcal{F}_e(S')$ with $\varphi^{-1}(F') \cap S = \varphi^{-1}(F'') \cap S$. Then

$$F' = \varphi\big(\varphi^{-1}(F') \cap S\big) = \varphi\big(\varphi^{-1}(F'') \cap S\big) = F''.$$

$\mathcal{F}_e(\varphi)$ is surjective: Let $F \in \mathcal{F}_e(S)$ and set $F' := \varphi(F)$. Then $\ker \varphi \subseteq T_F(S)$ and therefore $\varphi\big(L_F(S)\big)$ is closed. Hence

$$\begin{aligned}\varphi(T_F(S)) &= H\Big(\varphi\big(L_F(S)\big)\Big) = H\big(\overline{\varphi(\langle SF^{-1}\rangle)}\big) = H\big(\overline{\langle \varphi(S)\varphi(F)^{-1}\rangle}\big) \\ &= H\big(\overline{\langle S'F'^{-1}\rangle}\big) = T_{F'}(S').\end{aligned}$$

We conlcude that

$$S' \cap T_{F'}(S') = \varphi(S) \cap \varphi\big(T_F(S)\big) = \varphi\big(S \cap T_F(S)\big) = \varphi(F) = F',$$

thus $F' \in \mathcal{F}_e(S')$ and $F = \mathcal{F}_e(\varphi)F'$. If $\varphi(G) = G'$, $F \in \mathcal{F}_n(S)$ with $F = H(S_1) \cap S$ and $H(S_1)$ is normal, then $\ker \varphi \subseteq H(S_1)$ and thus $S_1' := \varphi(S_1)$ is a closed subsemigroup of G' containing S' such that $H(S_1') = \varphi\big(H(S_1)\big)$ is normal and

$$H(S_1') \cap S' = \varphi\big(H(S_1) \cap S\big) = \varphi(F) = F'.$$

We conclude that $F' \in \mathcal{F}_n(S')$. This completes the proof since the restriction to $\mathcal{F}_n(S')$ of $\mathcal{F}_n(\varphi)$ is injective, too. ■

Remark IV.19. We note that for faces $F \in \mathcal{F}_e(S)$ and $F' \in \mathcal{F}_e(S')$ we have that

$$F \subseteq \mathcal{F}_e(\varphi)(F') \Leftrightarrow \varphi(F) \subseteq F'.$$

Therefore the assignments $F' \mapsto \mathcal{F}_e(\varphi)F'$ and $F \mapsto \langle \varphi(F)\rangle_{\mathcal{F}_e(S')}$, the exposed face of S' generated by $\varphi(F)$, define a Galois connection between the complete lattices $\mathcal{F}_e(S)$ and $\mathcal{F}_e(S')$. ■

Example IV.20. (1) We refer to Example I.6 for an example, where $\mathcal{F}_e(S) \neq \mathcal{F}(S)$. This occurs even in the case of wedges in vector spaces.
(2) We give an example with $\mathcal{F}_e(S) \neq \mathcal{F}_n(S)$. Let $G = \mathrm{Sl}(2, \mathbb{R})^\sim$ and $\mathbf{L}(G) = \mathrm{sl}(2, \mathbb{R})$ (Example VI.23). The cone

$$W := \{hH + tT + xU : x \geq 0, h^2 + t^2 \leq x^2\}$$

is the tangent wedge of the invariant subgroup $S := \overline{\langle \exp W\rangle}$ ([HHL89, V.4], [Ne90, 1.5]). The half-line $F := \exp\big(\mathbb{R}^+(U + H)\big)$ is an exposed face of S

because the subsemigroup $\widetilde{S} := \overline{\langle \exp V \rangle}$ with $V = \overline{W - \mathbf{L}(F)}$ agrees with $L_F(S)$ and satisfies

$$F = S \cap H(\widetilde{S}) = S \cap T_F(S).$$

The simple Lie algebra $\mathbf{L}(G)$ contains no non-trivial ideals, hence F is not in $\mathcal{F}_n(S)$.

(3) We also give an example with $\mathcal{F}_e(S) \neq \mathcal{F}_c(S)$. This provides a second example for $\mathcal{F}_e(S) \neq \mathcal{F}_n(S)$. Let G denote the connected component of $\mathbf{1}$ in the affine group on $\mathbb{R}$. This group may be parametrized by $\mathbb{R}^+ \setminus \{0\} \times \mathbb{R}$ with

$$(a,b)(a',b') = (aa', ab' + b).$$

We set $S := \{(a,b) \in G : b \geq 0\}$. This is a Lie semigroup in G and every morphism $\varphi : S \to K$ into a compact monoid satisfies $\varphi(0,b) = \mathbf{1}$ for all $b \geq 0$ ([Rup88, p.328]). Hence $\varphi^{-1}(\mathbf{1}) \cap \operatorname{int} S \neq \emptyset$ and therefore the only compact exposed face of S is S itself (Lemma IV.31). In particular $H(S) \in \mathcal{F}_e(S) \setminus \mathcal{F}_c(S)$.

(4) To find an example with $\mathcal{F}_c(S) \neq \mathcal{F}_n(S)$ would be an interesting problem. As we will see in Section IX such an example does not exist in the category of invariant Lie semigroups. ∎

In the rest of this section we turn to the structure of faces of a Lie semigroup S and how they are related to faces of $\mathbf{L}(S)$.

Definition IV.21. Let $(W, \mathfrak{g})$ be a Lie wedge. A face $F \in \mathcal{F}(W)$ is called a *Lie face* if F is a Lie wedge. Since $H(W) = H(F)$, this is equivalent to the invariance of F under the group $\langle e^{\operatorname{ad} H(W)} \rangle$. Thus we write

$$\mathcal{F}_L(W) := \{F \in \mathcal{F}(W) : e^{\operatorname{ad} H(W)} F = F\}$$

for the set of all Lie faces of W.

Lemma IV.22. *The set $\mathcal{F}_L(W) \subseteq \mathcal{F}(W)$ is stable under arbitrary intersections and therefore a complete lattic in its own right.*

Proof. trivial ∎

Definition IV.23. Let $(W, \mathbf{L}(G))$ be a Lie wedge. We say that W is *global in* G if there exists a Lie semigroup (S, G) such that $\mathbf{L}(S) = W$. ∎

Lemma IV.24. *Let $(W, \mathbf{L}(G))$ a Lie wedge. Then the following assertions are equivalent:*

(1) *W is global in G.*

(2) $\mathbf{L}(\overline{\langle \exp W \rangle}) = W$.

Proof. If $\mathbf{L}(S) = W$ and $S \subseteq G$ is a closed Lie semigroup, then $\langle \exp W \rangle \subseteq S$ and therefore

$$W \subseteq \mathbf{L}(\overline{\langle \exp W \rangle}) \subseteq W.$$

If, conversely, $\mathbf{L}(\overline{\langle \exp W \rangle}) = W$, then $S := \overline{\langle \exp W \rangle})$ is a Lie subsemigroup of G with $\mathbf{L}(S) = W$. ∎

The following proposition is a useful tool to prove globality for Lie wedges W which sit nicely in global Lie wedges V.

Proposition IV.25. *Let G be a connected Lie group, $W \subseteq V \subseteq \mathbf{L}(G)$ Lie wedges, and suppose that*

1) $W \cap H(V) \subseteq H(W)$,
2) $\langle \exp H(W) \rangle$ *is closed, and*
3) V *is global in* G.

Then

$$\mathbf{L}(\overline{\langle \exp W \rangle}) = W.$$

Proof. Using [Ne89a, II.12] we find a V-positive function $f \in C^\infty(G)$ such that

$$df(\mathbf{1}) \in \operatorname{algint} V^*.$$

According to our assumption we have

$$\operatorname{algint} V^* \subseteq \operatorname{algint} W^*$$

(Proposition I.2) and therefore f is a W-positive function with $df(\mathbf{1}) \in \operatorname{algint} W$. Applying again [Ne89a, II.12] we conclude that

$$\mathbf{L}(\overline{\langle \exp W \rangle}) = W.$$

■

Lemma IV.26. *Let (S, G) be a Lie semigroup. Then every Lie face*

$$F \in \mathcal{F}_L\big(\mathbf{L}(S)\big)$$

is global in G.

Proof. We check the conditions of Proposition IV.25. If $F \in \mathcal{F}_L\big(\mathbf{L}(S)\big)$ then

$$F \subseteq \mathbf{L}(S) \quad \text{and} \quad H(F) = H\big(\mathbf{L}(S)\big).$$

Thus $F \cap H\big(\mathbf{L}(S)\big) = H\big(\mathbf{L}(S)\big) \subseteq H(F)$ and $\langle \exp H(F) \rangle = H(S)$ is closed in G. ■

The following result, whose proof rests on a certain amount of machinery, shows that faces of Lie semigroups are, at least locally around $\mathbf{1}$, not too far from Lie semigroups.

Theorem IV.27. *If (S, G) is a Lie semigroup and $F \in \mathcal{F}(S)$, then there exists a neighborhood U of $\mathbf{1}$ in G such that*

$$U \cap F \subseteq \overline{\langle \exp \mathbf{L}(F) \rangle}.$$

Proof. [Ne90a, III.2] ■

Remark IV.28. The Lemma IV.26 supplies us with Lie semigroups $F \subseteq S$ which are good candidates for faces of S, namely those generated by the Lie faces of $\mathbf{L}(S)$. That this idea is misleading in some cases will be shown by the following example, where there exists a face

$$F \in \mathcal{F}(S) \quad \text{with} \quad \mathbf{L}(F) \notin \mathcal{F}(\mathbf{L}(S)).$$

It is taken from [Su72] where it serves as an example of a control system where not all points are reachable with bang-bang controls. Exactly these points will be the face we are looking for. We define

$$B := \begin{pmatrix} 0 & 0 & 0 & 0 \\ 0 & 0 & 0 & 0 \\ 0 & 0 & 0 & 0 \\ 0 & 0 & 1 & 0 \end{pmatrix} \quad \text{and} \quad C := \begin{pmatrix} 0 & 0 & 0 & 0 \\ 1 & 0 & 0 & 0 \\ 0 & 1 & 0 & 0 \\ 0 & 0 & 0 & 0 \end{pmatrix}.$$

An easy computation shows that the Lie algebra generated by these matrices in $\mathrm{gl}(4, \mathbb{R})$ is isomorphic to $V \rtimes \mathbb{R}C$, where C acts on $V \cong \mathbb{R}^3$ as the matrix

$$\begin{pmatrix} 0 & 0 & 0 \\ 1 & 0 & 0 \\ 0 & 1 & 0 \end{pmatrix},$$

where B corresponds to the first base vector $(1,0,0)^\top$. This Lie algebra belongs to a nilpotent group of unipotent matrices. The Lie semigroup S in $\mathrm{Gl}(4, \mathbb{R})$ generated by the Lie wedge

$$W := \mathbb{R}^+(B + C) + \mathbb{R}^+(B - C)$$

has the property that every element $s \in S$ is reachable from $\mathbf{1}$ with an absolutely continuous S-monotone curve and every pair of points $s, s' \in S$ with $ss' \in \exp \mathbb{R}^+ B$ satisfies $s, s' \in \exp \mathbb{R}^+ B$. If $u : [0,T] \to [-1,1]$ is a measurable function the solution $\gamma_u : [0,T] \to G$ of the initial value problem

$$\gamma_u(0) = \mathbf{1} \quad \text{and} \quad \dot{\gamma}_u(t) = d\lambda_{\gamma_u(t)}(\mathbf{1})\big(B + u(t)C\big)$$

is an S-monotone curve $\gamma_u : ([0,T], \leq) \to (G, \leq_S)$ and

$$\gamma_u(t) = \begin{pmatrix} 1 & 0 & 0 & 0 \\ f(t) & 1 & 0 & 0 \\ \frac{1}{2} f(t)^2 & f(t) & 1 & 0 \\ g(t) & h(t) & t & 1 \end{pmatrix},$$

where

$$f(t) = \int_0^t u(\tau)\, d\tau, \quad h(t) = \int_0^t \tau u(\tau)\, d\tau, \quad \text{and} \quad g(t) = \int_0^t u(\tau) h(\tau)\, d\tau.$$

If $\gamma_{u_n}(T_n)$ is a sequence of points in S which converge to $s \in S$ then the explicit form of γ_u shows that T_n converges to a number $T \geq 0$ and thus $s = \gamma_u(T)$ for a measurable function $u : [0,T] \to [-1,1]$ because the set of all these functions is weak-$*$-compact, this topology has a countable base, and if u_n converges to u in this topology, this implies that $\gamma_{u_n}(T) \to \gamma_u(T)$ ([CL89]). Thus

$$S = \{\gamma_u(T) : T \in \mathbb{R}^+, u : [0,T] \to [-1,1] \ \text{measurable}\}.$$

It is clear that $F := \exp \mathbb{R}^+ B \subseteq S$ is closed. If $s = \gamma_u(T)$ and $s' = \gamma_{u'}(T')$ with $ss' \in F$, then $ss' = \gamma_{u''}(T + T')$ with

$$u''(\tau) := \begin{cases} u(\tau), & \text{for } 0 \leq \tau \leq T \\ u'(\tau - T), & \text{for } T \leq \tau \leq T + T'. \end{cases}$$

Thus the argument given in [Su72] shows that $u''(\tau) = 0$ almost everywhere on $[0, T + T']$ and therefore $s, s' \in \exp \mathbb{R}^+ B = F$. Consequently $F \in \mathcal{F}(S)$ and $\mathbf{L}(F) = \mathbb{R}^+ B$ is far from being a face of the two dimensional wedge W because it is a ray between the two extremal rays of W. ∎

In view of the preceding remark it is interesting that there exists a large class of Lie semigroups for which it is true that

$$\mathbf{L}\left(\mathcal{F}(S)\right) \subseteq \mathcal{F}\left(\mathbf{L}(S)\right).$$

Definition IV.29. A wedge W in a Lie algebra $\mathfrak{g}$ is called a *Lie semialgebra* if there exists a BCH-neighborhood B in $\mathfrak{g}$ such that

$$(W \cap B) * (W \cap B) \subseteq W.$$

∎

The crucial properties of semialgebras are the following.

Proposition IV.30. *If $W \subseteq \mathfrak{g}$ is a Lie semialgebra, then*

(1) *$W - W$ is a subalgebra of $\mathfrak{g}$,*

(2) *W is a Lie wedge, and*

(3) *every face of W is a Lie semialgebra.*

(4) *Every invariant wedge is a Lie semialgebra.*

Proof. (1) [HHL89, II.2.13].

(2) [HHL89, III.2.15].

(3) It is shown in [HHL89, II.2.20] that every exposed face of a Lie semialgebra is a Lie semialgebra. Using Proposition I.5 we find a finite chain

$$F = F_0 \subseteq F_1 \subseteq \ldots \subseteq F_n = W$$

of wedges $F_i \subseteq W$ such that $F_i \in \mathcal{F}_e(F_{i+1})$. Inductively, this implies that F_i is a Lie semialgebra for $i = n, \ldots, 0$.

(4) [HHL89, III.2.15]. ∎

Lemma IV.31. *Let F be a face of a LSg S and $F \cap \operatorname{int} S \neq \emptyset$. Then $F = S$.*

Proof. Let $f \in F \cap \operatorname{int} S$ and $x \in \mathbf{L}(S)$. Then there exists $t > 0$ such that $\exp(tx) \in fS^{-1}$ since fS^{-1} is a neighborhood of $\mathbf{1}$ in G. Hence $f \in \exp(tx)S \cap F$ and therefore $\exp(tx) \in F$. The fact that F is a face implies that $\exp(\mathbb{R}^+ x) \subseteq F$. Consequently $S \subseteq F$ because $x \in \mathbf{L}(S)$ was arbitrary. ∎

Proposition IV.32. *Let (S,G) be a LSg such that $\mathbf{L}(S)$ is a Lie semialgebra. Then*

$$\mathbf{L}\left(\mathcal{F}(S)\right) \subseteq \mathcal{F}\left(\mathbf{L}(S)\right).$$

Proof. Let $F \in \mathcal{F}(S)$ and $F' := T_{\mathbf{L}(F)}\left(\mathbf{L}(S)\right) \cap \mathbf{L}(S)$ the exposed face generated by $\mathbf{L}(F)$ in $\mathbf{L}(S)$. We consider two cases.
Case 1: $F' = \mathbf{L}(S)$. Then $\mathbf{L}(F) \cap \operatorname{algint} \mathbf{L}(S) \neq \emptyset$ because otherwise we find $\omega \in \mathbf{L}(S)^* \cap \mathbf{L}(F)^\perp \setminus \{0\}$ and $\ker \omega \cap \mathbf{L}(S)$ is an exposed face of $\mathbf{L}(S)$ containing $\mathbf{L}(F)$. Let $G_1 := \langle \exp\left(\mathbf{L}(S) - \mathbf{L}(S)\right)\rangle$ the analytic subgroup generated by $\exp \mathbf{L}(S)$ endowed with its Lie group topology. Then $S_1 := \overline{\langle \exp_{G_1} \mathbf{L}(S)\rangle} \subseteq G_1$ is a Lie semigroup and $\mathbf{L}(S_1) = \mathbf{L}(S)$ is generating in $\mathbf{L}(G_1) = \mathbf{L}(S) - \mathbf{L}(S)$. Moreover $F_1 := F \cap S_1$ is a face of S_1 with $\mathbf{L}(F) = \mathbf{L}(F_1)$. Now every $f \in \mathbf{L}(F) \cap \operatorname{int}_{\mathbf{L}(G_1)} \mathbf{L}(S_1)$ satisfies $\exp(tf) \in \operatorname{int}_{G_1}(S_1)$ for all $t > 0$. Therefore $\exp(f) \in F_1 \cap \operatorname{int}_{G_1} S_1$. Thus $F_1 = S_1$ (Lemma IV.31) and thus $F = S$ since S_1 is dense in S. In this case $\mathbf{L}(F) = \mathbf{L}(S) \in \mathcal{F}\left(\mathbf{L}(S)\right)$.
Case 2: $F' \neq \mathbf{L}(S)$. Then F' is a global Lie semialgebra (Lemma IV.26, Proposition IV.30) because semialgebras are Lie wedges (Proposition IV.30). Now $F' \in \mathcal{F}_e\left(\mathbf{L}(S)\right)$ and it suffices to show that $\mathbf{L}(F)$ is a face of F' (Proposition I.5). We iterate this construction inductively with

$$S_1 := \overline{\langle \exp F'\rangle}, \quad F_1 := F \cap S_1 \in \mathcal{F}(S_1), \quad \text{and} \quad \mathbf{L}(F) = \mathbf{L}(F_1).$$

This process stops because either Case 1 occurs or Case 2 diminishes the dimension of $\mathbf{L}(S)$. ∎

Remark IV.33. In view of the above counterexample, which shows that in general the tangent wedge of a face is not a face, one would like to see where the proof breaks down for general Lie wedges. This is the case if the smallest exposed face containing $\mathbf{L}(F)$ agrees with $\mathbf{L}(S)$. Then we find $f \in \mathbf{L}(F) \cap \operatorname{algint} \mathbf{L}(S)$ and also $a \in \mathbf{L}(G)$ with $f \pm a \subseteq \operatorname{algint} \mathbf{L}(S) \setminus \mathbf{L}(F)$. Then the Lie wedge

$$W := \mathbb{R}^+(f+a) + \mathbb{R}^+(f-a)$$

is global in G and $\mathbf{L}(F) \cap W := \mathbb{R}^+ a$ is the tangent wedge of the face $F \cap \overline{\langle \exp W\rangle}$. So every example of a face such that $\mathbf{L}(F)$ is not a face provides an example of the $(f \pm a)$-type as is Sussman's example above. ∎

It is noteworthy that the situation is much better for exposed faces.

Proposition IV.34. *Let S be a Lie semigroup. Then*

$$\mathbf{L}\left(\mathcal{F}_e(S)\right) \subseteq \mathcal{F}_e\big(\mathbf{L}(S)\big).$$

Proof. Suppose that $F \in \mathcal{F}_e(S)$ is exposed and set $V := \overline{\mathbf{L}(S) - \mathbf{L}(F)}$. Then

$$\mathbf{L}(F) \subseteq \mathbf{L}(S) \cap H(V) \subseteq \mathbf{L}(S) \cap \mathbf{L}\left(T_F(S)\right) = \mathbf{L}\left(S \cap T_F(S)\right) = \mathbf{L}(F).$$

Therefore

$$\mathbf{L}(F) = \mathbf{L}(S) \cap H(V) = \mathbf{L}(S) \cap H\big(\overline{\mathbf{L}(S) - \mathbf{L}(F)}\big)$$

and $\mathbf{L}(F)$ is an exposed face of $\mathbf{L}(S)$. ∎

Lemma IV.35. *Let $H(S) \neq F \in \mathcal{F}(S)$. Then $\mathbf{L}(F) \neq H\big(\mathbf{L}(S)\big)$.*

Proof. Assume that $F \neq H(S)$ is a face of S and that $\mathbf{L}(F) = H\big(\mathbf{L}(S)\big)$. Then Theorem IV.27 shows that $H(S)$ is isolated in F, i.e., there exists a neighborhood U of $\mathbf{1}$ in G such that

$$U \cap F \subseteq H(S).$$

Let $f \in F \setminus H(S)$. According to Proposition IV.4 in [Ne91d] we know that

$$F \setminus H(S) \subseteq S \setminus \text{reach}(S).$$

Thus we find $x \in F \setminus \text{reach}(S)$ and a monotone curve $\gamma : (\mathbb{R}^+, \leq) \to (G, \leq_S)$ such that $\gamma(0) = \mathbf{1}$, $\gamma(\mathbb{R}^+) \not\subseteq H(S)$ (Proposition IV.6), and $\gamma(t) \leq_S x$ for all $t \in \mathbb{R}^+$. Since $F \in \mathcal{F}(S)$, $\gamma(t) \leq_S x$ implies that $\gamma(t) \in F$ for every $t \in \mathbb{R}^+$. Hence $\gamma(\mathbb{R}^+) \subseteq F$. This leads to the contradiction $\gamma(\mathbb{R}^+) \subseteq H(S)$. ∎

Proposition IV.36. *Let $F \in \mathcal{F}_e(S)$. Then $L_F(S)$ is a Lie semigroup and there exists a sequence of wedges*

$$\mathbf{L}(S) = W_0 \subseteq W_1 \subseteq \ldots \subseteq W_n$$

such that

(1) $W_{i+1} = \mathbf{L}\left(\overline{\langle \exp L_{F_i}(W_i)\rangle}\right)$ *for an exposed face* $F_i \in \mathcal{F}_e(W_i)$ *and*

(2) $L_F(S) = \overline{\langle \exp W_n\rangle}$.

If F and S are invariant, then all the wedges W_i are invariant.

Proof. We set $W_0 := \mathbf{L}(S)$. If $F = H(S)$, then $L_F(S) = S = \overline{\langle \exp W_0\rangle}$ and we are done. If $F \neq H(S)$, then $\mathbf{L}(F) \neq H(W_0)$ (Lemma IV.35). We set

$$W_1 := \mathbf{L}\left(\overline{\langle \exp L_{\mathbf{L}(F)}(W_0)\rangle}\right) \supseteq L_{\mathbf{L}(F)}(W_0).$$

Then $\dim H(W_1) > \dim H(W_0)$ and $\exp(W_1) \subseteq L_F(S)$. According to Proposition IV.34 we know that $\mathbf{L}(F) \in \mathcal{F}_e(W_0)$. Suppose that $W_0, \ldots, W_i$ are constructed such that $\dim H(W_j) > \dim H(W_{j-1})$ and

$$W_j := \mathbf{L}\left(\overline{\langle \exp L_{F_{j-1}}(W_{j-1})\rangle}\right) \supseteq L_{F_{j-1}}(W_{j-1})$$

with $F_{j-1} \in \mathcal{F}_e(W_{j-1})$ for $j = 1, \ldots, i$ and $\exp W_i \subseteq L_F(S)$. There are two cases. If $\langle \exp W_i \rangle$ is dense in $L_F(S)$, the proof is complete. If not, we set

$$T := \overline{\langle \exp W_i \rangle} = \overline{\langle \exp L_{F_{j-1}}(W_{j-1}) \rangle}$$

and $F' := T_F(S) \cap T$. Then $S \subseteq T$ and therefore $F \subseteq F'$. This implies that $F' \not\subseteq H(T)$ because otherwise $F \subseteq H(T)$ and $L_F(S) \subseteq T$. Moreover

$$F' \subseteq T_{F'}(T) \cap T \subseteq T_F(S) \cap T = F'$$

is an exposed face of T. Consequently

$$F_i := \mathbf{L}(F') \in \mathcal{F}_e\big(\mathbf{L}(T)\big) = \mathcal{F}_e(W_i)$$

is different from $H(W_i)$ (Lemma IV.35). We conclude that $\exp W_{i+1} \subseteq L_F(S)$ for

$$W_{i+1} := \mathbf{L}\left(\overline{\langle \exp L_{F_i}(W_i) \rangle}\right) \supseteq L_{F_i}(W_i)$$

and that $\dim H(W_{i+1}) > \dim H(W_i)$. This procedure has to stop after finitely many steps since $\mathbf{L}(G)$ is a finite dimensional vector space and $\dim H(W_i)$ is enlarged in every step. Thus we find an $n \in \mathbb{N}$ such that $\overline{\langle \exp W_n \rangle} = L_F(S)$. Moreover, it is clear that all wedges F_i and W_i are invariant if this holds for F and S. ∎

Remark IV.37. It is not true for general semigroups that the unit group is not isolated in a face. It depends on the special structure of Lie semigroups. To find an example we take

$$G = \mathbb{R}^2 \quad \text{and} \quad S = \mathbb{R}^+ \times \mathbb{R}^+ \setminus \{(x,y) : 0 \leq x < y < 1\}.$$

Then

$$F := \{(0,0)\} \cup (\{0\} \times [1, \infty[)$$

is a face of S, i.e., $S \setminus F$ is an ideal and F is closed. Moreover $H(S) = \{0\}$ is isolated in F. The largest Lie semigroup contained in S is $S_L = \mathbb{R}^+(1,0) + \mathbb{R}^+(1,1) \neq S$. ∎

Problems IV.38. 1) Find an example of a Lie semigroup S with

$$\mathcal{F}_c(S) \neq \mathcal{F}_n(S)$$

or prove that $\mathcal{F}_c(S) = \mathcal{F}_n(S)$ holds in general.
2) Find an example of a face F of a Lie semigroup S which is not a Lie semigroup. Even if S is invariant, we do not know whether such an example exists or not.
3) Give a better and more general description of what happens in Sussmann's example (Remark IV.28). Maybe one can use these ideas to find an example in 2). The Lie semigroups S with $\mathbf{L}(S) = \mathbb{R}^+(B+C) + \mathbb{R}^+(B-C)$ seem to be sufficiently simple so that it should be possible to characterize those with $\mathbf{L}\left(\mathcal{F}(S)\right) \subseteq \mathcal{F}\big(\mathbf{L}(S)\big)$.

V. COMPACTIFICATIONS OF SUBSEMIGROUPS OF LOCALLY COMPACT GROUPS

In later sections we will consider compactifications of invariant subsemigroups of Lie groups. In this section we construct special compactifications of such semigroups. We give a criterion for the Alexandrov compactification to be a topological semigroup and give a more general construction if this criterion fails. The results of this section will be used to describe the Bohr compactification of invariant Lie semigroups but they are also interesting in their own right.

Definition V.1. Let S be a locally compact semigroup and $I \subseteq S$ a closed two sided ideal. We endow the Alexandrov compactification $S_0 := S \cup \{0\}$ with a semigroup structure by setting

$$0 \cdot s = s \cdot 0 = 0 \quad \text{for every} \quad s \in S_0.$$

We call this semigroup the *one-point compactification* of S. The set $I \cup \{(s,s) : s \in S\} \subseteq S \times S$ is a closed congruence on S. We write S/I for the quotient semigroup, the *Rees quotient semigroup of* S *mod* I. We identify $S \setminus I$ with the corresponding subset of S/I and endow S/I with the quotient topology.

The following theorem gives a criterion for the compatibility of the semigroup structure and the topology on S/I.

Theorem V.2. *Let S be a locally compact $\sigma-$compact semigroup.*

(1) *Let I be a closed ideal of S. Then S/I is a topological semigroup, i.e., the multiplication $S/I \times S/I \to S/I$ is continuous.*

(2) *Let $H \subseteq S$ be a compact normal subgroup, i.e., $sH = Hs$ for all $s \in S$. Then the set $S/H = \{sh : s \in S\}$ is a topological semigroup with respect to the quotient topology.*

Proof. (1) [CHK83, p. 50]
(2) In view of [CHK83, p.48] we have to ckeck that the prescription

$$a \sim b \quad :\Leftrightarrow \quad aH = bH$$

defines a closed congruence on S. Let $a \sim b$ and $a' \sim b'$. Then

$$aa'H = ab'H = aHb' = bHb' = bb'H.$$

Therefore $\sim$ is a congruence on S. Suppose that (a_i, b_i) is a net in $S \times S$ with $a_i \sim b_i$ and $(a,b) = \lim(a_i, b_i)$. Let $h \in H$. Then we find for every i an element

$h_i \in H$ such that $a_i h = b_i h_i$. We have to show that $a \sim b$. So we may assume that $h_1 = \lim h_i$ exists in H since H is compact. Now

$$ah = \lim a_i h = \lim b_i h_i = bh_1 \in bH.$$

The converse inclusion $bH \subseteq aH$ follows with the same argument. ∎

Lemma V.3. *Let S be a locally compact semigroup, then S_0 is a topological semigroup iff the multiplication function $m : S \times S \to S$ is proper.*

Proof. "$\Rightarrow$": Suppose that S_0 is a topological semigroup, $m_0 : S_0 \times S_0 \to S_0$ the multiplication function and let $K \subseteq S$ be a compact set. The set $m_0^{-1}(K) \subseteq S_0 \times S_0$ is closed and contained in $S \times S$, hence $m^{-1}(K) = m_0^{-1}(K)$ is compact in $S \times S$.
"$\Leftarrow$": Suppose that the mapping $m : S \times S \to S$ is proper and that (s_n, t_n) is a net in $S_0 \times S_0$ with $\lim(s_n, t_n) \notin S \times S$. Then the net $s_n t_n$ eventually leaves every compact subset of S. Therefore $\lim s_n t_n = 0$ holds in S_0. Hence S_0 is a topological semigroup. ∎

Definition V.4. We call a locally compact semigroup S *proper* if S_0 is a topological semigroup. ∎

Lemma V.5. *Suppose that S is a locally compact topological semigroup such that left- and right multiplications are proper. Let $I \subseteq S$ be the set of points s such that there exists a net $(s_n, t_n) \in S \times S$ with $\lim s_n t_n = s$ in S and $\lim(s_n, t_n) = (0, 0)$ in $S_0 \times S_0$. Then I is a closed two sided ideal in S.*

Proof. (1) I is an ideal in S: Let $s \in I$ and $t \in S$. We find a net (s_n, t_n) in $S \times S$ with $\lim s_n t_n = s$ and $\lim(s_n, t_n) = (0, 0)$. Then $\lim(ts_n, t_n) = (0, 0)$ because the left multiplciation $\lambda_t : s \mapsto ts$ is proper. In addition, we have that $\lim ts_n t_n = ts$, hence $ts \in I$. The proof of $IS \subseteq I$ is similar. Hence $IS \cup SI \subseteq I$ and I is a two sided ideal.
(2) I is closed: Let $\mathcal{U}(0)$ denote the filter base of neighborhoods of 0 in S_0. We show that

$$I = \bigcap_{V \in \mathcal{U}(0)} \overline{VV}.$$

For $s \in I$ with $s = \lim s_n t_n$ and $\lim(s_n, t_n) = (0, 0)$ we find for every $V \in \mathcal{U}(0)$ an n_0 such that $(s_n, t_n) \in V \times V$ for $n \geq n_0$, hence $s \in \overline{VV}$ for every $V \in \mathcal{U}(0)$. Conversely, suppose that $s \in \bigcap_{V \in \mathcal{U}(0)} \overline{VV}$. Then we find for every $V \in \mathcal{U}(0)$ and $U \in \mathcal{U}(s)$ an element $(s_{(V,U)}, t_{(V,U)}) \in V \times V$ with $s_{(V,U)} t_{(V,U)} \in U$. We endow the product $\mathcal{U}(0) \times \mathcal{U}(s)$ with the direction defined by

$$(U, V) \prec (U', V') \quad \text{if} \quad U' \subseteq U \quad \text{and} \quad V' \subseteq V.$$

Then it is clear that $\lim(s_{(V,U)}, t_{(V,U)}) = (0, 0)$ and $\lim s_{(V,U)} t_{(V,U)} = s$, hence $s \in I$. ∎

Definition V.6. Let S be a monoid. We define a quasi order on S by setting

$$s \prec s' \quad \text{if} \quad s' \in sS.$$

Note that this is the order dual of Green's $\mathcal{R}$-quasiorder.

If S is a submonoid of a group G, we extend this quasi order to G by setting

$$g \preceq_S g' \quad \text{if} \quad g' \in gS \quad \text{for} \quad g \in G.$$

We omit the subscript S if no confusion is possible. This order is closed if S is a compact semigroup or a closed subsemigroup of a group G. Let $(P, \leq)$ be a quasi ordered space. For a subset $M \subseteq P$ we write

$$\uparrow M := \{s \in P : (\exists m \in M) m \leq s\}, \quad \text{and} \quad \downarrow M := \{s \in P : (\exists m \in M) s \leq m\}.$$

For $M = \{s\}$ we set $\uparrow s := \uparrow\{s\}$, $\downarrow s := \downarrow\{s\}$, and $[s,t] := \uparrow s \cap \downarrow t$ for $s, t \in P$. ∎

Lemma V.7. *Suppose that G is a locally compact group and that $S \subseteq G$ is a closed submonoid with $\mathbf{1} \in \overline{\operatorname{int} S}$. Let*

$$\operatorname{comp}(S) := \{s \in S : [\mathbf{1}, s] \ \textit{is compact}\}.$$

Then $I = \overline{S \setminus \operatorname{comp}(S)}$, with I as in Lemma V.5.

Proof. We set $J := S \setminus \operatorname{comp}(S)$.
$J \subseteq I$: Let $s \in J$ and

$$\Lambda := \{K : K \subseteq [\mathbf{1}, s] \text{ is compact }\}.$$

Then $(\Lambda, \subseteq)$ is a directed set. For every $K \in \Lambda$ we find $s_K \in [\mathbf{1}, s] \setminus K$ because $[\mathbf{1}, s]$ is not compact. Hence there exists $\widetilde{s}_K \in S$ such that $s_K \widetilde{s}_K = s$. It is clear that $\lim s_K = 0$ in S_0, hence the net $\widetilde{s}_K = s_K^{-1} s$ eventually leaves every compact subset of G. Therefore $\lim \widetilde{s}_K = 0$ in S_0. We conclude that $s = \lim s_K \widetilde{s}_K \in I$.
$I \subseteq \overline{J}$: Let $s \in I$. For every neighborhood $U \in \mathcal{U}(\mathbf{1})$ we choose an element $s_U \in \operatorname{int} S \cap U$. Then $\lim s_U = \mathbf{1}$ and it sufficies to show that $s s_U \in J$ for every $U \in \mathcal{U}(\mathbf{1})$. It is clear that $s s_U \operatorname{int} S^{-1}$ is an open neighborhood of s. We choose a net $(t_n, \widetilde{t}_n) \in S \times S$ such that $\lim t_n \widetilde{t}_n = s$ and $\lim(t_n, \widetilde{t}_n) = (0,0)$. Then we find n_0 such that $t_n \widetilde{t}_n \in s s_U \operatorname{int} S^{-1}$ for $n \geq n_0$. We conclude that $t_n \in [\mathbf{1}, s s_U]$ for $n \geq n_0$, hence that $[\mathbf{1}, s s_U]$ is not compact. This shows that $s s_U \in J$. ∎

Theorem V.8. *Let G be a locally compact group and $S \subseteq G$ a closed submonoid with $\mathbf{1} \in \overline{\operatorname{int} S}$. Then the following assertions hold:*

(1) *S is proper iff $I = \emptyset$ iff* $\operatorname{comp}(S) = S$.
(2) *If $I \neq \emptyset$, then $I_0 := I \cup \{0\}$ is a closed ideal in S_0, and $S_I := S_0/I_0$ is a compact topological semigroup. It is homeomorphic to the one-point compactification of the space $S \setminus I$.*

(3) *Let* $\varphi : G \to H$ *be a homomorphism of locally compact groups such that* $N := \ker\varphi$ *is a compact subgroup of* $H(S)$. *Then*

$$\varphi\big(\mathrm{comp}(S)\big) = \mathrm{comp}\,\big(\varphi(S)\big) \quad \text{and} \quad \varphi(I_S) = I_{\varphi(S)}.$$

Proof. (1) Suppose that S is proper, then Lemma V.5 implies that $I = \emptyset$. Conversely, assume that $I = \emptyset$. Then Lemma V.5 shows that the multiplication of S_0 is continuous in the point $(0,0)$. The continuity on $S \times S \subseteq S_0 \times S_0$ is clear, and the continuity on $S \times \{0\} \cup \{0\} \times S$ follows from the fact that $\lim s_n = s \in S$ and $\lim t_n = 0$ with $s_n, t_n \in S$ implies that the net $s_n t_n$ eventually leaves every compact subset of G, hence that $\lim s_n t_n = \lim t_n s_n = 0$. Now (1) follows from Lemma V.7 since $I = \overline{S \setminus \mathrm{comp}(S)}$.
(2) It is clear that the mapping $S \setminus I \to S_0/I_0$ is an open embedding. The neighborhoods of I_0 in S_0/I_0 are the open sets containing I_0. These are the complements of compact subsets of $S \setminus I$. Hence S_I is homeomorphic to the Alexandrov compactification of $S \setminus I$. Suppose that S_I is not a topological semigroup, then we find a net $(s_n, \widetilde{s}_n) \in S_I \times S_I$ such that $\lim(s_n, \widetilde{s}_n) \notin S \setminus I \times S \setminus I$ and $s = \lim s_n \widetilde{s}_n \in S \setminus I$. We choose $t \in \mathrm{int}\, S$ such that $st \in S \setminus I$ (I is closed). Then the set $st \,\mathrm{int}\, S^{-1}$ is a neighborhood of s and we find n_0 such that $s_n \widetilde{s}_n \in st\,\mathrm{int}\, S^{-1}$ for $n \geq n_0$. We conclude that $s_n \in [\mathbf{1}, st]$ for $n \geq n_0$. But this set is compact because $st \in S \setminus I \subseteq \mathrm{comp}(S)$. Therefore $u := \lim s_n \in [\mathbf{1}, st]$. We conlude that $\lim u\widetilde{s}_n = s \in S$, hence that $\lim \widetilde{s}_n \notin I$, a contradiction.
(3) We contend that

$$\varphi([\mathbf{1}, s]) = [\mathbf{1}, \varphi(s)] \quad \text{for all} \quad s \in S. \tag{5.1}$$

It is trivial that

$$\varphi([\mathbf{1}, s]) = \varphi(S \cap sS^{-1}) \subseteq \varphi(S) \cap \varphi(s)\varphi(S)^{-1} = [\mathbf{1}, \varphi(s)].$$

Let $t' = \varphi(s') \in [\mathbf{1}, \varphi(s)]$. Then we find $t'' = \varphi(s'') \in \varphi(S)$ such that $\varphi(s) = t't'' = \varphi(s's'')$, hence

$$s' \in S \cap sNS^{-1} \subseteq S \cap sS^{-1} = [\mathbf{1}, s].$$

We conclude that $[\mathbf{1}, s]$ is compact if and only if $[\mathbf{1}, \varphi(s)]$ is compact since the homomorphism φ is a proper mapping. Whence $\varphi\big(\mathrm{comp}(S)\big) = \mathrm{comp}\,\big(\varphi(S)\big)$. Now we have

$$\begin{aligned}\varphi\big(S \setminus \mathrm{comp}(S)\big) &\subseteq \varphi(I_S) = \varphi\big(\overline{S \setminus \mathrm{comp}(S)}\big) \\ &\subseteq \overline{\varphi\big(S \setminus \mathrm{comp}(S)\big)} = \overline{\varphi(S) \setminus \mathrm{comp}\,\big(\varphi(S)\big)} = I_{\varphi(S)}.\end{aligned}$$

This shows that $\varphi(I_S)$ is dense in $I_{\varphi(S)}$. But it is also closed since $I_S N = I_S$. Consequently $\varphi(I_S) = I_{\varphi(S)}$. ■

Theorem V.9. *Suppose that (S,G) is a GLSg and $H(S)$ is compact. Then the following assertions hold:*

(1) *The ideal $S\setminus \operatorname{comp}(S)$ is closed and $S\setminus \operatorname{comp}(S)=I$.*

(2) *S is proper iff all order intervals $[\mathbf{1},s]$ are compact.*

(3) *If S is not proper, then the mapping $j: S\to S_I$ (Theorem V.8(2)) is a homomorphism of S into a compact monoid such that $j\mid_{S\setminus I}$ is an open embedding, the image of I consists of a single point and $H(S_I)=j\big(H(S)\big)$.*

Proof. (1) Let $H:=H(S)$, $M=G/H$ the associated homogeneous space and $\pi: G\to M, g\mapsto gH$ the canonical projection. We define a partial order on M by setting

$$\pi(g)\preceq \pi(g') \quad \text{if} \quad g\preceq g'$$

(cf. Definition IV.4). Applying [Ne91b, 1.26] we see that the set

$$\operatorname{comp}(M):=\{p\in M:[\pi(\mathbf{1}),p] \text{ is compact}\}$$

is closed. According to the definition of the order on M we have

$$\pi([\mathbf{1},s])=[\pi(\mathbf{1}),\pi(s)] \qquad \text{for every } s\in S.$$

Hence $[\mathbf{1},s]$ is compact if and only if $[\pi(\mathbf{1}),\pi(s)]$ is compact since π is a proper mapping. Therefore $\operatorname{comp}(S)=\pi^{-1}\big(\operatorname{comp}(M)\big)$ is closed in S. It follows from Lemma V.7 that $S\setminus\operatorname{comp}(S)=I$.
(2) This is a consequence of Theorem V.8(1) because a generating Lie subsemigroup S of a Lie group G has dense interior (Proposition IV.6).
(3) This follows from Theorem V.8(2). ∎

The following result is of crucial importance in later sections to descripe the structure of the Bohr compactification of a Lie semigroup.

Corollary V.10. *Let (S,G) be a LSg with $H(S)$ compact and $i_S: S\to S^\flat$ the Bohr compactification of S* (cf. Section IX). *Then the mapping i_S is injective on a neighborhood of $H(S)$. Moreover, it embeds a neighborhood of $H(S)$ topologically into $S^\flat$ and $H(S^\flat)=i_S\big(H(S)\big)$.*

Proof. Using the universal property of the compactification $i_S: S\to S^\flat$ and the homomorphism $j: S\to S_I$ from Theorem V.8, we find a continuous homomorphism $j^\flat: S^\flat\to S_I$ such that $j^\flat\circ i_S=j$. But the restriction of j to the neighborhhod $S\setminus I$ of $H(S)$ is an embedding. Therefore $i_S\mid_{S\setminus I}$ is an embedding, too. It is clear that $j^\flat\big(H(S^\flat)\big)\subseteq H(S_I)=H(S)$. If $s\in H(S^\flat)$ and $(s_n)_{n\in D}$ is a net in S with $s=\lim i_S(s_n)$, then

$$j^\flat(s)=\lim j(s_n)=\lim j^\flat\big(i_S(s_n)\big)\in H(S).$$

Consequently we find $h\in H(S)$ such that $s_n\to h$. Thus $s=i_S(h)$. This shows that $H(S^\flat)=i_S\big(H(S)\big)$. ∎

VI. INVARIANT SUBSEMIGROUPS OF LIE GROUPS

Section IV contains facts about general Lie semigroups. In this section we will heavily use the fact that the Lie semigroups (S, G) under consideration are invariant. In Lemma VI.1 we use the semialgebra property of $\mathbf{L}(S)$ and we often use the fact that the order $\leq_S$ is directed and that it satisfies

$$x, y \leq_S xy \quad \text{for} \quad x, y \in S.$$

To prove that $\mathbf{L}\left(L_F(S)\right)^*$ is a face of the dual wedge $\mathbf{L}(S)^*$ for every face of S, we need results on the closedness of normal subgroups of G if the fundamental group $\pi_1(G)$ has at most rank 1. All this information is not available for general Lie semigroups. At the end of this section we present the two most essential low dimensional examples of non-abelian invariant Lie semigroups: the invariant Lie semigroups in the group $\mathrm{Sl}(2,\mathbb{R})^\sim$ and an invariant Lie semigroup in the oscillator group. These examples are very instructive because we know that at least one of them is contained in every invariant Lie semigroup (S, G) if $\mathbf{L}(G)$ is not a compact Lie algebra and $H(S) = \{\mathbf{1}\}$ (Theorem VI.25). This permits us to conclude that exactly the invariant Lie semigroups with $\mathbf{L}(G)/\mathbf{L}\left(H(S)\right)$ compact and $H(S)$ compact have no non-compact order intervals, i.e., $\mathrm{comp}(S) = S$.

Lemma VI.1. *Let G be a connected Lie group, $W \subseteq \mathbf{L}(G)$ a semialgebra, C a compact convex symmetric CH-neighborhood in $\mathbf{L}(G)$. Then we find for every element $s = \exp(v_1) \cdot \ldots \cdot \exp(v_n) \in \langle \exp W \rangle$ a natural number $m \in \mathbb{N}$ and $w_1, \ldots, w_{m-1} \in W \cap \partial C$, $w_m \in C \cap W$ such that*

$$s = \exp(w_1) \cdot \ldots \cdot \exp(w_{m-1}) \cdot \exp(w_m).$$

Proof. We assume that $s \neq \mathbf{1}$. If $n = 1$, we write $v_1 = kw_1 + w_{k+1}$, where $w_1 \in \mathbb{R}^+ v_1 \cap \partial C$ and k is the largest integer with $[0,k]w_1 \subseteq [0,1]v_1$. We set $w_i := w_1$ for $i = 2, \ldots, k$ and get the desired representation of s because $w_{k+1} \in [0,1]w_1 \subseteq C$. Now we assume that $n \geq 2$, that $v_i \in W \cap C$ for $i = 1, \ldots, n$ (Case $n = 1$) and that the assertion holds for all products of less that n factors. We set

$$\widetilde{s} := \exp(v_1) \cdot \ldots \cdot \exp(v_{n-1}).$$

Then we find $w_1, \ldots, w_{m-1} \in \partial C \cap W$ and $\widetilde{w}_m \in C \cap W$ such that

$$\widetilde{s} = \exp(w_1) \cdot \ldots \cdot \exp(w_{m-1}) \cdot \exp(\widetilde{w}_m).$$

Then, since W is a Lie semialgebra (cf. Definition IV.29),

$$w := \widetilde{w}_m * v_n \in (W \cap C) * (W \cap C) \subseteq W.$$

We use the case $n = 1$ to write

$$\exp w = \exp(\widetilde{w}_m) \exp(v_n) = \exp(w_m) \cdot \ldots \cdot \exp(w_l)$$

with $w_m, \ldots, w_{l-1} \in W \cap \partial C$ and $w_l \in C \cap W$. We conclude that

$$s = \exp(w_1) \cdot \ldots \cdot \exp(w_{l-1}) \cdot \exp(w_l)$$

is the desired representation. ■

Proposition VI.2. *Let (S, G) be an invariant Lie semigroup with $H(S) = \{1\}$, $\varphi : S \to T$ a homomorphism of topological monoids and*

$$t \in \overline{\varphi(S)} \setminus \varphi(\operatorname{comp}(S)).$$

Suppose that the quasi order $\prec$ on T is closed. Then there exists $w \in \mathbf{L}(S) \setminus \{0\}$ such that

$$\varphi(\exp \mathbb{R}^+ w) \subseteq [1, t].$$

Proof. Let $S_0 := \langle \exp \mathbf{L}(S) \rangle$. This is a dense subsemigroup of S. Hence $\overline{\varphi(S)} = \overline{\varphi(S_0)}$. So we find a net $s_n \in S_0$ such that $t = \lim \varphi(s_n)$. The net s_n eventually leaves every compact subset of $\operatorname{comp}(S)$ since $t \notin \varphi(\operatorname{comp}(S))$. There are two cases:
Case 1): s_n eventually leaves every compact subset of S. Then we set $a_n := s_n$.
Case 2): There exists a subnet, which we again denote with s_n, such that $\lim s_n = s$ exists in S. Then $s \notin \operatorname{comp}(S)$ and $t = \varphi(s)$. Thus we find a sequence $a_n \in [1, s]$ which eventually leaves every compact subset of S.

In both cases we have found a net $a_n \in S$ which eventually leaves every compact subset of S and which has the following property. If $t_n \in T$ is a net with the same index set, $t_n \prec \varphi(a_n)$ for every n, and $t' = \lim \varphi(t_n)$, then

$$t' \in [1, t].$$

In Case 1) this follows from $t = \lim \varphi(a_n)$ and in Case 2) from

$$t_n \prec \varphi(a_n) \prec \varphi(s) = t.$$

Let $W := \mathbf{L}(S)$. Let us first assume that $a_n \in S_0$ holds for all $n \in \mathbb{N}$. Note that this is at least true in Case 1). According to Lemmas VI.1 and Proposition IV.30 we find representations

$$a_n = \exp w_1^n \cdot \ldots \cdot \exp w_{m_n}^n$$

with $w_i^n \in \partial C \cap W$ for $i = 1, \ldots, m_n - 1$ and $w_{m_n} \in C \cap W$ for a suitable compact convex neighborhood C of 0 in $\mathbf{L}(G)$. We set $w_i^n = 0$ for $i > m_n$. The fact that a_n eventually leaves every compact subset of S shows that $\lim_{n\to\infty} m_n = \infty$ because the sets $\left(\exp(C \cap W)\right)^k$ are compact for every $k \in \mathbb{N}$. The net $n \mapsto (w_i^n)_{i\in\mathbb{N}} \in C^{\mathbb{N}}$ has a convergent subnet because $C^{\mathbb{N}}$ is compact. Therefore we may assume that $w_i := \lim w_i^n$ exists in $\partial C \cap W$ for every $i \in \mathbb{N}$. Let $k \in \mathbb{N}$. Then

$$\exp w_1^n \cdot \ldots \cdot \exp w_k^n \leq_S \exp w_1^n \cdot \ldots \cdot \exp w_{m_n}^n = a_n,$$

consequently

$$\varphi(\exp w_1 \cdot \ldots \cdot \exp w_k) = \lim_{n\to\infty} \varphi(\exp w_1^n \cdot \ldots \cdot \exp w_k^n) \prec t$$

for every $k \in \mathbb{N}$. We have $w_i \in W \cap \partial C$ for every $i \in \mathbb{N}$ and therefore we find a converging subsequence $w_{i_l}, l \in \mathbb{N}$ with $w := \lim_{l\to\infty} w_{i_l}$. Let $k \in \mathbb{N}$, then

$$\exp kw = \lim_{l\to\infty} \exp(w_{i_l}) \exp(w_{i_{l+1}}) \cdot \ldots \cdot \exp(w_{i_{l+k-1}})$$

and

$$\begin{aligned} &\varphi\big(\exp(w_{i_l}) \exp(w_{i_{l+1}}) \cdot \ldots \cdot \exp(w_{i_{l+k-1}})\big) \\ &\prec \varphi\big(\exp(w_1) \exp(w_2) \cdot \ldots \cdot \exp(w_{i_{l+k-1}})\big) \prec \varphi(a_n). \end{aligned}$$

Here we use the invariance of S. We conclude that $\varphi(\exp kw) \in [\mathbf{1}, t]$ and hence that

$$\varphi(\exp \mathbb{R}^+ w) \subseteq [\mathbf{1}, t].$$

Now we drop the assumption that $a_n \in S_0$. Then there exists a sequence $b_m \in S_0$ with $\lim_{m\to\infty} b_m \to a_n$. Now the same arguments as above apply to the sequence b_m and show that there exists $0 \neq w \in W$ such that $\exp(\mathbb{R}^+ w) \subseteq [\mathbf{1}, a_n]$. Then, since $\varphi(a_n) \prec t$, we also have that

$$\varphi(\exp \mathbb{R}^+ w) \subseteq \varphi([\mathbf{1}, a_n]) \subseteq [\mathbf{1}, t].$$

∎

The following corollary is a remarkably sharper version of Proposition IV.6(4) in the case of invariant semigroups.

Corollary VI.3. *Let (S, G) be an invariant Lie semigroup with $H(S) = \{\mathbf{1}\}$ and $s \in S \setminus \operatorname{comp}(S)$. Then there exists $w \in \mathbf{L}(S) \setminus \{0\}$ such that*

$$\exp \mathbb{R}^+ w \subseteq [\mathbf{1}, s].$$

Proof. We apply Proposition VI.2 with $T = S$, $t = s$, and $\varphi = \mathrm{id}_S$. ∎

The preceding corollary means that every non-compact order interval even contains non-compact one-parameter semigroups. The following theorem, a recent result proved in [MN92], is something like a converse. Since the proof is rather involved, it uses methods from Lorentz geometry and optimal control theory, we only state the theorem.

Lemma VI.4. *Let (S,G) be an invariant Lie semigroup. Then*

$$\mathrm{comp}(S) \subseteq \exp \mathbf{L}(S).$$

It follows in particular that there exists a neighborhood U of $\mathbf{1}$ in G such that

$$S \cap U \subseteq \exp \mathbf{L}(S).$$

Proof. This follows from Theorem IV.8 in [MN92]. ∎

Lemma VI.5. *Let (S,G) be an ILSg and $s \in S \setminus H(S)$. Then there exists $w \in \mathbf{L}(S) \setminus H\big(\mathbf{L}(S)\big)$ with*

$$\exp w \in [\mathbf{1}, s].$$

Proof. First we set $W := \mathbf{L}(S)$ and assume that $H(S) = \{\mathbf{1}\}$. We have to consider two cases:
Case 1: The order interval $[\mathbf{1}, s]$ is non-compact. Then we find $w \in W \setminus \{0\}$ with $\exp \mathbb{R}^+ w \subseteq [\mathbf{1}, s]$ (Corollary VI.3).
Case 2: The order interval $[\mathbf{1}, s]$ is compact. Then, in view of Theorem VI.4, $s = \exp w \in \exp \mathbf{L}(S)$.

Suppose that $H(S) \neq \{\mathbf{1}\}$. We write $\varphi : G \to G/H(S)$ for the quotient homomorphism. Then $\varphi(s) \neq \mathbf{1}$ and we find $\widetilde{w} = d\varphi(\mathbf{1})w \in d\varphi(\mathbf{1})W \setminus \{0\}$ such that

$$\exp\big(d\varphi(\mathbf{1})w\big) = \varphi(\exp w) \in [\mathbf{1}, \varphi(s)] = \varphi([\mathbf{1}, s]).$$

Consequently $w \in W \setminus H(W)$ and $\exp w \in [\mathbf{1}, s]$. ∎

Lemma VI.6. *Let S be a subsemigroup of a connected topological group G. Then the following assertions hold:*

(1) $\mathrm{int}(S)$ *is a semigroup ideal.*

(2) *If S is closed, then $H(S) \in \mathcal{F}(S)$.*

(3) *If $\mathbf{1} \in \overline{\mathrm{int}\, S}$, then we have that*

$$S \subseteq \overline{\mathrm{int}\, S} \quad \textit{and} \quad \mathrm{int}\, S = \mathrm{int}\, \overline{S}.$$

(4) *Suppose that $\mathrm{int}\, S \neq \emptyset$ and $\overline{S} = G$. Then $S = G$.*

Proof. (1) Let $s \in \mathrm{int}\, S$, $t \in S$ and U a neighborhood of $\mathbf{1}$ in G such that $UsU \subseteq S$. Then $Ust \subseteq S$ and $tsU \subseteq S$. Thus $st, ts \in \mathrm{int}\, S$ and therefore $S(\mathrm{int}\, S), (\mathrm{int}\, S)S \subseteq \mathrm{int}\, S$.
(2) The fact that $H(S)$ is a closed subsemigroup of S, and the property that $S \setminus H(S)$ is a semigroup ideal of S, show that $H(S) \in \mathcal{F}(S)$ since $H(S)$ is closed.
(3) Let $s, s_n \in S$ with $\lim_{n\to\infty} s_n = \mathbf{1}$. Then $s = \lim_{n\to\infty} s s_n \in \overline{\mathrm{int}\, S}$ because $S \,\mathrm{int}\, S \subseteq \mathrm{int}\, S$. To see that $\mathrm{int}\, \overline{S} \subseteq \mathrm{int}\, S$ (the other inclusion is trivial), let $x \in \mathrm{int}\, \overline{S}$, U be a neighborhood of $\mathbf{1}$ in G with $xU^{-1} \subseteq \overline{S}$ and $W := \mathrm{int}\, S \cap U \neq \emptyset$. Then xW^{-1} is an open subset of $\overline{S}$ and contains an element $s \in S$, i.e., $s = xw^{-1}$ with $w \in \mathrm{int}\, S$. Therefore $x = sw \in S \,\mathrm{int}\, S \subseteq \mathrm{int}\, S$.
(4) Let $x \in (\mathrm{int}\, S)^{-1} \cap S$. Then, using (1), we find that

$$\mathbf{1} = xx^{-1} \in \overline{S} \cdot \mathrm{int}(S) \subseteq \overline{S\, \mathrm{int}(S)} \subseteq \overline{\mathrm{int}(S)}.$$

Now (1) shows that $\mathrm{int}\, S = \mathrm{int}\, \overline{S} = G$ and therefore $S = G$. ∎

Lemma VI.7. *Let $H, S \subseteq G$ be subsemigroups such that S normalizes H or vice versa. Then*

(1) *$SH = HS$ and $\overline{SH}$ are subsemigroups of G.*

(2) *Suppose that $\overline{S}$ and $\overline{H}$ are Lie semigroups. Then $\overline{SH}$ is a Lie semigroup and $H(\overline{SH})$ is a closed connected subgroup of G.*

If, in addition, $\operatorname{int} S \neq \emptyset$ the following assertions hold:

(3) *$SH = G$ iff $\overline{SH} = G$ iff $H^{-1} \cap \operatorname{int} S \neq \emptyset$.*

(4) *Suppose that H is a group and that there exists an $s \in S$ with $sH \subseteq S$ and $S \neq G$. Then $\operatorname{int} S \cap H = \emptyset$.*

Proof. (1) We may assume that H normalizes S. For $h \in H$ we have that $hS = Sh$, hence $SHSH = SSHH \subseteq SH$ and SH is a semigroup. Then its closure $\overline{SH}$ is also a semigroup.
(2) Let $W := \mathbf{L}(\overline{SH})$. Then $\mathbf{L}(\overline{S}) \cup \mathbf{L}(\overline{H}) \subseteq W$, and thus

$$\overline{SH} \subseteq \overline{\langle \exp W \rangle} \subseteq \overline{SH}.$$

The rest follows from Proposition IV.6.
(3) The facts that $\operatorname{int}(SH) \neq \emptyset$ and that G is connected imply, in view of Lemma VI.6, that $SH = G$ if and only if $\overline{SH} = G$. If $SH = G$ and $s \in \operatorname{int} S$, we find $t \in S$ and $h \in H$ such that $s^{-1} = th$, hence $h^{-1} = st \in \operatorname{int} S \cap H^{-1}$. Conversely, if $s \in \operatorname{int} S \cap H^{-1}$, then $\mathbf{1} \in \operatorname{int}(Ss^{-1}) \subseteq \operatorname{int}(SH)$. Therefore $SH = G$ since G is connected.
(4) Suppose that $\operatorname{int} S \cap H \neq \emptyset$. Then, by (2) above, $SH = G$. Choose $s_1 \in \operatorname{int} S$. Then $s_1 sH \subseteq s_1 S \subseteq \operatorname{int} S$. We find $s' \in S$ and $h \in H$ such that $hs' = (s_1 s)^{-1}$. Hence

$$s' = h^{-1}(s_1 s)^{-1} \in (s_1 sH)^{-1} \subseteq \operatorname{int} S^{-1}.$$

Therefore $\mathbf{1} \in \operatorname{int}(s'S) \subseteq \operatorname{int}(S)$ which contradicts that $S \neq G$ since G is connected. ∎

Remark VI.8. Let $S \subseteq G$ be an invariant subsemigroup and $H \subseteq G$ a subgroup. Then

$$\overline{SH} = \overline{S\widetilde{H}} \quad \text{for} \quad \widetilde{H} := H(\overline{SH}).$$

Proof. It is clear that $H \subseteq \widetilde{H}$, hence that $\overline{SH} \subseteq \overline{S\widetilde{H}}$. Conversely,

$$\overline{S\widetilde{H}} \subseteq \overline{S\overline{SH}} \subseteq \overline{SH}.$$

∎

Corollary VI.9. *If (S, G) is an ILSg and $F \subseteq S$ a subsemigroup, then*

$$L_F(S) = \overline{SF^{-1}}.$$

The normal exposed faces of S are exactly the invariant exposed faces.

Proof. This is a direct consequence of Lemma VI.7 because S is invariant and thus $\langle SF^{-1}\rangle = SF^{-1}$ is a semigroup. Let $F \in \mathcal{F}_n(S)$. Then F is an invariant subsemigroup of G since there exists a closed subsemigroup $S_1 \subseteq G$ such that $H(S_1)$ is normal and $F = H(S_1) \cap S$. Hence $L_F(S) \subseteq S_1$ is invariant and $T_F(S) \cap S = F$. Conversely, suppose that F is an invariant exposed face of S. Then $L_F(S)$ is invariant and $T_F(S) = H\big(L_F(S)\big)$ is normal with $F = S \cap T_F(S)$. ∎

Lemma VI.10. (Extension Lemma) *Let (S,G) be a GILSg, $F \subseteq S$ be a subsemigroup and $\varphi : S \to K$ a continuous homomorphism of S into a compact monoid K with $F \subseteq \varphi^{-1}\big(H(K)\big)$. Then there exists a unique continuous homomorphism $\widetilde{\varphi} : L_F(S) \to K$ such that $\widetilde{\varphi}|_S = \varphi$.*

Proof. (1) For $f \in F$ and $s \in S$ we have that

$$\varphi(fsf^{-1}) = \varphi(f)\varphi(s)\varphi(f)^{-1}.$$

Indeed, let $s \in S$ and $f \in F$. Then $\varphi(fsf^{-1})\varphi(f) = \varphi(fs)$ and therefore $\varphi(fsf^{-1}) = \varphi(f)\varphi(s)\varphi(f)^{-1}$ since $\varphi(f) \in H(K)$.
(2) Firstly we construct a continuous extension φ_1 of φ to SH. Let $sf^{-1} = s'f'^{-1}$. We claim that $\varphi(s)\varphi(f)^{-1} = \varphi(s')\varphi(f')^{-1}$. From $sf^{-1} = s'f'^{-1}$ we deduce that $fsf^{-1}f' = fs'$. Now (1) shows that

$$\varphi(fsf^{-1}f') = \varphi(f)\varphi(s)\varphi(f)^{-1}\varphi(f') = \varphi(f)\varphi(s'),$$

hence $\varphi(s)\varphi(f)^{-1} = \varphi(s')\varphi(f')^{-1}$. We set $\varphi_1(sf^{-1}) := \varphi(s)\varphi(f^{-1})$. Then, in view of the preceding argument, φ_1 is well defined.
(3) $\varphi_1 : SF^{-1} \to K$ is a semigroup homomorphism. For $s, s' \in S$ and $f, f' \in F$ we have, according to (1), that

$$\begin{aligned}\varphi_1(sf^{-1}s'f'^{-1}) &= \varphi_1\big(s(f^{-1}s'f)f^{-1}f'^{-1}\big)\\ &= \varphi_1\big(s(f^{-1}s'f)(f'f)^{-1}\big)\\ &= \varphi(s)\varphi(f^{-1}s'f)\varphi(f)^{-1}\varphi(f')^{-1}\\ &= \varphi(s)\varphi(f)^{-1}\varphi(s')\varphi(f)\varphi(f)^{-1}\varphi(f')^{-1}\\ &= \varphi(s)\varphi(f)^{-1}\varphi(s')\varphi(f')^{-1}\\ &= \varphi_1(sf^{-1})\varphi_1(s'f'^{-1}).\end{aligned}$$

(4) φ_1 is continuous on $(\operatorname{int} S)F^{-1}$: Let $s \in \operatorname{int} S$, $f \in F$ and s_n be a net in $\operatorname{int}(SF^{-1})$ with $\lim s_n = sf^{-1}$. Then $\lim \varphi_1(s_n)\varphi_1(f) = \lim \varphi_1(s_n f) = \varphi_1(s) = \varphi_1(sf^{-1})\varphi_1(f)$, hence $\lim \varphi_1(s_n) = \varphi_1(sf^{-1})$ since $s_n f$ eventually lies in $\operatorname{int} S$. This shows that φ_1 is continuous on $(\operatorname{int} S)F^{-1}$.
(5) $\operatorname{int} L_F(S) = (\operatorname{int} S)F^{-1}$: First we note that $\overline{(\operatorname{int} S)F^{-1}}$ is an open subsemigroup of G (Lemma VI.7). But $\mathbf{1} \in \overline{\operatorname{int} S} \subseteq \overline{(\operatorname{int} S)F^{-1}}$. Thus Lemmas VI.6 and VI.7 show that

$$\operatorname{int} L_F(S) = \operatorname{int} \overline{(\operatorname{int} S)F^{-1}} = (\operatorname{int} S)F^{-1}.$$

(6) Using (5) and Lemma VI.6 we find that $(\operatorname{int} S)F^{-1}$ is a dense ideal of $L_F(S)$. Then we use [CHK83, p.109] to find a unique continuous extension $\widetilde{\varphi}$ of φ_1 to $L_F(S)$. The uniqueness of $\widetilde{\varphi}$ implies that $\widetilde{\varphi}|_S = \varphi$. ∎

Proposition VI.11. *Lst S be a GILSg and $\varphi : S \to K$ be a morphism of S into a compact monoid. Then $F := \varphi^{-1}(H(K))$ is a compact exposed face of S.*

Proof. According to Lemma VI.10 we find a continuous homomorphism $\widetilde{\varphi} : L_F(S) \to K$ such that $\widetilde{\varphi}\mid_S = \varphi$. Let $H := T_F(S)$. Then it is clear that $\widetilde{\varphi}(H) \subseteq H(K)$, hence that

$$F \subseteq H \cap S \subseteq \varphi^{-1}(H(K)) = F.$$

This proves that F is an exposed face of S. ∎

Proposition VI.12. *Let (S, G) be an ILSg, $F \subseteq S$ a face and $H := FF^{-1}$. Then the following assertions hold:*

(1) *H is a group.*

(2) *$F = S \cap H$.*

(3) *$\mathbf{L}(F) = \mathbf{L}(S) \cap \mathbf{L}(H)$ is an invariant wedge in $\mathbf{L}(H)$.*

(4) *There exists a $\mathbf{1}$-neighborhood U in G such that*

$$U \cap F \subseteq \exp\left(\mathbf{L}(F)\right).$$

Proof. (1) Let $F \subseteq S$ be a face. Then, for every $f \in F$, we have that $S \cap Ff^{-1} \subseteq F$. Therefore

$$fFf^{-1} \subseteq fSf^{-1} \cap Ff^{-1} \subseteq S \cap Ff^{-1} \subseteq F \quad \text{for every} \quad f \in F.$$

This implies that $HH = FF^{-1}FF^{-1} = FF^{-1} = H$ and $H^{-1} = H$ is clear.

(2) It is clear that $F \subseteq H \cap S$. Conversely, let $s \in S \cap H$ and $s = ff'^{-1}$ with $f, f' \in F$. Then $f = sf' \in F \cap Sf' = Ff'$. Therefore $s \in F$. Consequently $F \cap H \subseteq F$.

(3) It follows from (2) that $\mathbf{L}(F) = \mathbf{L}(S) \cap \mathbf{L}(H)$ and from (1) that $\mathbf{L}(F)$ is invariant in $\mathbf{L}(H)$.

(4) We may choose U such that $S \cap U \subseteq \exp \mathbf{L}(S)$ (Theorem VI.4). Let $f = \exp w \in U \cap F$. Then the fact that F is a face shows that $\exp([0,1]w) \subseteq F$, thus $w \in \mathbf{L}(F)$. ∎

Proposition VI.13. *Let G be a connected Lie group and $N \subseteq G$ a dense analytic subgroup different from G. Then we find a torus $T \subseteq G$ such that*

$$G = NT, \quad \mathbf{L}(G) \cong \mathbf{L}(N) \rtimes \mathbf{L}(T), \quad \textit{and} \quad \operatorname{rank} \pi_1(G) \geq 2.$$

Proof. Let $p : \widetilde{G} \to G$ be the universal covering of G. Then $\widetilde{N} := \langle \exp_{\widetilde{G}} \mathbf{L}(N) \rangle$ contains $\widetilde{G}'$ ([Ho65, p.190]), hence $\widetilde{N} \subseteq \widetilde{G}$ is closed. Set $D := \ker p$. Then $D \cong D_1 \oplus D_2$, where D_1 is a free abelian group and D_2 is finite. Further, we find $d_1, \ldots, d_n \in D$ such that

$$D_1 = \mathbb{Z}d_1 \oplus \ldots \oplus \mathbb{Z}d_k \quad \text{and} \quad D_2 = \mathbb{Z}d_{k+1} \oplus \ldots \oplus \mathbb{Z}d_n.$$

According to [Ho65, p.180] there exists an abelian analytic subgroup $A \subseteq \widetilde{G}$ such that $D \subseteq A$, hence there are $x_1, \ldots, x_n \in \mathbf{L}(A)$ with $\exp x_i = d_i$. The density of $p^{-1}(N) = D\widetilde{N}$ shows that $D\widetilde{N}/\widetilde{N}$ is dense in the simply connected abelian group $\widetilde{G}/\widetilde{N}$. Note that $\widetilde{G}/\widetilde{N}$ is a vector space. Therefore we find $x_{i_1}, \ldots, x_{i_m}$ such that these elements span a subspace supplementary to $\mathbf{L}(N) \subseteq \mathbf{L}(G)$. Set $C := \langle \exp_{\widetilde{G}} \operatorname{span}\{x_{i_1}, \ldots, x_{i_m}\}\rangle$. Then

$$\widetilde{G} = \widetilde{N} \rtimes C, \quad \mathbf{L}(G) = \mathbf{L}(N) \rtimes \mathbf{L}(C)$$

and C is a vector group. Now $T := p(C)$ is a torus in G with $G = NT$, since $\widetilde{G} = C\widetilde{N}$. To prove the last assertion, we assume that $\operatorname{rank} \pi_1(G) \leq 1$. Then $\widetilde{N} \neq \widetilde{G}$ shows that $C \neq \{0\}$, hence that $\operatorname{rank} \pi_1(G) = \operatorname{rank} D = 1$, $D_1 = \mathbb{Z} d_1$, and $C = \mathbb{R} d_1$. Denote the quotient homomorphism $\widetilde{G} \to \widetilde{G}/\widetilde{N} \cong C$ with φ. Then $\varphi(D) = \varphi(\mathbb{Z} d_1)\varphi(D_2)$ is a finite union of closed sets and therefore closed and different from C. This contradicts the fact that $\varphi(D)$ is dense in $\widetilde{G}/\widetilde{N}$. ■

Proposition VI.14. *Let G be a connected Lie group with* $\operatorname{rank} \pi_1(G) \leq 1$ *and $\mathfrak{a} \subseteq \mathbf{L}(G)$ an ideal. Then $\langle \exp \mathfrak{a} \rangle \subseteq G$ is closed.*

Proof. Set $H := \overline{\langle \exp \mathfrak{a} \rangle}$. We consider the exact homotopy sequence of the fibre bundle $\pi : G \mapsto G/H$ ([Ste51, p.90]). The space G/H is a connected Lie group and therefore $\pi_2(G/H) = \{0\}$ since G/H is diffeomorphic to a product $\mathbb{R}^n \times K$, where K is a compact Lie group ([Ho65, p.180]) and $\pi_2(G/H) = \pi_2(K) = \{0\}$ ([BtD85, p.223]). We therefore have a short exact sequence

$$\{0\} = \pi_2(G/H) \to \pi_1(H) \to \pi_1(G) \to \pi_1(G/H) \to \pi_0(H) = \{0\},$$

since H is connected. Thus

$$1 \geq \operatorname{rank} \pi_1(G) = \operatorname{rank} \pi_1(H) + \operatorname{rank} \pi_1(G/H) \geq \operatorname{rank} \pi_1(H).$$

We conclude with Proposition IV.29 that H does not contain any dense analytic subgroup and consequently that $\langle \exp \mathfrak{a} \rangle = H$ is closed in G. ■

Corollary VI.15. *Let G be a connected Lie group with* $\operatorname{rank} \pi_1(G) \leq 1$, *$W \subseteq \mathbf{L}(G)$ an invariant wedge and $V \supseteq W$ a global Lie wedge such that*

$$W \cap H(V) \subseteq H(W).$$

Then W is global in G.

Proof. This follows from Proposition IV.25 and Proposition VI.14 since $H(W)$ is an ideal of $\mathbf{L}(G)$ and therefore $\langle \exp H(W) \rangle$ is closed in G. ■

The following proposition will also be useful in testing globality of invariant wedges.

Proposition VI.16. *Let G be a connected Lie group, $W \subseteq \mathbf{L}(G)$ a Lie wedge and $K \subseteq G$ a compact subgroup with $\mathfrak{k} := \mathbf{L}(K)$. Then the following assertions hold:*

(1) *If W is global in G, then $\mathfrak{k} \cap W \subseteq H(W)$.*

(2) *Suppose that $\mathfrak{k} \cap W \subseteq H(W)$ and that $\mathrm{Ad}(K)W = W$. Then*

  (a) *The Lie wedge W is global in G iff $V := W + \mathfrak{k}$ is global in G and $\langle \exp H(W) \rangle$ is closed in G.*

  (b) *If $\mathbf{L}(G) = \mathfrak{g} = \mathfrak{p} \rtimes \mathfrak{k}$ is a semidirect product, and $P := \langle \exp \mathfrak{p} \rangle$ is closed, then W is global in G iff $W \cap \mathfrak{p}$ is global in P and $\langle \exp H(W) \rangle$ is closed.*

Proof. (1) Let $S := \overline{\langle \exp W \rangle}$. Then, if W is global, we have that

$$\mathbf{L}(S \cap K) = \mathbf{L}(S) \cap \mathfrak{k} = W \cap \mathfrak{k} \subseteq H(W) = \mathbf{L}\left(H(S)\right)$$

because the compact subsemigroup $S \cap K$ of the compact group K has to be a group ([HHL89, V.0]).

(2a) First Proposition I.7 implies that V is a wedge and also a Lie wedge because $H(V) = \mathfrak{k} + H(W)$. That the globality of V implies the globality of W if $\langle \exp H(W) \rangle$ is closed, is a consequence of Proposition IV.25 since

$$W \cap H(V) \subseteq W \cap \left(H(W) + \mathbf{L}(K)\right) \subseteq H(W).$$

The converse is proved in [Ne89, III.5].

(2b) If W is global, then $W \cap \mathfrak{p}$, as the intersection of two global invariant wedges, is a global invariant wedge. If, conversely, $W \cap \mathfrak{p}$ is global in G, we set $V := (W \cap \mathfrak{p}) + \mathfrak{k}$. Then the invariance of $W \cap \mathfrak{p}$ and the fact that $H(V) = H(W \cap \mathfrak{p}) + \mathfrak{k}$ is a subalgebra, imply that V is a Lie wedge. Moreover

$$V \cap \mathfrak{k} = \mathfrak{k}, \ \mathrm{Ad}(K)V = V, \quad \text{and} \quad \langle \exp H(V) \rangle = \langle \exp H(W \cap \mathfrak{p}) \rangle K$$

is closed in G. Thus, by (2a), V is global in G. Now

$$W \cap H(V) = W \cap \left(H(W \cap \mathfrak{p}) + \mathfrak{k}\right) = H(W \cap \mathfrak{p}) + W \cap \mathfrak{k} \subseteq H(W)$$

and the globality of W follows from Proposition IV.25. ∎

We know from Section IV that $L_F(S)$ is infinitesimally generated for every exposed face $F \in \mathcal{F}_e(S)$ and we want to know more about its tangent wedge $\mathbf{L}\left(L_F(S)\right)$. When is its dual wedge a face of $\mathbf{L}(S)^*$? To answer this question we need the facts about non-closed normal analytic subgroups of Lie groups provided above.

Proposition VI.17. *Suppose that G is a connected Lie group with*

$$\operatorname{rank} \pi_1(G) \leq 1.$$

Let $V \subseteq \mathbf{L}(G)$ *be an invariant wedge with* $S := \overline{\langle \exp V \rangle} \neq G$. *Then*

$$\mathbf{L}(S)^* \in \mathcal{F}(V^*).$$

Proof. We prove this by induction on $\dim \mathbf{L}(G) - \dim H(V)$. If $\dim H(V) = \dim \mathbf{L}(G)$, then $S = G$ and $\mathbf{L}(S)^* = \{0\} \in \mathcal{F}(V^*) = \mathcal{F}(\{0\})$. Suppose that $\dim \mathbf{L}(G) > \dim H(V)$ and that the assertion holds for all wedges V' with $\dim H(V') > \dim H(V)$. If V is global, then $\mathbf{L}(S) = V$ and $\mathbf{L}(S)^* = V^* \in \mathcal{F}(V^*)$. Suppose that V is not global in G. The fact that $\operatorname{rank} \pi_1(G) \leq 1$ implies that the normal subgroup $\langle \exp H(V) \rangle$ is closed in G (Proposition VI.14). Then Proposition IV.25 shows that

$$F := V \cap H\big(\mathbf{L}(S)\big) \not\subseteq H(V)$$

and that $F \in \mathcal{F}_e(V)$. Then $L_F(V)$ is an invariant wedge between V and $\mathbf{L}(S)$ satisfying

$$\overline{\langle \exp L_F(V) \rangle} \subseteq S \neq G.$$

Moreover $\dim H\big(L_F(V)\big) > \dim H(V)$. Therefore

$$\mathbf{L}(S)^* = \mathbf{L}\left(\overline{\langle \exp L_F(V) \rangle}\right)^* \in \mathcal{F}\big(L_F(V)^*\big).$$

But we know that $L_F(V)^* = F^\perp \cap V^* \in \mathcal{F}_e(V^*)$ since $F \in \mathcal{F}_e(V)$ (Proposition I.4). Therefore $\mathbf{L}(S)^*$ is a face of the exposed face $L_F(V)^*$ of V^*. Hence it is a face (Proposition I.4(2)). ∎

Theorem VI.18. (The Face Theorem) *Let* (S, G) *be a GILSg and suppose that* $\operatorname{rank} \pi_1(G) \leq 1$ *and* $F \in \mathcal{F}_n(S)$. *Then*

$$\mathbf{L}\left(L_F(S)\right)^* \in \mathcal{F}\big(\mathbf{L}(S)^*\big).$$

Proof. We use induction on $\dim G - \dim H(S)$. If $G = L_F(S)$, the assertion is clear. If not, we use Lemma IV.35 and Proposition IV.34 to see that

$$\mathbf{L}\left(H(S)\right) \neq \mathbf{L}(F) \in \mathcal{F}_e(S).$$

Let $V := L_{\mathbf{L}(F)}\big(\mathbf{L}(S)\big)$. Then V is an invariant wedge in $\mathbf{L}(G)$ which is controllable since $\exp V \subseteq L_F(S) \neq G$. Set $T := \overline{\langle \exp V \rangle}$. Then (T, G) is a GILSg and

$$F' := T \cap T_F(S) = T \cap \left(H(L_F(S)\right) \in \mathcal{F}_n(T)$$

since $T \subseteq L_F(S)$. Moreover, $S \subseteq T$ and $F \subseteq F'$ entail that $L_F(S) = L_{F'}(T)$. The induction hypothesis now shows that

$$\mathbf{L}\left(L_F(S)\right)^* = \mathbf{L}\left(L_{F'}(T)\right)^* \in \mathcal{F}\big(\mathbf{L}(T)^*\big).$$

But Proposition VI.17 implies that $\mathbf{L}(T)^*$ is a face of V^*. According to the definition of V we have that

$$V^* = \mathbf{L}(F)^\perp \cap \mathbf{L}(S)^* \in \mathcal{F}_e\big(\mathbf{L}(S)^*\big)$$

(Proposition I.4). Therefore $\mathbf{L}\left(L_F(S)\right)^*$ is a face of $\mathbf{L}(S)^*$ (Proposition I.4). ∎

Example VI.19. This example shows that the assumption on the fundamental group of G in Proposition VI.17 is really necessary. Let

$$W := \mathbb{R}^+(1,1,0) + \mathbb{R}^+(1,-1,0) + \mathbb{R}(0,1,\sqrt{2}) \subseteq \mathbb{R}^3$$

and

$$Z := \mathbb{Z}(0,1,0) + \mathbb{Z}(0,0,1).$$

We set $G := \mathbb{R}^3/Z$. Note that $\pi_1(G) \cong Z$ has rank 2. Then $W \subseteq \mathbf{L}(G) = \mathbb{R}^3$ is a generating invariant wedge and $\exp : \mathbb{R}^3 \to G, x \mapsto x + Z$ is a covering homomorphism of Lie groups. We have that

$$V := \mathbf{L}(\overline{\langle \exp W \rangle}) = \mathbf{L}(\overline{W + Z}) = \mathbb{R}^+ \oplus \mathbb{R}^2$$

and $V^* = \mathbb{R}^+ \times \{0\} = \mathbb{R}^+(1,0,0) \notin \mathcal{F}(W^*)$ because W^* is a two-dimensional wedge spanned by $(\sqrt{2},\sqrt{2},-1)$ and $(\sqrt{2},-\sqrt{2},1)$ which contains $(1,0,0)$ in its algebraic interior.

Example VI.20. This example shows that the assumption on the fundamental group of G in Theorem VI.18 is necessary. To construct an example we have to go up to dimension 4. We denote with $e_1 = (1,0,0,0),\ldots,e_4 = (0,0,0,1)$ the canonical base vectors in $\mathbb{R}^4$ and set $f_1 := e_1 - e_2$ and $f_2 := e_3 - e_4$. Let

$$W := \sum_{i=1}^{4} \mathbb{R}^+ e_i \subseteq \mathbb{R}^4 \quad \text{and} \quad Z := \mathbb{Z}(f_1 + \sqrt{2}f_2) + \mathbb{Z}f_2.$$

We set $G := \mathbb{R}^4/Z$. Then, as above, $W \subseteq \mathbf{L}(G) = \mathbb{R}^4$ is a generating invariant wedge and $\exp : \mathbb{R}^4 \to G, x \mapsto x + Z$ is a covering homomorphism of Lie groups. We notice that the vector $\omega := (1,1,1,1)$ is orthogonal to f_1, f_2 and lies in the interior of W^*. Therefore $V := W + \operatorname{span}\{f_1, f_2\}$ is a wedge in $\mathbb{R}^4$ which satisfies

$$W \cap H(V) = W \cap \operatorname{span}\{f_1, f_2\} = \{0\} \subseteq H(W).$$

The subgroup $\exp H(V) = \exp(\mathbb{R}f_1 + \mathbb{R}f_2)$ is closed in G. Therefore Proposition IV.25 shows that W is global in G, i.e., $\mathbf{L}(S) = W$ for $S := \exp W$ (Definition IV.23). Hence (S, G) is a GILSg. Let $x := \exp a \in S$ with $a = (1,1,0,0)$ and denote the exposed face generated by x with F. Then

$$F = T_x(S) \cap S.$$

We compute $L_x(S) = L_F(S)$. First we note that

$$L_a(W) = \mathbb{R}e_1 + \mathbb{R}e_2 + \mathbb{R}^+e_3 + \mathbb{R}^+e_4 \subseteq \mathbf{L}\left(L_F(S)\right).$$

But

$$\exp(\operatorname{span} Z) \subseteq \exp\left(\overline{(T_a(W) \cap \operatorname{span} Z) + Z}\right) = \exp\left(\overline{\mathbb{R}f_1 + Z}\right) \subseteq L_x(S).$$

Consequently

$$V_1 := \mathbb{R}e_1 + \mathbb{R}e_2 + \mathbb{R}f_2 + \mathbb{R}^+e_4 \subseteq \mathbf{L}\left(L_F(S)\right).$$

This wedge is global in G since it contains Z and therefore $V_1 = \mathbf{L}(L_F(S))$. We have that

$$\mathbf{L}\left(L_F(S)\right)^* = V_1^* = \mathbb{R}^+(0,0,1,1) \notin \mathcal{F}(W^*) = \mathcal{F}(W).$$

■

Remark VI.21. This example shows in particular, in view of the results of Section XI, that the idempotents in the Bohr compactification of a GILSg can only be visualized as faces of the wedge $\mathbf{L}(S)^*$ if the fundamental group of G has at most rank 1. One should also note that these difficulties even occur in abelian Lie groups, as the above examples show. ∎

Compact order intervals and low dimensional examples

Example VI.22. Let $G = \mathbb{R}^n$ and $W \subseteq G$ a pointed wedge. Then all sets $[0,w] = W \cap (w - W)$ are compact. To see this, let $\omega \in \operatorname{int} W^*$. Then the sets $\omega^{-1}([0,\alpha]) \cap W$ are compact for every $\alpha \geq 0$. For $w \in W$ we have $[0,w] \subseteq \omega^{-1}\big([0,\omega(w)]\big) \cap W$, hence $[0,w]$ is compact. ∎

Example VI.23. Let $G = \mathrm{Sl}(2,\mathbb{R})^\sim$ and $\mathbf{L}(G) = \mathrm{sl}(2,\mathbb{R})$. We use the same notations as in [HHL89] for the elements of L:

$$H = \begin{pmatrix} 1 & 0 \\ 0 & -1 \end{pmatrix}, \quad T = \begin{pmatrix} 0 & 1 \\ 1 & 0 \end{pmatrix}, \quad \text{and} \quad U = \begin{pmatrix} 0 & 1 \\ -1 & 0 \end{pmatrix}.$$

These matrices satisfy the relations

$$[U,T] = 2H, \quad [U,H] = -2T, \quad \text{and} \quad [H,T] = 2U.$$

We use the invariant quadratic form k on L defined by

$$k(hH + tT + xU) = -h^2 - t^2 + x^2.$$

The cone

$$W := \{hH + tT + xU : x \geq 0, h^2 + t^2 \leq x^2\}$$

is invariant under the adjoint action and global in G ([Ne90, 1.5], [HHL89, V.4]), hence $S := \langle \exp W \rangle$ is an invariant subsemigroup of G with $\mathbf{L}(S) = W$, $H(S) = \{\mathbf{1}\}$ and $\operatorname{int} S \neq \emptyset$. We show that

$$\operatorname{comp}(S) = \{\exp w : w \in W, k(w) < \pi^2\}.$$

In the following we refer to the explicit description of S given in [HHL89, V.4].

Proof. "$\subseteq$": Suppose that $s \in S$ and that $[\mathbf{1}, s]$ is compact. Then Theorem VI.4 provides an element $w \in \mathbf{L}(W)$ with $\exp w = s$. If $k(w) \geq \pi^2$, then w is conjugate to $\sqrt{k(w)}U$, hence $s \geq \exp(\pi U)$. To prove that, in this case, $s \notin \operatorname{comp}(S)$, it suffices to show that

$$\exp\big(\mathbb{R}^+(U + H)\big) \subseteq [\mathbf{1}, \exp(\pi U)].$$

We define

$$\gamma_\lambda : \mathbb{R}^+ \to G, t \mapsto \exp\big(t(\cosh \lambda\, U + \sinh \lambda\, H)\big).$$

For every $\lambda > 0$ we have $\cosh\lambda\, U + \sinh\lambda\, H \in W$ and $\gamma_\lambda(\pi) = \exp(\pi U)$, whence $\gamma_\lambda([0,\pi]) \subseteq [\mathbf{1}, \exp(\pi U)]$. Let $t > 0$. Then

$$\begin{aligned}\exp\big(t(H+U)\big) &= \lim_{\lambda\to\infty} \exp\big(\frac{t}{\cosh\lambda}(\cosh\lambda U + \sinh\lambda H)\big)\\ &= \lim_{\lambda\to\infty} \gamma_\lambda(\frac{t}{\cosh\lambda}) \in [\mathbf{1}, \exp(\pi U)].\end{aligned}$$

"$\supseteq$": Let $w \in W$ with $k(w) < \pi^2$. We have to consider two cases:
Case 1: $k(w) > 0$. Then s is conjugate to $\exp\big(\sqrt{k(w)}U\big)$ and we may assume that $s = \exp(tU)$ with $t < \pi$. Then

$$[\mathbf{1}, s] = S \cap sS^{-1} = \exp(\partial W)\exp(\mathbb{R}^+U) \cap \exp(]-\infty, t]U)\exp(-\partial W)$$

and this set is compact because we have that

$$0 \leq \arcsin(\tanh h) \leq x \leq t - \arcsin(\tanh h)$$

for $g = \exp(xU)\exp(hH) \in [\mathbf{1}, s]$. This leads to $|h| \leq \operatorname{artanh}(\sin\frac{t}{2})$.
Case 2: $k(w) = 0$: We may assume that $w = h(H+U)$. Then $[\mathbf{1}, s] \subseteq \partial S = \exp(\partial W)$ because $\operatorname{int} S$ is an ideal. The half space $\widetilde{S} := \exp\big(\mathbb{R}^-(H+U)\big)S$ is a subsemigroup of G with $\partial S \cap \partial\widetilde{S} = \exp\big(\mathbb{R}^+(H+U)\big)$. The same argument as above proves that $[\mathbf{1}, s] \subseteq \partial\widetilde{S}$, hence

$$[\mathbf{1}, s] \subseteq \partial S \cap \partial\widetilde{S} = \exp\big(\mathbb{R}^+(H+U)\big).$$

Now it is clear that

$$[\mathbf{1}, s] = \exp\big([0,t](H+U)\big),$$

and this set is compact. ∎

Example VI.24. Let $G = \mathcal{O}_1$ be the 4-dimensional oscillator group. We represent G as $\mathbb{C} \times \mathbb{R} \times \mathbb{R}$ with the multiplication

$$(v,z,r)(v',z',r') = (e^{-r'i}v + e^{ri}v', z + z' + \frac{1}{2}\operatorname{Im}\langle e^{-r'i}v, e^{ri}v'\rangle, r + r').$$

Using $d\exp(0)$, we identify $\mathbf{L}(G) = A_4$ with $\mathbb{C} \times \mathbb{R} \times \mathbb{R}$. The cone

$$W := \{(v,z,r) : r - z \geq 0, 2rz + \frac{1}{2}|v|^2 \leq 0\}$$

is invariant in $\mathbf{L}(G)$ and global in G. According to [Dö92] we know that

$$S := \overline{\langle \exp W\rangle} = S_1 \cup S_2,$$

where

$$S_1 := \{0\} \times \mathbb{R}^- \times \{0\} \bigcup \{(v,z,r) : z \leq \frac{\sin(2r)}{4(\cos(2r) - 1)}|v|^2, 0 < r < 2\pi\}$$

and

$$S_2 := \mathbb{C} \times \mathbb{R} \times [\pi, \infty[.$$

We claim that $\operatorname{comp}(S) = S_1$.

Let $s = (v,z,r) \in S_2$. Then $\{v\}\times]-\infty, z] \times \{r\} \subseteq [\mathbf{1}, s]$, hence $[\mathbf{1}, s]$ is not compact. Let $s = (v,z,r) \in S_1$. Suppose that $[\mathbf{1}, s]$ is not compact. Then we find $w = (v',z',r') \in W$ such that $\exp\mathbb{R}^+w \subseteq [\mathbf{1}, s]$ (Corollary VI.3). Then we must have $r' = 0$ and therefore $v' = 0$ (definition of W). We conclude that $\{v\}\times]\infty, z] \times \{r\} \subseteq [\mathbf{1}, s] \subseteq S_1$, a contradiction. ∎

Theorem VI.25. (Compact Order Intervals) *Let (S,G) be a GILSg with $H(S)=\{\mathbf{1}\}$. Then $S=\mathrm{comp}(S)$ iff $\mathbf{L}(G)$ is a compact Lie algebra.*

Proof. Let $\mathfrak{g} := \mathbf{L}(G)$ and $W := \mathbf{L}(S)$. Then $W \subseteq \mathfrak{g}$ is an invariant generating pointed cone in $\mathfrak{g}$. Suppose that $\mathfrak{g}$ is not compact. Then Proposition II.18 applies and shows that $\Omega_P^+ \neq \emptyset$. Let $\omega \in \Omega_P^+$ and $0 \neq x \in \mathfrak{g}^\omega$. Then $Q(x) \in W$. We distinguish two cases:
Case 1: $\omega(Q(x)) < 0$. Then $\langle x \rangle \cong \mathrm{sl}(2,\mathbb{R})$ and $Q(x) \in V := W \cap \langle x \rangle$. We conclude that V is a non-zero invariant cone in $\mathrm{sl}(2,\mathbb{R})$, hence is generating. Now Example VI.23 shows that there are elements $s \in \exp V$ such that $[\mathbf{1},s]$ is not compact.
Case 2: $\omega(Q(x)) = 0$. Then $A(x)$ is isomorphic to the four dimensional oscillator algebra, where $A(x) = \mathbb{R}h \oplus \langle x \rangle$ and $h \in \operatorname{int} W \cap \mathfrak{h}$. We conclude that $V := A(x) \cap W$ is a generating invariant cone in $A(x)$ and Example VI.24 shows that there are elements in $s \in \exp V$ such that $[\mathbf{1},s]$ is not compact. Putting these facts together, we have proved that $\mathfrak{g}$ is compact if $\mathrm{comp}(S)=S$.

Now we assume that $\mathfrak{g}$ is compact. Let $K \subseteq G$ be a maximal compact subgroup and set $S' := SK$. The connectedness of G implies that K is connected [Ho65, p.180]. K is normal since it contains the commutator subgroup. We conclude that (S',G) is a GILsg (Lemma IV.11) and it clearly suffices to prove that $\mathrm{comp}(S') = S'$. Using [Ne89, III.5], we see that $\mathbf{L}(K) \cap W = \{0\}$, since W is pointed. Therefore

$$\mathbf{L}\left(H(S')\right) = H\left(W + \mathbf{L}(K)\right) = \mathbf{L}(K) \quad \text{and} \quad H(S') = K.$$

Let $\varphi : G \to G/K$ be the quotient mapping. Then $G/K \cong \mathbb{R}^n$ ([Ho65, p.180]) and $(\varphi(S'), G/K)$ is a GILSg. Consequently $\varphi(S')$ is a pointed wedge in a vector space and Example VI.22 shows that $\mathrm{comp}\left(\varphi(S')\right) = \varphi(S')$. The mapping φ is proper. Therefore $\varphi\left(\mathrm{comp}(S')\right) = \mathrm{comp}\left(\varphi(S')\right)$ (Theorem V.8). This shows that $\mathrm{comp}(S') = S'$. ∎

Corollary VI.26. *Let (S,G) be a GILSg. Then $\mathrm{comp}(S) = S$ iff the Lie algebra $\mathbf{L}(G)/\mathbf{L}\left(H(S)\right)$ is compact and $H(S)$ is compact. If this is true, then the following assertions hold:*

(a) *$Z\left(H(S)\right)_0$ is a central subgroup.*
(b) *$\mathbf{L}(G) \cong \mathfrak{r} \oplus \mathfrak{k}$, where $\mathfrak{k}$ is a compact semisimple ideal.*
(c) *$\mathfrak{r}$ is nilpotent of degree $\mathbf{1}$, i.e., $[\mathfrak{r},\mathfrak{r}] \subseteq Z(\mathfrak{r})$.*
(d) *$S = \exp \mathbf{L}(S)$.*

Proof. Suppose that $\mathrm{comp}(S) = S$. Then $H(S) = [\mathbf{1},\mathbf{1}]$ is compact. Moreover, in view of Theorem V.8, Theorem VI.25 shows that $\mathbf{L}(G)/\mathbf{L}\left(H(S)\right)$ is a compact Lie algebra. Conversely, suppose that these conditions are satisfied. Then Theorem VI.25 implies that all order intervals in $S/H(S)$ are compact. Now we apply Theorem V.8 to see that all order intervals in S are compact.

Next we assume that $S = \mathrm{comp}(S)$ holds.
(a) Let

$$\mathfrak{k} := [H(\mathbf{L}(S)), H(\mathbf{L}(S))].$$

This is a compact semisimple ideal of $\mathbf{L}(G)$. Therefore $\mathbf{L}(G) \cong \mathfrak{g}_1 \oplus \mathfrak{k}$, as is easily seen with Levi's Theorem. Hence we may assume that $H(S)$ is abelian, i.e., a torus. The automorphism group of the torus is discrete and consequently $[\mathfrak{g}_1, \mathbf{L}(H(S))] = \{0\}$ because G acts on $H(S)$ as a discrete group, hence trivially. This shows that $Z(H(S))_0 \subseteq Z(G)$.

(b) Theorem VI.25 shows that $\mathbf{L}(G)/\mathbf{L}(H(S))$ is a compact Lie algebra because all order intervals in $S/H(S)$ are compact (Theorem V.8). Let $\pi : \mathbf{L}(G) \to \mathfrak{g}_1 := \mathbf{L}(G)/H(\mathbf{L}(S)) = \mathfrak{z} \oplus \mathfrak{k}_1$ denote the quotient homomorphism, $\mathfrak{z} = Z(\mathfrak{g}_1)$ and $\mathfrak{k}_1 = [\mathfrak{g}_1, \mathfrak{g}_1]$. Then there exists a lift $\varphi : \mathfrak{k}_1 \to \mathbf{L}(G)$. Let $\mathfrak{r} \subseteq \mathbf{L}(G)$ denote the radical and $\mathfrak{k} := \varphi(\mathfrak{k}_1)$. Then $\pi(\mathfrak{r}) = \mathfrak{z}$ and therefore $\pi(\mathfrak{r} + \mathfrak{k}) = \mathfrak{g}_1$. Thus $\mathbf{L}(G) = \mathfrak{r} \rtimes \mathfrak{k}$ since $\ker \pi \subseteq Z(\mathbf{L}(G)) \subseteq \mathfrak{r}$. We choose a vectorspace $V \subseteq \mathfrak{r}$ which is $\mathfrak{k}$-invariant and complementary to $Z(\mathbf{L}(G))$. Then $[\mathfrak{k}, V] \subseteq V$ and $\pi([\mathfrak{k}, V]) \subseteq [\mathfrak{k}_1, \mathfrak{z}] = \{0\}$. We conclude that $[\mathfrak{k}, V] = \{0\}$ since $\ker \pi \cap V = \{0\}$, and that $[\mathfrak{r}, \mathfrak{k}] = \{0\}$.

(c) Using the fact that $\pi(\mathfrak{r})$ is abelian, we see that $[\mathfrak{r}, \mathfrak{r}] \subseteq \mathbf{L}(H(S)) \subseteq Z(\mathfrak{r})$.

(d) Theorem VI.4. ∎

Example VI.27. To see that, in general, one cannot sharpen the conclusion in Corollary VI.26(c), we consider a quotient G of the three dimensional Heisenberg group modulo a non-trivial discrete central subgroup. Then $Z(G) \cong \mathbb{R}/\mathbb{Z}$ is a circle and $G/Z(G) \cong \mathbb{R}^2$. Let $C \subseteq \mathbb{R}^2$ be a pointed cone, and $S \subseteq G$ its inverse image. Then S is an invariant Lie semigroup in G with $H(S) = Z(G)$. Thus Corollary VI.26 implies that $\operatorname{comp}(S) = S$ and $\mathfrak{g}$ is nilpotent, but not abelian. ∎

Corollary VI.28. *Let (S, G) be a generating invariant Lie semigroup such that $\mathbf{L}(G)$ is compact. Then*

$$S = \exp \mathbf{L}(S).$$

Proof. We consider the connected Lie group $H(S)$. Since $\mathbf{L}(H(S))$ is a compact Lie algebra, there exists a vector group $V \subseteq H(S)$ and a compact subgroup $K \subseteq H(S)$ such that $H(S) \cong V \times K$. The Lie algebra $\mathbf{L}(V)$ of V is an ideal of $\mathbf{L}(G)$. Hence

$$\mathbf{L}(V) = \big(\mathbf{L}(V) \cap Z(G)\big) \oplus \big(\mathbf{L}(V) \cap [\mathbf{L}(G), \mathbf{L}(G)]\big).$$

Since the subgroup $\exp[\mathbf{L}(G), \mathbf{L}(G)] \subseteq G$ is compact, we conclude that $V \subseteq Z(\mathbf{L}(G))$, so that $V \subseteq Z(G)$. Then $(S/V, G/V)$ is a generating invariant Lie semigroup and $H(S/V) \cong K$ is compact. Thus $\exp \mathbf{L}(S/V) = S/V$ follows from Corollary VI.26(d). Let $s \in S$ and $\pi : G \to G/V$ the quotient homomorphism. Then there exists $w \in \mathbf{L}(S)$ such that

$$\pi(s) = \exp_{G/V}\big(d\pi(\mathbf{1})w\big) = \pi(\exp w).$$

Therefore we find $v \in \mathbf{L}(V)$ such that $s = \exp(w)\exp(v)$. Since $v \in Z(\mathbf{L}(G))$, this leads to $s = \exp(w + v)$. To complete the proof we only have to note that $V \subseteq \mathbf{L}(S)$ implies that $w + v \in \mathbf{L}(S)$. ∎

Pathologies in ILSGs

Example VI.29. (Disconnected order intervals) Let G be the 4-dimensional oscillator group from Example VI.24 and

$$S := S_1 \cup S_2,$$

where

$$S_1 := \{0\} \times \mathbb{R}^- \times \{0\} \bigcup \Big\{(v,z,r) : z \leq \frac{\sin(2r)}{4\big(\cos(2r)-1\big)}|v|^2, 0 < r < 2\pi\Big\}$$

and

$$S_2 := \mathbb{C} \times \mathbb{R} \times [\pi, \infty[.$$

Set $s := (0,1,\pi)$. We show that

$$[\mathbf{1}, s] = \{0\} \times \mathbb{R}^- \times \{0\} \bigcup \{0\} \times [1,\infty[\times \{\pi\}.$$

Proof. Suppose that $t = (v,z,r), t' = (v',z',r') \in S$ with $tt' = s$. Then $r + r' = \pi$ and there are three cases:
Case 1: $r = 0$. Then $v = 0$ and $z \leq 0$. We conclude that

$$t \in \{0\} \times \mathbb{R}^- \times \{0\} \quad \text{and} \quad t' \in \{0\} \times [1,\infty[\times \{\pi\}.$$

Case 2: $r' = 0$. Then $v' = 0$ and $z' \leq 0$. We conclude that

$$t' \in \{0\} \times \mathbb{R}^- \times \{0\} \quad \text{and} \quad t \in \{0\} \times [1,\infty[\times \{\pi\}.$$

Case 3: $r \in]0,\pi[$. From $tt' = s$ we obtain the relations:

$$r = \pi - r' \quad \text{and} \quad v = -e^{(r+r')i}v' = v'.$$

This shows that $\operatorname{Im}\langle e^{-r'i}v, e^{ri}v'\rangle = -\operatorname{Im}|v|^2 = 0$, hence that

$$z + z' = 1.$$

The fact that $s \in \partial S$ shows that $t, t' \in \partial S$. Thus

$$z = \frac{\sin(2r)}{4\big(\cos(2r)-1\big)}|v|.$$

Putting all these fact together we get

$$z' = \frac{\sin(2r')}{4\big(\cos(2r')-1\big)}|v'|^2 = \frac{-\sin(2r)}{4\big(\cos(2r)-1\big)}|v|^2 = -z.$$

This contradicts $z + z' = 1$ and proves our assertion. ∎

Another example of this type is the "parking-ramp" in the group $G/(\{0\} \times \mathbb{R} \times \{0\})$ (cf. [HoRu91]).

Proposition VI.30. (Peripheral one-parameter semigroups entering the interior) *Let (S,G) be a GILSg and $x \in C^1(W)$, i.e., $V := \overline{W + \mathbb{R}x}$ is a half space, such that ∂V is no subalgebra. Then there exists $T > 0$ such that*

$$\exp(\mathbb{R}^+ x) \cap \operatorname{int} S = \exp(]T, \infty[x).$$

Proof. Let $B \subseteq 0$ be a neighborhood of 0 such that $\exp|_B$ is a diffeomorphism onto $\exp(B)$ and $\exp(B) \cap S \subseteq \exp W$ (Lemma VII.4). Then $\exp(tx) \in \partial S$ for all sufficiently small $t > 0$. Therefore $T := \inf\{t \in \mathbb{R}^+ : \exp(tx) \in \operatorname{int} S\} > 0$. We show that $T < \infty$. If this is false, the Lie semigroup $T := \overline{S \exp \mathbb{R}x}$ does not agree with G (Lemma VI.7). Then $\mathbf{L}(T)$ is a Lie wedge which contains the half space V, hence $\mathbf{L}(T) = V$. This implies that V is a Lie wedge, in particular that ∂V is a subalgebra. This is a contradiction. Now the fact that $\operatorname{int} S$ is an ideal shows that

$$\exp(\mathbb{R}^+ x) \cap \operatorname{int} S = \exp(]T, \infty[x).$$

∎

We have already seen in Chapter IV that the interior of a generating Lie semigroup is a dense ideal. If a Lie semigroup (S,G) is not generating, then the semigroup $\langle \exp \mathbf{L}(S) \rangle$ is contained in an analytic subgroup A of G. So, if S has non-empty interior, then the analytic subgroup A must be dense in G. In Theorem VI.31 and the following example we show how to construct an example of a non-generating invariant Lie semigroup having interior points.

Theorem VI.31. *Let (S,A) be a GILSg and $W := \mathbf{L}(S)$ such that*

(1) *there exists $x \in Z(\mathbf{L}(A))$ with $\mathbb{R}x \cap W = \{0\}$ such that* $\exp(\mathbb{R}x)$ *is closed,*

(2) *$V := \mathbb{R}x + W$ is global in A,*

(3) *and there exists $s \in S$ such that $s \exp \mathbb{R}x \subseteq S$.*

Then there exists a Lie group G and a homomorphism $i : A \to G$ such that

(a) *$i(A) \subseteq G$ is dense,*

(b) *$S_1 := \overline{i(\langle \exp W \rangle)}$ has non-empty interior,*

(c) $\mathbf{L}(S_1) \subseteq \mathbf{L}\big(i(A)\big) \neq \mathbf{L}(G)$.

Proof. Let $T := \mathbb{R}^2/\mathbb{Z}^2$ be the 2-dimensional torus and $y \in \mathbf{L}(T)$ such that $\exp_T(\mathbb{R}y)$ is dense in T. We set

$$G := (A \times T)/D \quad \text{with} \quad D = \exp\big(\mathbb{R}(x, -y)\big),$$

denote the quotient morphism $A \times T \to G$ with p, and define $i : A \to G$ by $i(a) := p(a, \mathbf{1})$. Note that D is closed since $\exp(\mathbb{R}x)$ is closed. Now (a) is clearly satisfied.

(b) We have that

$$\begin{aligned} p(T) &= p(\overline{\exp \mathbb{R}(0,y)}) \subseteq \overline{\exp \mathbb{R} dp(\mathbf{1})(0,y)} \\ &= \overline{\exp \mathbb{R} dp(\mathbf{1})(x,0)} = \overline{i(\exp \mathbb{R}x)} \subseteq i(s)^{-1} S_1 \end{aligned}$$

and therefore

$$i(s)p(T)i(\operatorname{int}_A S) = i(s)p(\operatorname{int}_A S \times T) \subseteq \operatorname{int} S_1.$$

(c) According to our assumption, the Lie wedge $V := \mathbb{R}x + W$ is global in A, hence $V \oplus \mathbf{L}(T)$ is global in $A \times T$ and Lie generating with

$$\mathbb{R}(x, -y) \subseteq H\big(V \oplus \mathbf{L}(T)\big) = \mathbb{R}x \oplus \mathbf{L}(T).$$

Hence $\widetilde{V} := dp(\mathbf{1})\big(V \oplus \mathbf{L}(T)\big)$ is global in G ([Ne89, III.8]). We have

$$di(\mathbf{1})W \cap H(\widetilde{V}) \subseteq di(\mathbf{1})W \cap dp(\mathbf{1})\big(\mathbb{R}x \oplus \mathbf{L}(T)\big) \subseteq di(\mathbf{1})W \cap \mathbb{R}di(\mathbf{1})(x, 0) = \{0\}.$$

Therefore Proposition IV.25 shows that

$$\mathbf{L}(S_1) = \mathbf{L}\left(\overline{\langle \exp_G di(\mathbf{1})W\rangle}\right) = di(\mathbf{1})W \subseteq di(\mathbf{1})\,\mathbf{L}(A) = \mathbf{L}\left(i(A)\right) \neq \mathbf{L}(G).$$

■

Example VI.32. We set $A := \mathcal{O}_1 \times \mathcal{O}_1$, where $\mathcal{O}_1$ is the 4-dimensional Oscillator group from Example VI.24. Let $W_1 \subseteq \mathbf{L}(\mathcal{O}_1)$ be the invariant pointed generating cone from Example VI.24 and $z \in \mathbf{L}\left(Z(\mathcal{O}_1)\right)$ such that $\mathbb{R}z \cap W_1 = \mathbb{R}^+ z$. Then Example VI.24 shows that W_1 is global in $\mathcal{O}_1$ and that there exists $s_1 \in S_1 := \overline{\langle \exp W_1\rangle}$ such that

$$s_1 Z(\mathcal{O}_1) \subseteq S_1.$$

We set

$$W := W_1 \oplus W_1 \subseteq \mathbf{L}(A) \quad \text{and} \quad x := (z, -z).$$

The following assertions hold:

(1) $\mathbb{R}x \cap W \subseteq W \cap Z\big(\mathbf{L}(A)\big) \cap \mathbb{R}x = \big(\mathbb{R}^+(z, 0) \oplus \mathbb{R}^+(0, z)\big) \cap \mathbb{R}x = \{0\}$.

(2) $\exp \mathbb{R}x \subseteq A$ is closed and central because $Z(A)_0$ is a 2-dimensional vector group and $x \in \mathbf{L}\left(Z(A)\right)$.

(3) Let $s := (s_1, s_1)$. Then

$$sZ(A) = s_1 Z(\mathcal{O}_1) \times s_1 Z(\mathcal{O}_1) \subseteq \overline{\langle \exp W\rangle}.$$

(4) $V := W + \mathbb{R}x$ is global in A because it is a generating invariant wedge (Section VIII).

So we may apply Theorem VI.31 to find an example of a non-generating Lie semigroup having interior points. ■

In view of the preceding example, one may think that every Lie semigroup S lying in a dense analytic subgroup A such that $\mathbf{L}(S)$ is Lie generating in $\mathbf{L}(A)$ has non-void interior. That this is false in general, is shown in the following example.

Example VI.33. (Thin Lie semigroups) Let $\mathbb{T}^2 := \mathbb{R}^2/\mathbb{Z}^2$ denote the 2-torus. Set $G := \mathbb{T}^2 \times \mathbb{R}$ and pick a generator of a dense one-parameter subgroup $x_0 \in \mathbf{L}(\mathbb{T}^2)$. We set

$$W := \{(\lambda x_0, y) \in \mathbf{L}(\mathbb{T}^2) \times \mathbb{R} : |\lambda| \leq y\}.$$

Then $V := W + \mathbf{L}(\mathbb{T}_2)$ is global in G and

$$W \cap H(V) = W \cap \mathbf{L}(\mathbb{T}^2) = \{0\} \subseteq H(W).$$

Therefore $\mathbf{L}(\overline{\langle \exp W \rangle}) = W$ (Proposition IV.25) and $\overline{\langle \exp W \cup \exp(-W) \rangle} = G$ since

$$\mathbb{R}x \times \{0\} + \{0\} \times \mathbb{R} \subseteq W - W.$$

■

VII. CONTROLLABILITY OF INVARIANT WEDGES

The global theory of invariant Lie semigroups is divided into two main parts: the existence theory and the structure theory. The following two sections, Section VII and VIII, are devoted to the existence theory. In the next section we consider the problem of deciding whether a given invariant wedge is the tangent wedge of a Lie semigroup or not (the globality problem) and in this section we consider a weaker version of this problem, the controllability problem: When does the exponential image $\exp W$ of an invariant wedge generate the whole group? Recall that a Lie wedge $W \subseteq \mathbf{L}(G)$, where G is a connected Lie group, is said to be *controllable in* G if $\langle \exp W \rangle = G$. We will characterize those invariant wedges which are controllable in the associated simply connected group. Moreover, one can easily reduce the problem to the case where W is pointed and generating, and then give a criterion for its intersection $C = W \cap \mathfrak{h}$ with a compactly embedded Cartan algebra $\mathfrak{h}$ which is checkable only by consideration of the root data $(\mathfrak{h}, \Omega)$. First we introduce a new notion which is appropriate in this context and which links the controllability problem to the globality problem.

Definition VII.1. A Lie wedge $\big(W, \mathbf{L}(G)\big)$ is said to be *maximal global in* G if $W \neq \mathbf{L}(G)$ is Lie generating and W is the largest global Lie wedge containing W which is different from $\mathbf{L}(G)$. ∎

Lemma VII.2. *Let G be a connected Lie group. For every Lie generating Lie wedge W which is not controllable in G we find a maximal global Lie wedge V containing W. Moreover, V is the tangent wedge of a maximal subsemigroup M of G which is closed.*

Proof. (cf. [Ne91a, II.3]) Suppose that W is not controllable. Then $S := \overline{\langle \exp W \rangle} \neq G$ since W is Lie generating (Proposition IV.6 and Lemma VI.6). Consequently $W' := \mathbf{L}(S) \neq \mathbf{L}(G)$ is global in G. Let $\mathcal{K}$ be a maximal tower of global Lie wedges which contain W' and are different from $\mathbf{L}(G)$. We claim that $V := \overline{\bigcup \mathcal{K}}$ is maximal global. Let $C \in \mathcal{K}$. Then $S_C := \overline{\langle \exp C \rangle}$ is a subsemigroup of G with $S \subseteq S_C \neq G$. Therefore $S_C \cap \operatorname{int} S^{-1} = \emptyset$. This leads to $S_V \cap \operatorname{int} S^{-1} = \emptyset$ for $S_V := \overline{\langle \exp V \rangle}$. Thus $\mathbf{L}(S_V) = V \neq \mathbf{L}(G)$ because $\mathcal{K}$ is maximal and hence V is maximal global. Now we consider a maximal tower $\mathcal{T}$ of subsemigroups of G which contain S_V and are different from G. Set $M := \bigcup \mathcal{T}$. Then M is a subsemigroup of G with $M \cap \operatorname{int} S^{-1} = \emptyset$. Now maximality implies that M is closed and $V \subseteq \mathbf{L}(M) \neq \mathbf{L}(G)$. Finally, the definition of V shows that $\mathbf{L}(M) = V$. ∎

Theorem VII.3. (First Reduction Theorem) *Every maximal global Lie wedge* $W \subseteq \mathbf{L}(G)$ *contains the commutator algebra* $[\mathfrak{n}, \mathfrak{n}]$ *of the nilradical* $\mathfrak{n}$ *of* $\mathbf{L}(G)$.

Proof. According to Lemma VII.2 we find a maximal subsemigroup M of G with $\mathbf{L}(M) = W$ and $\operatorname{int} M \neq \emptyset$. Then we apply Proposition V.5.31 in [HHL89] to see that

$$[\mathfrak{n}, \mathfrak{n}] = \mathbf{L}\left([\exp \mathfrak{n}, \exp \mathfrak{n}]\right) \subseteq \mathbf{L}(M) = W$$

since $\exp \mathfrak{n}$ is a connected normal subgroup of G. ∎

Now we apply the theory of Lie algebras with cone potential to get a better reduction in this case.

Proposition VII.4. *Let* $\mathfrak{g}$ *be a Lie algebra with cone potential,* $\mathfrak{h}$ *a compactly embedded Cartan algebra and* Ω_R^+ *as in* Definition II.14. *Then* $\mathfrak{n} = \mathfrak{z} \oplus \mathfrak{n}_{\text{eff}}$, *with* $\mathfrak{n}_{\text{eff}} := \bigoplus_{\omega \in \Omega_R^+} \mathfrak{g}^\omega$, *is the nilradical of* $\mathfrak{g}$. *Suppose that* $[\mathfrak{n}, \mathfrak{n}] \subseteq I \subseteq Z(\mathfrak{g})$. *Denote the quotient homomorphism* $\mathfrak{g} \to \mathfrak{g}/I, x \mapsto x + I$ *with* φ *and set* $\widetilde{\mathfrak{n}} := [\mathfrak{n}, \mathfrak{n}] + \mathfrak{n}_{\text{eff}}$. *Then the following assertions hold:*

(1) $\widetilde{\mathfrak{n}}$ *is an ideal of* $\mathfrak{g}$.

(2) $\mathfrak{g}/(I + \mathfrak{n}_{\text{eff}})$ *is reductive.*

(3) $\varphi(\widetilde{\mathfrak{n}}) = \varphi(\mathfrak{n}_{\text{eff}})$ *is an abelian ideal of* $\mathfrak{g}/I$.

(4) *Every invariant wedge* $V \subseteq \mathfrak{g}/I$ *with* $V - V = \varphi(\widetilde{\mathfrak{n}})$ *agrees with* $\varphi(\widetilde{\mathfrak{n}})$.

Proof. The facts about the real root decomposition of $\mathfrak{g}$ are contained in Proposition II.10.

(1) We choose a Levi algebra $\mathfrak{s} \subseteq \mathfrak{g}$ as in Proposition II.11. It follows from Proposition II.11(4) that

$$\begin{aligned}[\mathfrak{g}, \widetilde{\mathfrak{n}}] &= [\mathfrak{s} + \mathfrak{h} \cap \mathfrak{r} + \mathfrak{n}_{\text{eff}}, \widetilde{\mathfrak{n}}] \\ &= [\mathfrak{s}, \mathfrak{n}_{\text{eff}}] + [\mathfrak{h} \cap \mathfrak{r}, \mathfrak{n}_{\text{eff}}] + [\mathfrak{n}_{\text{eff}}, \mathfrak{n}_{\text{eff}}] \\ &\subseteq \mathfrak{n}_{\text{eff}} + \mathfrak{n}_{\text{eff}} + [\mathfrak{n}, \mathfrak{n}] \subseteq \widetilde{\mathfrak{n}}.\end{aligned}$$

Thus $\widetilde{\mathfrak{n}}$ is an ideal of $\mathfrak{g}$.

(2) We have

$$\begin{aligned}[\mathfrak{g}, \mathfrak{r}] &= [\mathfrak{s} + \mathfrak{h} \cap \mathfrak{r} + \mathfrak{n}_{\text{eff}}, \mathfrak{h} \cap \mathfrak{r} + \widetilde{\mathfrak{n}}] \\ &= [\mathfrak{s}, \mathfrak{n}_{\text{eff}}] + [\mathfrak{h} \cap \mathfrak{r}, \mathfrak{n}_{\text{eff}}] + [\mathfrak{n}_{\text{eff}}, \mathfrak{n}_{\text{eff}}] \\ &\subseteq \widetilde{\mathfrak{n}} \subseteq I + \mathfrak{n}_{\text{eff}}.\end{aligned}$$

Therefore

$$[\operatorname{Rad}\left(\mathfrak{g}/(I + \mathfrak{n}_{\text{eff}})\right), \mathfrak{g}/(I + \mathfrak{n}_{\text{eff}})] = [\mathfrak{r}/(I + \mathfrak{n}_{\text{eff}}), \mathfrak{g}/(I + \mathfrak{n}_{\text{eff}})] = \{0\}$$

which implies that the radical in $\mathfrak{g}/(I + \mathfrak{n}_{\text{eff}})$ is central, i.e., $\mathfrak{g}/(I + \mathfrak{n}_{\text{eff}})$ is reductive.

(3) To see that $\varphi(\mathfrak{n}_{\text{eff}})$ is abelian, we compute

$$[\varphi(\widetilde{\mathfrak{n}}), \varphi(\widetilde{\mathfrak{n}})] = \varphi([\widetilde{\mathfrak{n}}, \widetilde{\mathfrak{n}}]) \subseteq \varphi(I) = \{0\}.$$

(4) Suppose that $V \subseteq \mathfrak{g}/I$ is an invariant wedge with $\varphi(\widetilde{\mathfrak{n}}) = V - V$. Then $W := \varphi^{-1}(V)$ is an invariant wedge in $\mathfrak{g}$ with

$$I \subseteq H(W) \quad \text{and} \quad W - W = \mathfrak{n}_{\text{eff}} + I.$$

We apply Theorem I.1 to the module $M = \mathfrak{n}_{\text{eff}} + I$ of the compact group $K := \mathrm{INN}_{\mathfrak{g}}(\mathfrak{h})$. We have

$$M_{\text{fix}} = I, \qquad M_{\text{eff}} = \mathfrak{n}_{\text{eff}},$$

and the wedge W is K-invariant. Suppose that $W \neq M$. Then W is not a vector space and therefore algint $W \cap M_{\text{eff}} = \emptyset$ because $M_{\text{fix}} = I \subseteq H(W)$. This contradicts Theorem I.10. ∎

Theorem VII.5. (Second Reduction Theorem) *Suppose that $L(G)$ has cone potential. Then every maximal global Lie wedge $W \subseteq \mathbf{L}(G)$ contains $\widetilde{\mathfrak{n}}$.*

Proof. According to Lemma VII.2 we find a maximal semigroup $M \subseteq G$ with $\mathbf{L}(M) = W$ and $\operatorname{int} M \neq \emptyset$. From the First Reduction Theorem we know that $[\mathfrak{n}, \mathfrak{n}] \subseteq H(W)$. Set $A := \overline{\exp[\mathfrak{n}, \mathfrak{n}]}$. This is a closed central subgroup of G since $[\mathfrak{n}, \mathfrak{n}] \subseteq Z(\mathbf{L}(G))$ (Proposition II.10) and $Z(G)$ is closed. Let $\varphi : G \to G/A$ denote the quotient homomorphism. Now [HHL89, V.5.11] implies that $\varphi(M) \subseteq G/A$ is maximal. Setting $\mathfrak{a} := \mathbf{L}(A)$, we conclude with Proposition VII.4 that $d\varphi(\mathbf{1})\widetilde{\mathfrak{n}} \subseteq \mathbf{L}(G/A) = \mathbf{L}(G)/\mathfrak{a}$ is an abelian ideal and contains no invariant wedges V of $\mathbf{L}(G)/\mathfrak{a}$ with $V - V = d\varphi(\mathbf{1})\widetilde{\mathfrak{n}}$ which are different from $d\varphi(\mathbf{1})\widetilde{\mathfrak{n}}$. We apply [HHL89, V.5.3] to see that

$$V := \{x \in d\varphi(\mathbf{1})\widetilde{\mathfrak{n}} : \exp_{G/A}(x) \in \varphi(M)\}$$

is a closed invariant wedge in $d\varphi(\mathbf{1})\widetilde{\mathfrak{n}}$ which satisfies $V - V = d\varphi(\mathbf{1})\widetilde{\mathfrak{n}}$. This leads to $V = d\varphi(\mathbf{1})\widetilde{\mathfrak{n}}$ (Proposition VII.4). We find that

$$\exp_{G/A}\big(d\varphi(\mathbf{1})\widetilde{\mathfrak{n}}\big) = \varphi\big(\exp_G \widetilde{\mathfrak{n}}\big) \subseteq \varphi(M).$$

Therefore $\widetilde{\mathfrak{n}} \subseteq \mathbf{L}(M) = W$. ∎

Corollary VII.6. *Let W be a Lie generating Lie wedge in $\mathbf{L}(G)$ and suppose that $\mathbf{L}(G)$ has cone potential. Then W is controllable in G iff the wedge $V := \overline{d\varphi(\mathbf{1})W}$ is controllable in G/A, where $A := \overline{\exp \widetilde{\mathfrak{n}}}$ and $\varphi : G \to G/A$ denotes the quotient homomorphism.*

Proof. The controllability of W in G clearly implies that V is controllable in G/A. Suppose that W is not controllable in G. Then there exists a maximal global Lie wedge $W_1 \subseteq \mathbf{L}(G)$ such that $W \subseteq W_1$ (Lemma VII.2). Now the Second Reduction Theorem shows that $\widetilde{\mathfrak{n}} \subseteq W_1$, hence that $A \subseteq M$, where M is a maximal closed subsemigroup of G with $\mathbf{L}(M) = W_1$ (Lemma VII.2). Therefore

$$G/A \neq \varphi(M) = \overline{\varphi(M)} \supseteq \langle \exp \overline{d\varphi(\mathbf{1})W} \rangle = \langle \exp V \rangle.$$

Again Lemma VII.2 applies and shows that V is not controllable in G/A. ∎

Corollary VII.6 and Proposition VII.4(2) reduce the controllability problem in Lie groups whose Lie algebras have cone potential to the reductive case. Before we turn our attention to the controllability problem in reductive Lie groups, we have to do some preliminary work on invariant subsemigroups in a special class of solvable Lie groups. One should notice that the methods used here to get the main result about controllability in reductive Lie groups are completely different from those used in [Ne90] to get an analogeous result in the semisimple case. We also note that Hilgert has some results on controllablity of invariant wedges which overlap with our Controllability Theorems in the reductive case (cf. [Hi92]).

Definition VII.7. We say that a solvable Lie algebra $\mathfrak{t}$ is of *real type* if there exists a Cartan algebra $\mathfrak{h} \subseteq \mathfrak{t}$ and a set $\Lambda \subseteq \widehat{\mathfrak{h}}$ such that

$$\mathfrak{t} = \mathfrak{h} \oplus \bigoplus_{\alpha \in R+} \mathfrak{t}^{\alpha},$$

where Λ is contained in an open half space of $\widehat{\mathfrak{h}}$ and

$$\mathfrak{t}^{\alpha} = \{x \in \mathfrak{t} : [h, x] = \alpha(h)x \quad \text{for all} \quad h \in \mathfrak{h}\}.$$

Lemma VII.8. *Suppose that $\mathfrak{s}$ is a quasihermitean semisimple Lie algebra, $\mathfrak{k} \subseteq \mathfrak{s}$ a maximal compactly embedded subalgebra. Then there exists an Iwasawa decomposition $\mathfrak{s} = \mathfrak{k} + \mathfrak{t}$ such that $\mathfrak{t}$ is a solvable Lie algebra of real type. Moreover, there exists a subset $\Pi \subseteq \Omega_P^+$* (Definition II.17) *and $x_\nu \in \mathfrak{g}^\nu$ for $\nu \in \Pi$ such that*

$$\mathfrak{h} := \operatorname{span}\{Ix_\nu : \nu \in \Pi\}$$

is a Cartan algebra of $\mathfrak{t}$ and a system $\Lambda \subseteq \widehat{\mathfrak{h}}$ which is contained in an open half space such that

$$x_\nu + Q(x_\nu) \in \mathfrak{t}^{\alpha_\nu} \quad \textit{for} \quad \nu \in \Pi,$$

where $\alpha_\nu \in \Lambda$. The subalgebra $E := \bigoplus_{\nu \in \Pi} \langle x_\nu \rangle$ is isomorphic to $\mathrm{sl}(2, \mathbb{R})^n$.

Proof. From [HHL89, A.2.21] we deduce that any two maximal compactly embedded subalgebras of S are conjugate under an inner automorphism of L. Therefore we may choose a fixed $\mathfrak{k}$. The rest follows from [Ne90, 2.11, 2.18]. ∎

Lemma VII.9. *Let $\mathfrak{t}$ be a solvable Lie algebra of real type and $h \in \mathfrak{h}$ such that $\alpha(h) \neq 0$ whenever $\alpha \in \Lambda$. Then the following assertions hold:*

(1) $\mathfrak{n} := \bigoplus_{\alpha \in \Lambda} \mathfrak{t}^\alpha$ *is the nilradical and the commutator algebra of* $\mathfrak{t}$.

(2) *The mapping* $\Phi : \mathfrak{n} \to \mathfrak{n}, x \mapsto (-x) * e^{\operatorname{ad} h} x$ *is one-to-one, where* $* : \mathfrak{n} \times \mathfrak{n} \to \mathfrak{n}$ *denotes the Campbell-Hausdorff-multiplication on* $\mathfrak{n}$.

Proof. (cf. [Hel84, p.183]) (1) This follows from the fact that

$$[\mathfrak{t}^\alpha, \mathfrak{t}^{\alpha'}] \subseteq \mathfrak{t}^{\alpha+\alpha'} \quad \text{for} \quad \alpha, \alpha' \in \Lambda$$

and $0 \notin \Lambda + \Lambda$, since Λ is contained in an open half space.

(2) (cf. [Hel84, p.183]) We use induction on the dimension of $\mathfrak{n}$. If $\dim \mathfrak{n} = \{0\}$, there is nothing to prove. Suppose that the assertions holds for all solvable Lie algebras $\mathfrak{t}_1 = \mathfrak{h}_1 + \mathfrak{n}_1$ of real type with $\dim \mathfrak{n}_1 < \dim \mathfrak{n}$. Firstly we take $\beta \in \Lambda \setminus (\Lambda + \Lambda)$, $x \in \mathfrak{t}^\beta$ and choose a subspace $\mathfrak{t}_1^\beta \subseteq \mathfrak{t}^\beta$ such that $\mathfrak{t}^\beta = \mathbb{R}x \oplus \mathfrak{t}_1^\beta$. Then we set

$$\mathfrak{t}_1 := \mathfrak{h} + \mathfrak{n}_1 \quad \text{with} \quad \mathfrak{n}_1 := \mathfrak{t}_1^\beta + \bigoplus_{\alpha \in \Lambda \setminus \{\beta\}} \mathfrak{t}^\alpha.$$

It is clear that $\mathfrak{t}_1 \subseteq \mathfrak{t}$ is a subalgebra of real type and that $\Phi(\mathfrak{n}_1) \subseteq \mathfrak{n}_1$ since $e^{\operatorname{ad} h} |_{\mathfrak{t}^\alpha} = e^{\alpha(h)} \operatorname{id}_{\mathfrak{t}^\alpha}$. Therefore $\dim \mathfrak{n}_1 < \dim \mathfrak{n}$ and the induction hypothesis shows that $\Phi |_{\mathfrak{n}_1}$ is one-to-one. But $\mathfrak{n} = \mathfrak{n}_1 * \mathbb{R}x$ since $\mathfrak{n}_1$ is normal in $\mathfrak{n}$. For $t \in \mathbb{R}$ and $y \in \mathfrak{n}_1$ we have that

$$\begin{aligned}\Phi(y * tx) &= -tx * -y * e^{\operatorname{ad} h}(y * tx) \\ &= -tx * -y * e^{\operatorname{ad} h} y * e^{\operatorname{ad} h} tx \\ &= I_{-tx}\big(\Phi(y)\big) * (-tx) * e^{\alpha(h)} tx \\ &= I_{-tx}\big(\Phi(y)\big) * (e^{\alpha(h)} - 1)tx.\end{aligned}$$

Given $y' \in \mathfrak{n}_1$ and $t'x \in \mathbb{R}x$ we set

$$t := (e^{\alpha(h)} - 1)^{-1} t' \quad \text{and} \quad y := \Phi |_{\mathfrak{n}_1}^{-1}\big(I_{tx}(y')\big).$$

Then $\Phi(y * tx) = y' * t'x$ and it follows that Φ is injective and onto, hence one-to-one. ■

Proposition VII.10. *Let T be a connected Lie group such that $\mathfrak{t} := \mathbf{L}(G)$ is solvable of real type, and $S \subseteq T$ a closed invariant subsemigroup with dense interior. Then*

$$S = S[T,T] = [T,T]S.$$

Moreover, we have for every closed subset $M \subseteq T$ which is invariant under conjugation and contains a Cartan algebra that $M = T$.

Proof. According to the assumption that $\mathfrak{t}$ is solvable of real type we find a Cartan algebra $\mathfrak{h} \subseteq \mathfrak{t}$ and $\Lambda \subseteq \widehat{\mathfrak{h}}$ such that

$$\mathfrak{t} = \mathfrak{h} \oplus \bigoplus_{\alpha \in \Lambda} \mathfrak{t}^\alpha,$$

where Λ is contained in an open half space. The commutator algebra is $\mathfrak{n} = \bigoplus_{\alpha \in \Lambda} \mathfrak{t}^\alpha$ (Lemma VII.9). We endow $\mathfrak{n}$ with the Campbell-Hausdorff multiplication $*$. Then $\widetilde{T} = \mathfrak{n} \rtimes \mathfrak{h}$ is the universal covering group of T, where $\mathfrak{h}$ acts on $\mathfrak{n}$ by $h.n = e^{\operatorname{ad} h} n$. The product in $\widetilde{T}$ is given by

$$(n,h)(n',h') = (n * e^{\operatorname{ad} h} n', h + h') \quad \text{for} \quad n, n' \in \mathfrak{n},\ h, h' \in \mathfrak{h}.$$

We write $q : \widetilde{T} \to T$ for the universal covering morphism, $p : \widetilde{T} \to \{0\} \times \mathfrak{h}$ for the projection onto $H := \{0\} \times \mathfrak{h}$ along $N := \mathfrak{n} \times \{0\}$, and set $\widetilde{S} = q^{-1}(S)$.

This is a closed invariant subsemigroup of $\widetilde{T}$ with dense interior. We claim that $N\widetilde{S} = \widetilde{S}$.
Step 1) $p(\widetilde{S}) = \widetilde{S} \cap H$: It is clear that $\widetilde{S} \cap H \subseteq p(\widetilde{S})$. Let $(n,h) \in \widetilde{S}$. We have to show that $(0,h) \in \widetilde{S}$. To see this we choose an element $x \in \mathfrak{h}$ such that $\alpha(x) < 0$ for all $\alpha \in \Lambda$. Then

$$I_{(0,tx)}(n,h) = (0,tx)(n,h)(0,-tx) = (e^{t\,\mathrm{ad}\,x}n, h) \in I_{(0,tx)}(\widetilde{S}) = \widetilde{S}.$$

We write $n = \sum_{\alpha \in R^+} n_\alpha$. Then

$$\lim_{t\to\infty} e^{t\,\mathrm{ad}\,x} n = \lim_{t\to\infty} \sum_{\alpha\in R^+} e^{t\alpha(h)} n_\alpha = 0.$$

Thus $(0,h) = \lim_{t\to\infty} I_{(0,tx)}(n,h) \in \widetilde{S} \cap H$.
Step 2) For every regular element $h \in \mathfrak{h}$ we have that $\{I_{(x,0)}(0,h) : x \in \mathfrak{n}\} = N(0,h)$: To see this, we compute

$$\begin{aligned} I_{(-x,0)}(0,h) &= (-x,0)(0,h)(x,0) = (-x,h)(x,0) \\ &= (-x * e^{\mathrm{ad}\,h}x, h) = (-x * e^{\mathrm{ad}\,h}x, 0)(0,h). \end{aligned}$$

Now we use the Lemma VII.9 to see that

$$\{I_{(x,0)}(0,h) : x \in \mathfrak{n}\} = \{(-x * e^{\mathrm{ad}\,h}x, 0) : x \in \mathfrak{n}\}(0,h) = N(0,h).$$

Step 3) $N\widetilde{S} = \widetilde{S}$: If $(n,h) \in \widetilde{S}$ such that h is a regular element, then $(0,h) \in \widetilde{S}$ (Step 1) and $N(0,h) \subseteq \widetilde{S}$ (Step 2). Therefore $N(n,h) \subseteq \widetilde{S}$ whenever h is regular. But the set of those elements is dense in $\widetilde{S}$ because it is dense in its interior. This proves that $N\widetilde{S} = \widetilde{S}$.
Using Step 3 we see that

$$[T,T]S = q([\widetilde{T},\widetilde{T}])q(\widetilde{S}) = q(N\widetilde{S}) = q(\widetilde{S}) = S.$$

The last assertion is a consequence of Step 2. ■

Lemma VII.11. *Let E be a Lie group such that $\mathbf{L}(E) \cong \mathrm{sl}(2,\mathbb{R})$ and H,T,U in $\mathbf{L}(E)$ such that*

$$[H,T] = 2U, \quad [U,T] = 2H, \quad \text{and} \quad [U,H] = -2T.$$

Moreover, suppose that $V \subseteq \mathbf{L}(E)$ is an invariant generating cone. Then there exists an element $z \in Z(E)$ such that

$$z\exp\big(\mathbb{R}H + \mathbb{R}(T+U)\big) \subseteq \langle \exp(V) \rangle.$$

The above relations are satisfied for the elements

$$H = 2Ix_\nu, \quad T = -2x_\nu, \quad \text{and} \quad U = 2Q(x_\nu) = 2[Ix_\nu, x_\nu]$$

in Lemma VII.8 *provided x_ν is suitably normalized.*

Proof. Assume the notation of Lemma VII.8. We choose $x_\nu \in \mathfrak{g}^\nu$ for $\nu \in \Pi$ such that $\nu\big(Q(x_\nu)\big) = 1$ (Proposition II.8). Then

$$Q(x_\nu) = [Ix_\nu, x_\nu], \quad [Q(x_\nu), Ix_\nu] = -x_\nu, \quad \text{and} \quad [Q(x_\nu), x_\nu] = Ix_\nu.$$

This proves the last assertion. That the semigroup $\langle \exp V \rangle$ contains cosets $z\exp\big(\mathbb{R}x + \mathbb{R}(y+z)\big)$ can easily be seen with [HHL89, V.4.46] since it is sufficient to prove this for the universal covering group $\widetilde{E} \cong \mathrm{Sl}(2,\mathbb{R})^\sim$. ■

Lemma VII.12. *Let G be a reductive Lie group such that $\mathfrak{g} := \mathbf{L}(G)$ is quasihermitean, $\mathbf{L}(G) = Z(\mathfrak{g}) \oplus [\mathfrak{g}, \mathfrak{g}]$ and $[\mathfrak{g}, \mathfrak{g}] = \mathfrak{k} + \mathfrak{t}$ an Iwasawa decomposition as in* Lemma VII.8. *Suppose that (S, G) is a GILSg with $S \neq G$. Then there exists an $s \in S \cap Z(G)$ with*

$$sT \subseteq S, \quad \operatorname{int} S \cap T = \emptyset, \quad \text{and} \quad \overline{ST} \neq G,$$

where $T := \langle \exp \mathfrak{t} \rangle = \exp \mathfrak{t}$.

Proof. Using Lemma VI.7 we see that the second and third assertion are consequences of the first one. We write $q : \widetilde{G} \to G$ for the universal covering of G and we set

$$\widetilde{S} := \overline{\langle \exp_{\widetilde{G}} \mathbf{L}(S) \rangle} \quad \text{and} \quad \widetilde{T} := \exp_{\widetilde{G}} \mathfrak{t}.$$

Then $\overline{q(\widetilde{S})} = S$, $q(\operatorname{int} \widetilde{S}) \subseteq \operatorname{int} S$, $q\big(Z(\widetilde{G})\big) \subseteq Z(G)$ and $q(\widetilde{T}) = T$. Hence it suffices to prove the assertion in the simply connected case. So let us assume that $\pi_1(G) = \{0\}$. If $I \subseteq \mathfrak{g}$ is an ideal such that $\mathfrak{g} = I \oplus H\big(\mathbf{L}(S)\big)$ we have that $G \cong A \times H(S)$, where $A = \langle \exp I \rangle$ and T, S, and $Z(G)$ are adjusted to the decomposition. Thus we may also assume that $H(S) = \{\mathbf{1}\}$. Then $\mathbf{L}(S)$ is a pointed generating invariant wedge in $\mathfrak{g}$. We choose $\Pi \subseteq \Omega_P^+$ as in Lemma VII.8. For every $\nu \in \Pi$ the intersection of $\langle x_\nu \rangle$ with $\mathbf{L}(S)$ is a generating invariant cone since $Q(x_\nu) \in \mathbf{L}(S) \cup -\mathbf{L}(S)$ (Proposition III.15). Lemma VII.11 implies that we find $z_\nu \in Z(\langle \exp \langle x_\nu \rangle \rangle)$ such that

$$z_\nu \exp(\mathbb{R} I x_\nu) \subseteq \big\langle \exp\big(\mathbf{L}(S) \cap \langle x_\nu \rangle\big)\big\rangle \subseteq S.$$

According to Lemma VII.8 we have $[\langle x_\nu \rangle, \langle x_\mu \rangle] = \{0\}$ for $\nu \neq \mu$. Then we find that

$$\prod_{\nu \in \Pi} \big(z_\nu \exp(\mathbb{R} I x_\nu)\big) = \Big(\prod_{\nu \in \Pi} z_\nu\Big) \exp(\mathfrak{h}) \subseteq S.$$

Next we show that there exists an element $s \in \operatorname{int} S \cap Z(G)$. To see this, we choose a compactly embedded Cartan algebra $\widetilde{\mathfrak{h}}_1 \subseteq \mathfrak{g}' = [\mathfrak{g}, \mathfrak{g}]$. Then $\widetilde{\mathfrak{h}} = Z(\mathfrak{g}) \oplus \widetilde{\mathfrak{h}}_1$ is a compactly embedded Cartan algebra of $\mathfrak{g}$. Moreover, we choose a nice positive system Ω^+ of real roots such that

$$C_{\min} \subseteq C := \mathbf{L}(S) \cap \widetilde{\mathfrak{h}} \subseteq C_{\max}$$

(Proposition III.15). If $\mathfrak{g}' = \bigoplus_{i=1}^m \mathfrak{s}_i$ is a decomposition into simple ideals, we have that

$$C_{\min} = \sum_{i=1}^m (\mathfrak{s}_i \cap C_{\min}).$$

In addition, we have that $C_{\min} \cap \mathfrak{s}_i \neq \{0\}$ if and only if $\mathfrak{s}_i$ is non-compact. We assume that $\mathfrak{s}_{k+1}, \ldots, \mathfrak{s}_m$ are precisely the compact ideals. Viewing $\mathfrak{g}'$ as an $\bigoplus_{i=k+1}^m \mathfrak{s}_i$-module, Theorem I.10 implies that

$$\operatorname{algint} C_{\min} \subseteq \operatorname{int}_{\mathfrak{g}'}(W).$$

Now we consider $\mathfrak{h} \cap \mathfrak{s}_i$ for $i = 1, \ldots, k$ as a $\mathcal{W}$-module, where $\mathcal{W}$ is the Weyl group of $\mathfrak{s}_i$ (Definition III.2). Then, using Theorem I.10 and the invariance of $C_{\min}$ under $\mathcal{W}$, we find $z_i \in \mathfrak{z}_{\mathfrak{k}} \cap \operatorname{algint}(C_{\min} \cap \mathfrak{s}_i)$. In view of Proposition II.19 we may also assume that $e^{\operatorname{ad} z_i}|_{\mathfrak{s}_i} = \operatorname{id}_{\mathfrak{s}_i}$ since $e^{\operatorname{ad} tz_i}$ acts as multiplication with a complex scalar of modulus 1 on a $\mathfrak{k}$-invariant complement of $\mathfrak{k} \cap \mathfrak{s}_i$ in $\mathfrak{s}_i$. Setting

$$z := \sum_{i=1}^{k} z_i \in C_{\min}$$

we have found an element $z \in \operatorname{int}_{\mathfrak{g}'}(W \cap \mathfrak{g}')$ with $e^{\operatorname{ad} z} = \operatorname{id}_{\mathfrak{g}}$, i.e., $\exp(z) \in Z(G)$.

Suppose that $\operatorname{int} S \cap Z(G) = \emptyset$. Then $\exp\big(z + Z(\mathfrak{g})\big) \cap \operatorname{int} S = \emptyset$. In particular $\big(z + Z(\mathfrak{g})\big) \cap \operatorname{int} W = \emptyset$ and we find a functional $\omega \in \widehat{\mathfrak{g}} \setminus \{0\}$ such that

$$\omega\big(z + Z(\mathfrak{g})\big) \subseteq]-\infty, 0] \quad \text{and} \quad \omega(\operatorname{int} W) \subseteq]0, \infty[.$$

Hence ω vanishes on $Z(\mathfrak{g})$ and thus

$$\omega \in W^* \cap Z(\mathfrak{g})^{\perp} = \overline{W + Z(\mathfrak{g})}^{\,*}$$

(Proposition I.2.4). This contradicts the fact that

$$z \in \operatorname{int}_{\mathfrak{g}'} W \subseteq \operatorname{int} \overline{W + Z(\mathfrak{g})}$$

because $\omega(z) \leq 0$. Consequently we find $z' \in Z(\mathfrak{g})$ such that

$$s := \exp(z + z') \in \exp(\operatorname{int} W) \cap Z(G) \subseteq \operatorname{int} S \cap Z(G).$$

Using [Ne91b, 2.11] we find an element $s' \in \operatorname{int} S$ and $m \in \mathbb{N}$ with

$$\prod_{\nu \in \Pi} z_\nu = s^m s'^{-1} = s'^{-1} s^m.$$

So $s^m \in Z(G) \cap \operatorname{int} S$ and

$$s^m \exp(\mathfrak{h}) = s'\big(\prod_{\nu \in \Pi} z_\nu\big) \exp(\mathfrak{h}) \subseteq s'S \subseteq \operatorname{int} S.$$

For every $t \in T$ this leads to

$$I_t\big(s^m \exp(\mathfrak{h})\big) = s^m I_t(\exp \mathfrak{h}) \subseteq \operatorname{int} S.$$

But, according to Proposition VII.10, we know that

$$T = \overline{\bigcup_{t \in T} I_t\big(\exp \mathfrak{h}\big)}.$$

Therefore $s^m T \subseteq \operatorname{int} S$. ∎

Proposition VII.13. *Let G be a connected Lie group and $M \subseteq G$ be a closed subsemigroup such that $\mathbf{L}(M)$ is no subalgebra of $\mathbf{L}(G)$. Then*

$$\mathbf{L}(M) \cap \mathbf{L}(K) \subseteq H(\mathbf{L}(M)) \quad \text{and} \quad \mathbf{L}(M)^* \cap \mathbf{L}(K)^\perp \neq \{0\}$$

for every compact subgroup $K \subseteq G$.

Proof. Let $S := \overline{\langle \exp \mathbf{L}(M) \rangle} \subseteq M$ be the largest Lie semigroup contained in M. Then $\mathbf{L}(S) = \mathbf{L}(M)$ is not a subalgebra of $\mathbf{L}(G)$ and, according to [Ne89, II.12], we find a function $f \in C^\infty(G)$ such that $df(\mathbf{1}) \in \operatorname{algint} \mathbf{L}(M)^*$. Moreover, Proposition III.3 in [Ne89] permits us to assume that $f(Kg) = \{f(g)\}$ for all $g \in G$. Therefore

$$df(\mathbf{1}) \in \mathbf{L}(K)^\perp \cap \operatorname{algint} \mathbf{L}(M)^*.$$

We conclude that

$$\mathbf{L}(M) \cap \mathbf{L}(K) \subseteq \ker df(\mathbf{1}) \cap \mathbf{L}(M) \subseteq H(\mathbf{L}(M))$$

and

$$\{0\} \neq df(\mathbf{1}) \in \mathbf{L}(M)^* \cap \mathbf{L}(K)^\perp = \big(\mathbf{L}(M) + \mathbf{L}(K)\big)^*.$$

∎

Proposition VII.14. *Suppose that $\mathbf{L}(G)$ is reductive and that $W \subseteq \mathbf{L}(G)$ is a generating invariant wedge which is not controllable in G. Then*

$$W^* \cap \mathfrak{t}^\perp \cap [\mathfrak{k}, \mathfrak{k}]^\perp \neq \{0\}, \quad \text{i.e.,} \quad W + \mathfrak{t} + [\mathfrak{k}, \mathfrak{k}] \neq \mathbf{L}(G)$$

where $\mathbf{L}(G) = Z(\mathbf{L}(G)) \oplus (\mathfrak{t} + \mathfrak{k})$ is an Iwasawa decomposition as in Lemma VII.8.

Proof. Set $S := \overline{\langle \exp W \rangle}$ and $T := \langle \exp \mathfrak{t} \rangle$. Then $S \neq G$ (Lemma VII.2). Using Lemma VII.12 we see that $ST \neq G$. Now we find a maximal subsemigroup $M \subseteq G$ with $ST \subseteq M$. The subgroup $K := \exp[\mathfrak{k}, \mathfrak{k}] \subseteq G$ is compact and therefore

$$\mathbf{L}(M)^* \cap [\mathfrak{k}, \mathfrak{k}]^\perp \neq \{0\}$$

by Lemma VII.13. Now $W + \mathfrak{t} \subseteq \mathbf{L}(M)$ implies that

$$\{0\} \neq \mathbf{L}(M)^* \cap [\mathfrak{k}, \mathfrak{k}]^\perp \subseteq W^* \cap \mathfrak{t}^\perp \cap [\mathfrak{k}, \mathfrak{k}]^\perp.$$

∎

Before we prove the first two Controllability Theorems we justify the restriction to pointed generating invariant wedges.

Proposition VII.15. *Let G be a connected Lie group and $W \subseteq \mathbf{L}(G)$ an invariant wedge. Then the following assertions hold:*

(1) *Suppose that W is not generating, then W is not controllable in G.*

(2) *Suppose that W is generating and that $H := \overline{\langle \exp H(W) \rangle}$. Write $\pi : G \to G/H$ for the canonical projection. Then W is controllable in G iff the pointed generating invariant wedge $d\pi(\mathbf{1})W$ is controllable in G/H. The group $\langle \exp H(W) \rangle$ is closed if G is simply connected.*

Proof. (1) Set $I := W - W$. Then the analytic subgroup generated by I is different from G and contains $\langle \exp W \rangle$. Therefore W is not controllable in G.
(2) It is clear that the controllability of W in G implies the controllability of $d\pi(\mathbf{1})W$ in G/H because

$$\langle \exp d\pi(\mathbf{1})W \rangle = \pi(\langle \exp W \rangle).$$

Suppose that $W_1 := d\pi(\mathbf{1})W$ is controllable in G/H. Then $\pi(\langle \exp W \rangle) = G/H$ and therefore $HS = G$ for $S := \langle \exp W \rangle$. Set $H_1 := \langle \exp H(W) \rangle$. Then the semigroup $H_1 S$ is dense in G and it has dense interior (Lemma IV.11). Hence $H_1 S = G$ (Lemma IV.10). Since $H_1 \subseteq S$, this leads to $H_1 S = S = G$, and W is controllable in G. The last assertion is a consequence of Proposition IV.30. ∎

Theorem VII.16. (First Controllability Theorem) *Let W be a pointed generating invariant wedge in $\mathbf{L}(G)$ and suppose that W is not controllable in G. Then*

$$W^* \cap \widetilde{\mathfrak{n}}^\perp \cap \mathfrak{t}^\perp \cap [\mathfrak{k}, \mathfrak{k}]^\perp \neq \{0\},$$

with $\widetilde{\mathfrak{n}}$ *as in* Proposition VII.4 *and* $\mathfrak{t}, \mathfrak{k}$ *as in* Lemma VII.8.

Proof. First we note that $\mathbf{L}(G)$ has cone potential (Proposition II.7). Set $A := \overline{\exp \widetilde{\mathfrak{n}}}$ and $\varphi : G \to G/A, g \mapsto gA$. Then $V := \overline{d\varphi(\mathbf{1})W}$ is an invariant generating wedge in $\mathbf{L}(G/A)$ which is not controllable in G/A (Corollary VII.6) and $\mathbf{L}(G/A)$ is a reductive quasihermitean Lie algebra (Proposition VII.4). Now Proposition VII.14 implies the existence of a non-zero functional

$$\omega \in V^* \cap d\varphi(\mathbf{1})(\mathfrak{t} + [\mathfrak{k}, \mathfrak{k}])^\perp.$$

Thus

$$\omega \circ d\varphi(\mathbf{1}) \in W^* \cap \widetilde{\mathfrak{n}}^\perp \cap \mathfrak{t}^\perp \cap [\mathfrak{k}, \mathfrak{k}]^\perp.$$

∎

Theorem VII.17. (Second Controllability Theorem) *Let G be a simply connected Lie group and $W \subseteq \mathbf{L}(G)$ a pointed generating invariant wedge. Then W is not controllable in G if and only if*

(Con) $$V := W^* \cap \widetilde{\mathfrak{n}}^\perp \cap \mathfrak{t}^\perp \cap [\mathfrak{k}, \mathfrak{k}]^\perp \neq \{0\}.$$

Moreover,

$$\mathbf{L}\left(H(\overline{\langle \exp W \rangle})\right) \subseteq \bigcap_{\omega \in V} \ker \omega.$$

Proof. In view of Theorem VII.16 we only have to prove that the above condition is sufficient for W not to be controllable in G. To see this, we use the method of adapted functions from [Ne89]. Let $\mathfrak{a} := \mathfrak{r} + \mathfrak{t}$ and $\mathfrak{q} := \mathfrak{k}$, where $\mathfrak{r}$ is the radical of $\mathbf{L}(G)$ and $A := \langle \exp \mathfrak{a} \rangle$, $Q := \langle \exp \mathfrak{q} \rangle$. Then $\mathfrak{a}$ and $\mathfrak{q}$ are subalgebras of $\mathbf{L}(G)$. It follows from [Hel78, p.270] that the mapping

$$m : A \times Q \to G, \ (a, q) \mapsto a.q$$

is a diffeomorphism. Let $\omega \in W^* \cap \widetilde{\mathfrak{n}}^\perp \cap \mathfrak{t}^\perp \cap [\mathfrak{k}, \mathfrak{k}]^\perp \setminus \{0\}$. Then

$$\begin{aligned} \omega([\mathfrak{a}, \mathfrak{a}] + [\mathfrak{q}, \mathfrak{q}]) &\subseteq \omega([\mathfrak{g}, \mathfrak{r}] + [\mathfrak{t}, \mathfrak{t}] + [\mathfrak{k}, \mathfrak{k}]) \\ &\subseteq \omega([\mathfrak{g}, \mathfrak{r}]) \subseteq \omega(\widetilde{\mathfrak{n}}) = \{0\}. \end{aligned}$$

From $\{0\} = \pi_1(G) = \pi_1(A) \times \pi_1(Q)$ we deduce that the groups A and Q are both simply connected. Now [Ne89, IV.10, IV.11] apply and we get a function $f \in C^\infty(G)$ with $df(\mathbf{1}) = \omega$ such that

$$df(g) = \omega \circ \mathrm{Ad}(b) \in W^* \quad \text{for all } g \in G \quad \text{with} \quad g \in Ab.$$

Thus f is a non-constant W-positive function and therefore W is not controllable in G ([Ne89, II.13]) and

$$\mathbf{L}((H(\overline{\langle \exp W \rangle})) \subseteq \ker \omega = df(\mathbf{1})$$

since $f\big(H(\overline{\langle \exp W \rangle})\big) \subseteq f(\{\mathbf{1}\})$. ■

Since every pointed generating wedge $W \subseteq \mathbf{L}(G)$ is determined uniquely by its intersection with a compactly embedded Cartan algebra $\mathfrak{h} \subseteq \mathbf{L}(G)$ (Proposition III.34), it should be possible to formulate the condition (Con) for controllability only with the root data associated to H, i.e., the real root system Ω (Section II) and the wedge $C := W \cap \mathfrak{h}$. This is what we do next. We will need some of the results in [Ne90], where the controllability problem is solved for invariant wedges in semisimple Lie algebras.

First we consider the reduction from Corollary VII.6 on the Cartan algebra level.

Lemma VII.18. *Let W be a pointed generating invariant wedge in $\mathfrak{g}$, $\widetilde{\mathfrak{n}} = \mathfrak{n}_{\mathrm{eff}} + [\mathfrak{n}, \mathfrak{n}]$ the ideal defined in* Proposition VII.4, *and $W_1 := \overline{W + \widetilde{\mathfrak{n}}}$. Then*

$$W_1 \cap \mathfrak{h} = \overline{W \cap \mathfrak{h} + [\mathfrak{n}, \mathfrak{n}]}.$$

Proof. Let $p : \mathfrak{g} \to \mathfrak{h}$ denote the projection onto the fixed point set of the action of the compact group $\overline{e^{\mathrm{ad}\, \mathfrak{h}}}$. Then, according to Theorem I.10,

$$W_1 = p(W_1) = p\big(L_{\widetilde{\mathfrak{n}}}(W)\big) = L_{p(\widetilde{\mathfrak{n}})}\big(p(W)\big) = L_{[\mathfrak{n},\mathfrak{n}]}(W \cap \mathfrak{h}) = \overline{W \cap \mathfrak{h} + [\mathfrak{n}, \mathfrak{n}]}$$

since $p(\widetilde{\mathfrak{n}}) = \widetilde{\mathfrak{n}} \cap \mathfrak{h} = [\mathfrak{n}, \mathfrak{n}]$. ■

This lemma provides the reduction to the reductive case which we consider now.

Definition VII.19. Let $\mathfrak{g}$ be a simple quasihermitean Lie algebra, $\mathfrak{h} \subseteq \mathfrak{g}$ a compactly embedded Cartan algebra and $\Omega^+ \subseteq \Omega$ a positive system such that $\omega(z) > 0$ for all $\omega \in \Omega_P^+$ and an element $z \in \mathfrak{z}_\mathfrak{k}$. According to [Ne90, 2.4] we may choose $z \in \mathfrak{z}_\mathfrak{k}$ such that $\omega(z) = 1$ for all $\omega \in \Omega_P^+$. Assuming this normalization, we set

$$l := z - \sum_{\nu \in \Pi} Q(x_\nu),$$

where $\Pi \subseteq \Omega_P^+$ is as in Lemma VII.8 and $x_\nu \in \mathfrak{g}^\nu$ with $\nu\big(Q(x_\nu)\big) = 1$ for all $\nu \in \Pi$. Moreover, we define the linear functional $\widehat{l}$ on $\mathfrak{g}$ by

$$\widehat{l}(x) = -B(x, l),$$

where B is the Cartan Killing form of $\mathfrak{g}$. ∎

Proposition VII.20. *Let $\mathfrak{g}$ be a quasihermitean reductive Lie algebra, $W \subseteq \mathfrak{g}$ a pointed generating invariant wedge, $\Omega^+ \subseteq \Omega$ a positive system such that $C := W \cap \mathfrak{h} \subseteq C_{\max}(\Omega^+)$, $[\mathfrak{g}, \mathfrak{g}] = \mathfrak{k} + \mathfrak{t}$ an Iwasawa decomposition of $\mathfrak{g}'$ as in* Lemma VII.8, *$\mathfrak{g}' = \bigoplus_{i=1}^n \mathfrak{s}_i$, where $\mathfrak{s}_i$ are the simple ideals and $\mathfrak{s}_i$ is compact for $i = m+1, \ldots, n$, and $l_i \in \mathfrak{s}_i$ as defined in* Definition VII.19. *Then the following assertions are equivalent:*

(1) $W^* \cap \mathfrak{t}^\perp \cap \mathfrak{k}'^\perp \neq \{0\}$,

(2) $W + \mathfrak{t} + \mathfrak{k}' \neq \mathfrak{g}$, *and*

(3) $(C \cap [\mathfrak{g}, \mathfrak{g}])^* \cap \sum_{i=1}^m \mathbb{R}^+ \widehat{l}_i \neq \{0\}$ *or* $C^* \cap [\mathfrak{g}, \mathfrak{g}]^\perp \neq \{0\}$.

Proof. (1) $\Leftrightarrow$ (2): Since $(W + \mathfrak{t} + \mathfrak{k}')^* = W^* \cap \mathfrak{t}^\perp \cap \mathfrak{k}'^\perp$ this follows from the fact that $\operatorname{int} W \neq \emptyset$ and the Hahn-Banach theorem.

Let $V := W \cap [\mathfrak{g}, \mathfrak{g}]$. This is a pointed generating invariant wedge in the semisimple ideal $\mathfrak{g}_1 := V - V$ of $\mathfrak{g}$. This ideal contains $\bigoplus_{i=1}^m \mathfrak{s}_i$ because $C \subseteq C_{\max}(\Omega^+)$ implies that

$$Q(\mathfrak{g}^\omega) \subseteq \omega(C) Q(\mathfrak{g}^\omega) \subseteq \mathcal{C}(C) \subseteq C \quad \text{for} \quad \omega \in \Omega_P^+.$$

Thus $C_{\min} \subseteq V$ and the smallest ideal of $\mathfrak{g}$ containing $C_{\min}$ is the sum of all hermitean ideals.

(2) $\Rightarrow$ (3): Suppose that

$$(C \cap [\mathfrak{g}, \mathfrak{g}])^* = (V \cap \mathfrak{h})^* = {\mathfrak{g}_{\text{eff}}}^\perp \cap V^*$$

does not intersect the cone $\sum_{i=1}^m \mathbb{R}^+ \widehat{l}_i$ non-trivially. Then Proposition 2.24 in [Ne90] applies and shows, in view of the equivalence of (1) and (2), that $V + (\mathfrak{k}' \cap \mathfrak{g}_1) + \mathfrak{t} = \mathfrak{g}_1$. Thus

$$V + \mathfrak{k}' + \mathfrak{t} = [\mathfrak{g}, \mathfrak{g}] \subseteq W + \mathfrak{k}' + \mathfrak{t}.$$

Consequently we find a functional

$$0 \neq \omega \in (W + \mathfrak{k}' + \mathfrak{t})^* \subseteq [\mathfrak{g}, \mathfrak{g}]^\perp.$$

This functional is contained in $C^\star$ because it vanishes on $\mathfrak{g}_{\text{eff}} \subseteq [\mathfrak{g},\mathfrak{g}]$ and $\omega(C) = \omega(W) \subseteq \mathbb{R}^+$.
(3) $\Rightarrow$ (2): First we assume that $C^\star \cap [\mathfrak{g},\mathfrak{g}]^\perp \neq \{0\}$. We take $\omega \in C^\star \cap [\mathfrak{g},\mathfrak{g}]^\perp$. Then $\omega \in W^* \cap \mathfrak{t}^\perp \cap \mathfrak{k}'^\perp$ because $\mathfrak{t} + \mathfrak{k}' \subseteq [\mathfrak{g},\mathfrak{g}]$. If

$$(C \cap [\mathfrak{g},\mathfrak{g}])^\star \cap \sum_{i=1}^{m} \mathbb{R}^+ \widehat{l}_i \neq \{0\}$$

we conclude with [Ne90, II.24] that

$$V + \mathfrak{t} + (\mathfrak{k}' \cap \mathfrak{g}_1) \neq \mathfrak{g}_1.$$

Thus $V + \mathfrak{t} + \mathfrak{k}' \neq [\mathfrak{g},\mathfrak{g}]$. This leads to $W + \mathfrak{t} + \mathfrak{k}' \neq \mathfrak{g}$ because

$$(W + \mathfrak{t} + \mathfrak{k}') \cap [\mathfrak{g},\mathfrak{g}] = V + \mathfrak{t} + \mathfrak{k}' \neq [\mathfrak{g},\mathfrak{g}].$$

■

Now we give a translation of the controllability criterion (Con) from Theorem VII.17 which involves only the root data and the intersection $C = W \cap \mathfrak{h}$.

Theorem VII.21. (The Controllability Criterion) *Let $\mathfrak{g}$ be a finite dimensional Lie algebra containing a pointed generating invariant wedge W, $\mathfrak{h}$ a compactly embedded Cartan algebra in $\mathfrak{g}$, $\mathfrak{s}$ an $\mathfrak{h}$-invariant Levi algebra, $\mathfrak{h}_\mathfrak{s} := \mathfrak{h} \cap \mathfrak{s}$, $C = W \cap \mathfrak{h}$, $\Omega^+ \subseteq \Omega$ a positive system such that $C \subseteq C_{\max}(\Omega^+)$ and $C_0 := L_{[\mathfrak{n},\mathfrak{n}]}(C)$, where $\mathfrak{n}$ is the nilradical of $\mathfrak{g}$. Then the controllability condition*

(Con) $$W^* \cap \widetilde{\mathfrak{n}}^\perp \cap \mathfrak{t}^\perp \cap \mathfrak{k}^\perp \neq \{0\}$$

is equivalent to

$$(C_0 \cap \mathfrak{h}_\mathfrak{s})^\star \cap \sum_{i=1}^{m} \mathbb{R}^+ \widehat{l}_i \neq \{0\} \quad \textit{or} \quad C_0^\star \cap \mathfrak{h}_\mathfrak{s}^\perp \neq \{0\}.$$

Proof. Let $W_0 := L_{\widetilde{\mathfrak{n}}}(W)$. Then (Con) is equivalent to $W_0^* \cap \mathfrak{t}^\perp \cap \mathfrak{k}'^\perp \neq \{0\}$ and we have that $C_0 = W_0 \cap \mathfrak{h}$ (Lemma VII.18). Let $\mathfrak{g}_1 := \mathfrak{s} + \mathfrak{h}_1$, where $\mathfrak{h}_1 \subseteq \mathfrak{h}$ is a vector space complement to $\mathfrak{h}_\mathfrak{s} + \mathfrak{h} \cap [\mathfrak{n},\mathfrak{n}]$. Then $\mathfrak{g}_1$ is a reductive subalgebra of $\mathfrak{g}$ and $\mathfrak{g} = \mathfrak{g}_1 + H(W_0)$ (cf. Proposition VII.4). Since $W_0^* \cap \mathfrak{t}^\perp \cap \mathfrak{k}'^\perp \neq \{0\}$ is equivalent to $W_0 + \mathfrak{t} + \mathfrak{k}' \neq \mathfrak{g}$ (Proposition VII.20) we find, setting $W_1 := W_0 \cap \mathfrak{g}_1$ and $C_1 = W_1 \cap \mathfrak{h}$, that this is equivalent to $W_1 + \mathfrak{t} + \mathfrak{k}' \neq \mathfrak{g}_1$. Now Proposition VII.20 shows that this is equivalent to

$$(C_1 \cap \mathfrak{s})^\star \cap \sum_{i=1}^{m} \mathbb{R}^+ \widehat{l}_i \neq \{0\} \quad \text{or} \quad C_1^\star \cap [\mathfrak{n},\mathfrak{n}]^\perp \cap \mathfrak{s}^\perp \neq \{0\},$$

because $C \subseteq C_{\max}$ implies that $C_0 \subseteq C_{\max}$ since $[\mathfrak{n},\mathfrak{n}] \subseteq Z(\mathfrak{g}) \subseteq H(C_{\max})$. The intersection $C_0 \cap \mathfrak{s}$ equals $C_1 \cap \mathfrak{s}$ because $\mathfrak{s} \subseteq \mathfrak{g}_1$ and

$$C_0^\star \cap \mathfrak{h}_\mathfrak{s}^\perp = (C_1 + [\mathfrak{n},\mathfrak{n}])^\star \cap \mathfrak{h}_\mathfrak{s}^\perp = [\mathfrak{n},\mathfrak{n}]^\perp \cap C_1^\star \cap \mathfrak{h}_\mathfrak{s}^\perp = [\mathfrak{n},\mathfrak{n}]^\perp \cap (C_1^\star \cap \mathfrak{s}^\perp).$$

This proves the equivalence of the two conditions stated in the theorem above.■

VIII. GLOBALITY OF INVARIANT WEDGES

In the previous section we considered the controllability problem for invariant wedges which is, in some sense, a weakened version of the globality problem: Given an invariant wedge W in the Lie algebra $\mathbf{L}(G)$ of the Lie group G, does there exist a Lie semigroup $S \subseteq G$ such that $\mathbf{L}(S) = W$? Again the main idea is to reduce the problem to the reductive case. This is much harder than it was for the controllability problem. The main result in this direction is the Reduction Theorem VIII.7 which gives only sufficient conditions for globality but which permits one to prove the globality of all pointed generating invariant wedges in some large classes of Lie algebras. Another main result is the existence of at least one pointed generating invariant wedge which is global in a simply connected group G whenever $\mathbf{L}(G)$ contains pointed generating invariant wedges. This leads to a characterization of those simply connected Lie groups which possess non-degenerate group orders in terms of algebraic properties of the Lie algebra.

We want to use the results of Hofmann [Hof90b] on the singularities of the exponential function and therefore we start with the necessary definitions.

Definition VIII.1. Let $\mathfrak{g}$ be a finite dimensional Lie algebra. We define a function $\sigma : \mathfrak{g} \to \mathbb{R}^+$ by

$$\sigma(x) := \max\{|\operatorname{Im}(\lambda)| : \lambda \in \operatorname{Spec}(\operatorname{ad} x)\}.$$

■

Lemma VIII.2. *The function σ vanishes exactly on all $x \in \mathfrak{g}$ such that*

$$\operatorname{Spec}(\operatorname{ad} x) \in \mathbb{R}$$

and its restriction to any solvable subalgebra of $\mathfrak{g}$ is a seminorm.

Proof. [Hof90b, 4.33] ■

Proposition VIII.3. *Let $\exp : \mathbf{L}(G) \to G$ denote the exponential function of a simply connected Lie group G. Define*

$$U := \{x \in \mathbf{L}(G) : \sigma(x) < \pi\} \quad \text{and} \quad V := \exp(U).$$

Then the following assertions hold:

(1) *U contains the nilradical of* $\mathbf{L}(G)$,

(2) *U is star-shaped and symmetric, and*

(3) *the function* $\exp|_U : U \to V$ *is a diffeomorphism onto an open symmetric neighborhood of* $\mathbf{1}$ *in* G.

Let

$$U^{(2)} := \{(x, x') \in U \times U : \exp(x)\exp(x') \in V\}.$$

Then there is an analytic function $(x, y) \mapsto x \circ y : U^{(2)} \to U$ *with* $\exp(x \circ y) = \exp(x)\exp(y)$ *for* $(x, y) \in U^{(2)}$ *which extends the Campbell-Hausdorff multiplication defined near* $(0, 0)$ *in* $\mathbf{L}(G) \times \mathbf{L}(G)$.

Proof. [Hof90b, 4.6, 5.2] ■

Proposition VIII.4. *Let* $\mathfrak{g}$ *be a finite dimensional Lie algebra,* $\mathfrak{n} \subseteq \mathfrak{g}$ *a nilpotent ideal, and endow* $\mathfrak{g}$ *with a norm* $\|\cdot\|$ *such that*

$$\|[x, y]\| \leq \|x\|\|y\| \quad \text{for all} \quad x, y \in \mathfrak{g}.$$

Suppose that C is an open convex neighborhood of 0 *in* $\mathfrak{g}$ *such that*

$$\|x * y\| < \log 2 \quad \text{for all} \quad x, y \in C,$$

and U as in Proposition VIII.3.

Then the following assertions hold:

(1) *The BCH-series converges for* $x \in \mathfrak{n}$ *and* $y \in \mathfrak{g}$ *with* $\|y\| < \log 2$.

(2) $C * C \subseteq U$ *and* $\mathfrak{n} * (C * C) \subseteq U$.

(3) $(\mathfrak{n} * C) \times (\mathfrak{n} * C) \subseteq U^{(2)}$.

(4) $[0, 1](\mathfrak{n} * C) \subseteq \mathfrak{n} * C$.

Proof. (1) Choose $n \in \mathbb{N}$ with $\mathfrak{n}^n = \{0\}$. The ideals $\mathfrak{n}^i$ in $\mathfrak{n}$ are characteristic, hence ideals of $\mathfrak{g}$. Let

$$x * y = H(x, y) = \sum_{r,s \geq 0} H_{rs}(x, y),$$

where $H_{rs}(x, y)$ is homogeneous of degree r in x and s in y. Then $H_{rs}(x, y) = 0$ for $r > n$ because $H_{rs}(x, y) \in \mathfrak{n}^r$ and $\mathfrak{n}^n = \{0\}$. Therefore it remains to show that $\sum_{s \geq 0} H_{rs}(x, y)$ converges for every $x \in \mathfrak{n}$ and $\|y\| < \log 2$. This follows from the fact that this sum is homogeneous in x and converges for $\|x\| + \|y\| < \log 2$ ([HHL89, A.1.3]).

(2) This follows from

$$\sigma(x * y) \leq \|x * y\| < \log 2 < \pi \quad \text{for} \quad x, y \in C$$

and from

$$\sigma(n * x) \in \sigma(x + \mathfrak{n}) \subseteq \{\sigma(x)\} + \sigma(\mathfrak{n}) = \{\sigma(x)\} \quad \text{for} \quad x \in C,$$

which is a consequence of Lemma VIII.2.
(3) For $n, n' \in \mathfrak{n}$ and $c, c' \in C$ we compute

$$\begin{aligned}\exp(n * c)\exp(n' * c') &= \exp(n)\exp(c)\exp(n')\exp(c') \\ &= \exp(n)\exp(e^{\operatorname{ad} c}n')\exp(c)\exp(c') \\ &= \exp(n * e^{\operatorname{ad} c}n')\exp(c * c') \\ &\subseteq \exp\big(\mathfrak{n} * (C * C)\big) \\ &\subseteq \exp(U) = V\end{aligned}$$

because $\mathfrak{n} * (C * C) \subseteq U$. Thus $(\mathfrak{n} * C) \times (\mathfrak{n} * C) \subseteq U^{(2)}$.
(4) Let $n \in \mathfrak{n}, c \in C$ and $t \in [0,1]$. We claim that

$$\exp\big(t(n * c)\big)\exp(-tc) \in \exp(\mathfrak{n}). \tag{8.1}$$

According to [Ho65, p.135] the analytic normal subgroup $\exp \mathfrak{n} \subseteq G$ is closed, where G is the simply connected Lie group with $\mathbf{L}(G) = \mathfrak{g}$. Firstly we find $\varepsilon > 0$ such that the BCH-series $t(n * c) * (-tc)$ converges for $|t| < \varepsilon$. For those t we have

$$t(n * c) * (-tc) \in t(n * c) - tc + [\mathfrak{g}_1, \mathfrak{g}_1],$$

where $\mathfrak{g}_1 := \mathbb{R}c + \mathfrak{n}$ is the solvable subalgebra of $\mathfrak{g}$ generated by $\mathfrak{n}$ and c. Hence $[\mathfrak{g}_1, \mathfrak{g}_1] = [\mathbb{R}c + \mathfrak{n}, \mathbb{R}c + \mathfrak{n}] \subseteq \mathfrak{n}$ since $\mathfrak{n}$ is an ideal.We conclude that

$$t(n * c) * (-tc) \in t(n * c) - tc + \mathfrak{n} \subseteq t\mathfrak{n} + tc - tc + \mathfrak{n} = \mathfrak{n}.$$

This prove that (8.1) holds for $|t| < \varepsilon$. But the mapping

$$t \mapsto \gamma(t) := \exp\big(t(n * c)\big)\exp(-tc), \quad \mathbb{R} \to G$$

is analytic and therefore $\gamma(\mathbb{R}) \subseteq \exp(\mathfrak{n})$. Thus

$$\exp\big(t(n * c)\big) \in \exp(\mathfrak{n})\exp(tc) = \exp(\mathfrak{n} * tc) \subseteq \exp(U) = V$$

for all $t \in [0,1]$. Suppose that $t(n * c) \notin \mathfrak{n} * C$. Then we find a minimal $t_0 > 0$ with $t_0(n * c) \notin \mathfrak{n} * C$. Then $t_0(n * c) \in U$ because

$$\sigma\big(t_0(n * c)\big) = t_0\sigma(n * c) = t_0\sigma(c) \leq \log 2 < \pi$$

and consequently

$$t_0(n * c) = \exp|_U^{-1}\Big(\exp\big(t_0(n * c)\big)\Big) \in \exp|_U^{-1}\exp(\mathfrak{n} * tc) \subseteq \mathfrak{n} * C.$$

∎

For the next proposition we recall the definition of a Lie semialgebra (Definition IV.29).

Proposition VIII.5. *Let $W \subseteq \mathfrak{g}$ be a Lie semialgebra in $\mathfrak{g}$ and $\mathfrak{n} \subseteq \mathfrak{g}$ be a nilpotent ideal. Choose C as in* Proposition VIII.4. *Then*

$$\big(W \cap (\mathfrak{n} * C)\big) \circ \big(W \cap (\mathfrak{n} * C)\big) \subseteq W.$$

Proof. Let $x, y \in W \cap (\mathfrak{n} * C)$. We want to apply [HHL89, II.2.40] and have to check the requirements of this theorem with $D = U^{(2)}$.
(1) $[0,1]x \times [0,1]y \in [0,1](\mathfrak{n} * C) \times [0,1](\mathfrak{n} * C) \subseteq (\mathfrak{n} * C) \times (\mathfrak{n} * C) \subseteq U^{(2)}$ (Proposition VIII.4).
(2) $\big([0,1]x \circ [0,1]y\big) \times \{0\} \subseteq U^{(2)}$ since

$$\exp([0,1]x \circ [0,1]y)\exp(0) = \exp([0,1]x)\exp([0,1]y) \subseteq \exp(\mathfrak{n} * C)\exp(\mathfrak{n} * C) \subseteq V$$

because $(\mathfrak{n} * C) \times (\mathfrak{n} * C) \subseteq U^{(2)}$ (Proposition VIII.4).
(3) $[0,1](U^{(2)} \cap (\mathfrak{g} \times \{0\})) = [0,1](U \times \{0\}) = U \times \{0\} \subseteq U^{(2)}$. Now [HHL89, II.2.40] shows that $x \circ y \in W$. ∎

Lemma VIII.6. *Let G be a connected Lie group, (S, G) a Lie semigroup and $f : (G, \leq_S) \to (\mathbb{R}, \leq)$ a monotone function with $df(\mathbf{1}) \in \operatorname{algint} W^*$ and $f(\mathbf{1}) = 0$. Then the sets*

$$I_\varepsilon := f^{-1}([\varepsilon, \infty[) \cap S$$

are right ideals in S and for every neighborhood U of $\mathbf{1}$ in G we find $\varepsilon > 0$ with

$$S \subseteq UH(S) \cup I_\varepsilon.$$

Proof. Let $s \in I_\varepsilon$ and $t \in S$. Then $f(st) \geq f(s) \geq \varepsilon$, hence $st \in I_\varepsilon$. This shows that I_ε is a right ideal in S. Suppose that there exists a neighborhood U of $\mathbf{1}$ in G such that $UH(S) \cup I_\varepsilon$ does not contain S for an $\varepsilon > 0$. It is clear that f is constant on the sets

$$gH(S) = \{g' \in G : g \leq_S g' \leq_S g\},$$

hence there exists a smooth function $\widetilde{f} : G/H(S) \to \mathbb{R}$ which is monotone with respect to the quotient ordering on $G/H(S)$ and $d\widetilde{f}\big(\pi(\mathbf{1})\big) \in \operatorname{algint} d\pi(\mathbf{1})W$, where $\pi : G \to G/H(S), g \mapsto gH(S)$ denotes the canonical projection. Now we have a contradiction to [Ne91b, 1.8]. ∎

Theorem VIII.7. (The Reduction Theorem) *Let G be a simply connected Lie group, $\mathfrak{g} := \mathbf{L}(G)$, $\mathfrak{n} \subseteq \mathfrak{g}$ a nilpotent ideal, $W_1 \subseteq W_2 \subseteq \mathfrak{g}$ Lie wedges, such that W_2 is global in G, $H(W_2) = \mathfrak{n}$ and W_1 is invariant. Then W_1 is global in G.*

Proof. Let $S := \overline{\langle \exp W_1 \rangle}$, $T := \overline{\langle \exp W_2 \rangle}$ and $V := \mathbf{L}(S)$. Then $V \subseteq W_2 = \mathbf{L}(T)$ and therefore $H(V) \subseteq H(W_2) = \mathfrak{n}$. We find a function $f \in C^\infty(G)$ such that $f : (G, \leq_T) \to (\mathbb{R}, \leq)$ is monotone and $df(\mathbf{1}) \in \operatorname{algint} W_2^*$ ([Ne89, II.12]).

We choose $C \subseteq \mathfrak{g}$ as in Lemma VIII.4. With Lemma VIII.6 we find an $\varepsilon > 0$ such that

$$T \subseteq H(T)\exp(C) \cup I_\varepsilon \subseteq \exp(\mathfrak{n})\exp(C) \cup I_\varepsilon = \exp(\mathfrak{n} * C) \cup I_\varepsilon. \tag{8.2}$$

We claim that $S \subseteq \exp\big(W_1 \cap (\mathfrak{n} * C)\big) \cup I_\varepsilon$. We proceed in three steps.
Step 1) $\big(W_1 \cap (\mathfrak{n}*C)\big) \circ \big(W_1 \cap (\mathfrak{n}*C)\big) \subseteq W$. This follows directly from Proposition VIII.5 and the fact that every invariant wedge is a semialgebra (Lemma IV.30).
Step 2) Let $w, w' \in W_1$. Then there are three cases:

Case 1) $w, w' \in \mathfrak{n} * C$: Then $\exp(w)\exp(w') \in \exp(W_1)$ by Step 1). Let $\exp(w)\exp(w') = \exp(w_1)$ and suppose that $w_1 \notin \mathfrak{n} * C$. Then there exists a minimal $t_0 > 0$ such that $t_0 w_1 \notin \mathfrak{n} * C$. Then $t_0 \leq 1$ and $t_0 w \in U$ (Proposition VIII.4). Therefore, in view of (8.2),

$$\exp(t_0 w_1) \in S \setminus \exp(\mathfrak{n} * C) \subseteq T \setminus \big(H(T)\exp(\mathfrak{n} * C)\big) \subseteq I_\varepsilon.$$

Hence $\exp w_1 = \exp(t_0 w_1)\exp\big((1 - t_0)w_1\big) \in I_\varepsilon$ because $I_\varepsilon \subseteq S$ is a right semigroup ideal.

Case 2) $w \notin \mathfrak{n} * C$: Then the above argument shows that $\exp(w) \in I_\varepsilon$ and therefore $\exp(w)\exp(w') \in I_\varepsilon$ because I_ε is a right ideal.

Case 3) $w' \notin \mathfrak{n} * C$: Then the same argument as in Case 2) applies and shows that $\exp(w') \in I_\varepsilon$. Here we have to use the fact that S is invariant which implies that I_ε is a two sided ideal. Therefore, in all cases, we find that

$$\exp(w)\exp(w') \in I_\varepsilon.$$

Putting these facts together, we see with an obvious induction that

$$\langle \exp W_1 \rangle \subseteq \exp\big(W_1 \cap (\mathfrak{n} * C)\big) \cup I_\varepsilon.$$

Step 3) We know from Propositions VIII.3 and VIII.4 that $\exp|_{\mathfrak{n}*C}$ is a diffeomorphism. Therefore

$$\mathbf{L}(S) = L_1(S) = L_1\Big(\exp\big(W_1 \cap (\mathfrak{n} * \widetilde{C})\big)\Big) = L_0\big(W_1 \cap (\mathfrak{n} * \widetilde{C})\big) = W_1,$$

where L_0 and L_1 denote the subtangent cones in 0 and $\mathbf{1}$ respectively (cf. [HHL89, IV.1]). This proves that W_1 is global in G. ∎

Lemma VIII.8. *Let W be a pointed generating invariant wedge in the Lie algebra $\mathfrak{g}$, $\mathfrak{k}_1 \subseteq \mathfrak{g}$ the sum of all compact simple ideals and $\mathfrak{n}$ the nilradical of $\mathfrak{g}$. Then there exists an ideal $\mathfrak{g}_1 \subseteq \mathfrak{g}$ such that*

$$\mathfrak{g} \cong \mathfrak{g}_1 \oplus \mathfrak{k}_1$$

and a reductive subalgebra $\mathfrak{g}_2 \subseteq \mathfrak{g}_1$ such that $\mathfrak{g}_1 \cong \mathfrak{n} \rtimes \mathfrak{g}_2$. The wedge

$$W_{\mathrm{red}} := L_{\mathfrak{n}}(W) \cap \mathfrak{g}_2$$

is pointed and invariant in $\mathfrak{g}_2$. *The intersection of* W_{red} *with the compactly embedded Cartan algebra* $\mathfrak{g}_2 \cap \mathfrak{h}$ *of* $\mathfrak{g}_2$ *is*

$$W_{\text{red}} \cap \mathfrak{h} = L_{Z(\mathfrak{g})}(W \cap \mathfrak{h}) \cap \mathfrak{g}_2.$$

The wedge W_{red} *is the projection of the wedge* W *along the ideal* $\mathfrak{n} + \mathfrak{k}_1$ *onto the reductive subalgebra* $\mathfrak{g}_2$.

Proof. Since $\mathfrak{k}_1$ is a compact ideal of $\mathfrak{g}$, it is contained in every Levi algebra $\mathfrak{s}$. We fix a compactly embedded Cartan algebra $\mathfrak{h}$ of $\mathfrak{g}$ and an $\mathfrak{h}$-invariant Levi algebra $\mathfrak{s}$ (Proposition II.10). We denote the radical of $\mathfrak{g}$ with $\mathfrak{r}$. Then $[\mathfrak{k}_1, \mathfrak{r}] \subseteq \mathfrak{k}_1 \cap \mathfrak{r} = \{0\}$. If $\mathfrak{s}_1 \subseteq \mathfrak{s}$ is an ideal of $\mathfrak{s}$ such that $\mathfrak{s} = \mathfrak{s}_1 \oplus \mathfrak{k}_1$, we conclude that $\mathfrak{g}_1 := \mathfrak{r} + \mathfrak{s}_1 \cong \mathfrak{r} \rtimes \mathfrak{s}_1$ is an ideal of $\mathfrak{g}$ because $[\mathfrak{k}_1, \mathfrak{g}_1] = \{0\}$. Moreover, $\mathfrak{g} = \mathfrak{k}_1 \oplus \mathfrak{g}_1$. We choose a vector space complement $\mathfrak{h}_1 \subseteq \mathfrak{h}$ to the center $Z(\mathfrak{g}) = \mathfrak{n} \cap \mathfrak{h}$ (Proposition II.10). Then $\mathfrak{g}_2 := \mathfrak{h}_1 + \mathfrak{s}_1$ is a reductive subalgebra of $\mathfrak{g}_1$ because $[\mathfrak{h}_1, \mathfrak{s}] = \{0\}$ (Proposition II.10). It follows from $\mathfrak{g}_2 \cap \mathfrak{n} = \{0\}$ that $\mathfrak{g}_1 \cong \mathfrak{n} \rtimes \mathfrak{g}_2$. The invariance of W_{red} follows from the invariance of

$$L_{\mathfrak{n}}(W) = \overline{W - \mathfrak{n}}$$

in $\mathfrak{g}$. We find a positive system $\Omega^+ \subseteq \Omega$ of positive real roots with respect to $\mathfrak{h}$ such that

$$C_{\min}(\Omega^+) \subseteq C := W \cap \mathfrak{h} \subseteq C_{\max}(\Omega^+)$$

(Proposition III.15). Using Theorem I.10 we get that

$$L_{\mathfrak{n}}(W) \cap \mathfrak{h} = L_{\mathfrak{n} \cap \mathfrak{h}}(W \cap \mathfrak{h}) = L_{Z(\mathfrak{g})}(C).$$

Now $Z(\mathfrak{g}) \subseteq H(C_{\max})$ and therefore

$$T_{\mathfrak{n}}(W) \cap \mathfrak{h} = T_{Z(\mathfrak{g})}(C) \subseteq H(C_{\max}) = Z(\mathfrak{g}) \oplus (\mathfrak{h} \cap \mathfrak{k}_1)$$

(Proposition III.20). The group $\text{INN}_{\mathfrak{g}} \mathfrak{k}_1$ is compact and acts on $\mathfrak{g}$ such that

$$\mathfrak{g}_{\text{fix}} = \mathfrak{g}_1 \quad \text{and} \quad \mathfrak{g}_{\text{eff}} = \mathfrak{k}_1.$$

Thus Theorem I.10 implies that

$$\text{int}\,\big(L_{\mathfrak{n}}(W)\big) \cap \mathfrak{g}_1 \neq \emptyset$$

since the wedge $L_{\mathfrak{n}}(W)$ is in particular invariant under $\text{INN}_{\mathfrak{g}} \mathfrak{k}_1$. But

$$L_{\mathfrak{n}}(W) \cap \mathfrak{g}_1 = \mathfrak{n} + \big(L_{\mathfrak{n}}(W) \cap \mathfrak{g}_2\big),$$

which implies that $L_{\mathfrak{n}}(W) \cap \mathfrak{g}_2$ is generating in $\mathfrak{g}_2$. To see that this wedge is pointed, we use that

$$H(W_{\text{red}}) \cap \mathfrak{h} = T_{\mathfrak{n}}(W) \cap \mathfrak{g}_2 \cap \mathfrak{h} \subseteq \big(Z(\mathfrak{g}) + (\mathfrak{h} \cap \mathfrak{k}_1)\big) \cap \mathfrak{g}_2 = \{0\},$$

the fact that $\mathfrak{h} \cap \mathfrak{g}_2$ is a compactly embedded Cartan algebra of $\mathfrak{g}_2$, and Theorem III.23 (Reconstruction of Ideals). Consequently W_{red} is a pointed generating invariant wedge in $\mathfrak{g}_2$. ∎

Theorem VIII.9. (The First Globality Theorem) *Let G be a simply connected Lie group, $W \subseteq \mathbf{L}(G)$ a pointed generating invariant wedge. Then W is global in G if W_{red} is global in the closed normal subgroup*

$$G_{\text{red}} := \langle \exp(W_{\text{red}} - W_{\text{red}}) \rangle .$$

Proof. Let $\mathfrak{g}_1, \mathfrak{k}_1$ and $\mathfrak{g}_2$ as in Lemma VIII.8, and set $G_1 := \langle \exp \mathfrak{g}_1 \rangle$. Then

$$G \cong G_1 \times \exp \mathfrak{k}_1 \quad \text{and} \quad G_1 \cong \exp \mathfrak{n} \rtimes \langle \exp \mathfrak{g}_2 \rangle = \exp \mathfrak{n} \rtimes G_{\text{red}} .$$

Therefore G_{red} is closed in G and simply connected. We assume that W_{red} is global in G_{red} and set $W_1 := L_{\mathfrak{n}}(W) \cap \mathfrak{g}_1$. Now $\mathfrak{g}_1 = \mathfrak{n} + \mathfrak{g}_2$ and $W_1 = \mathfrak{n} + W_{\text{red}}$. This is the tangent wedge of the the closed subsemigroup

$$\exp \mathfrak{n} \cdot \overline{\langle \exp W_{\text{red}} \rangle} \quad \text{in} \quad G_1$$

and therefore global. Let $W' := W \cap \mathfrak{g}_1$. Then

$$L_{\mathfrak{n}}(W') \subseteq L_{\mathfrak{n}}(W) \cap \mathfrak{g}_1 = W_1 ,$$

thus $T_{\mathfrak{n}}(W') = \mathfrak{n} = H(W_1)$ since W_{red} is pointed (Lemma VIII.8). The Reduction Theorem (Theorem VIII.7) now implies that W' is global in G_1. We conclude that the wedge $W' + \mathfrak{k}_1$ is global in G because it is tangent to the semigroup $\overline{\langle \exp W' \rangle \exp \mathfrak{k}_1}$. Now we consider $\mathfrak{g}$ as a module of the compact group $e^{\operatorname{ad} \mathfrak{k}_1}$. Then

$$\mathfrak{g}_{\text{eff}} = \mathfrak{k}_1 \quad \text{and} \quad \mathfrak{g}_{\text{fix}} = \mathfrak{g}_1 .$$

Hence $W' = W \cap \mathfrak{g}_1 = (W + \mathfrak{k}_1) \cap \mathfrak{g}_1$ (Theorem I.10). This proves that $W' + \mathfrak{k}_1 = W + \mathfrak{k}_1$. The wedge $W \cap \mathfrak{k}_1$ is an ideal of $\mathfrak{g}$ (Theorem I.10) since it is an invariant subspace, hence $\{0\}$, because W is pointed. Thus

$$W \cap H(W + \mathfrak{k}_1) = W \cap \mathfrak{k}_1 = \{0\} \subseteq H(W)$$

and Corollary IV.31 proves that W is global in G. ∎

In view of the First Globality Theorem we now have to find sufficient conditions for a pointed generating invariant wedge in a reductive Lie algebra to be global. Such conditions can be found by using the same functionals $\widehat{l_i}$ as in the Controllability Criterion whose definition we recall from Definition VII.19.

Theorem VIII.10. (The Second Globality Theorem) *Let G be a simply connected Lie group, $W \subseteq \mathfrak{g} := \mathbf{L}(G)$ a pointed generating invariant wedge and $\Omega^+ \subseteq \Omega$ a nice positive system such that*

$$C := W \cap \mathfrak{h} \subseteq C_{\max}(\Omega^+).$$

Then W is global in G if

$$C^\star \cap \sum_{i=1}^{m}]0, \infty[\, \widehat{l_i} \neq \emptyset .$$

Proof. We want to apply Lemma VIII.8 to reduce the problem to the semisimple case. With the notation of this Lemma we have that

$$\mathfrak{n} = Z(\mathfrak{g}), \quad \mathfrak{g}_2 = \bigoplus_{i=1}^{m} \mathfrak{s}_i, \quad \text{and} \quad \mathfrak{k}_1 = \bigoplus_{i=m+1}^{n} \mathfrak{s}_i,$$

where $\mathfrak{s}_1, \ldots, \mathfrak{s}_m$ are the hermitean ideals of $\mathfrak{g}$ and $\mathfrak{s}_{m+1}, \ldots, \mathfrak{s}_n$ are the compact ideals. Using Lemma VIII.8 and Theorem VIII.9 we see that we only have to show that

$$W_{\rm red} := L_{Z(\mathfrak{g})}(W) \cap \mathfrak{g}_2$$

is global. Let $\mathfrak{h}_i := \mathfrak{h} \cap \mathfrak{g}_i$. Then, according to Lemma VIII.8,

$$C_{\rm red} := W_{\rm red} \cap \mathfrak{h} = L_{Z(\mathfrak{g})}(C) \cap \bigoplus_{i=1}^{m} \mathfrak{h}_i.$$

We choose $\delta_i \in]0, \infty[$ such that

$$\nu := \sum_{i=1}^{m} \delta_i \widehat{l_i} \in C^\star.$$

Then

$$\nu \in C^\star \cap Z(\mathfrak{g})^\perp = \big(L_{Z(\mathfrak{g})}(C)\big)^\star \subseteq C^\star_{\rm red}.$$

Lemma VIII.8 asserts that the wedge $W_{\rm red}$ is pointed and generating. Thus Proposition III.24 in [Ne89] shows that that there exists $\omega \in W^*_{\rm red}$ such that $\omega(\mathfrak{s}_i) \neq \{0\}$ for $i = 1, \ldots, m$ and $\omega(\mathfrak{k}' + \mathfrak{t}) = \{0\}$, where $\mathfrak{g}' = \mathfrak{k} + \mathfrak{t}$ is an Iwasawa decomposition of $\mathfrak{g}$ as in Section VII. Using Proposition III.4 in [Ne89] this leads to

$$\mathfrak{s}_i \not\subseteq H\big(\mathbf{L}(\overline{\langle \exp W_{\rm red} \rangle})\big) \quad \text{for} \quad i = 1, \ldots, n.$$

Since this is an ideal of $\mathfrak{g}_2$, we conclude that $\mathbf{L}(\overline{\langle \exp W_{\rm red}})$ is pointed, and with Proposition IV.28 that $W_{\rm red}$ is global. Now an application of Theorem VIII.9 concludes the proof. ∎

The Second Globality Theorem motivates the following definition:

Definition VIII.11. Let $\mathfrak{g}$ be a quasihermitean Lie algebra. Then $\mathfrak{g}$ is said to be of *globality type* if $l_i = 0$ for all hermitean ideals $\mathfrak{s}_i$ of $\mathfrak{g}/\operatorname{Rad}(\mathfrak{g})$. Note that all surjective images, i.e., all quotients of Lie algebras of globality type are of globality type. To see this, let $\alpha : \mathfrak{g} \to \mathfrak{g}_1$ be a surjective morphism of Lie algebras and $\mathfrak{g}$ be of globality type. Then $\alpha\big(\operatorname{Rad}(\mathfrak{g})\big) = \operatorname{Rad}(\mathfrak{g}_1)$ and therefore α induces a surjective homomorphism

$$\widetilde{\alpha} : \mathfrak{g}/\operatorname{Rad}(\mathfrak{g}) \to \mathfrak{g}_1/\operatorname{Rad}(\mathfrak{g}_1).$$

Consequently $\mathfrak{g}_1/\operatorname{Rad}(\mathfrak{g}_1)$ is isomorphic to an ideal of $\mathfrak{g}/\operatorname{Rad}(\mathfrak{g})$ and the assertion follows. ∎

Theorem VIII.12. (The Main Theorem on Globality) *Let G be a simply connected Lie group such that $\mathbf{L}(G)$ is of globality type, then every generating invariant wedge $W \subseteq \mathbf{L}(G)$ is global in G.*

Proof. First we suppose that W is pointed. Since $l_i = 0$, we find that the corresponding functionals $\widehat{l_i}$ vanish, and therefore

$$C^\star \cap \sum_{i=1}^{m}]0, \infty[\widehat{l_i} = C^\star \cap \{0\} \neq \emptyset.$$

Now the assertion follows from the Second Globality Theorem. If W is not pointed we set $H := \langle \exp H(W) \rangle$ and write $\varphi : G \to G/H$ for the associated quotient homomorphism. Note that H is closed and that G/H is simply connected ([Ho65, p.135]). Therefore $d\varphi(\mathbf{1})W$ is a pointed generating invariant wedge in the Lie algebra $\mathbf{L}(G)/H(W)$ of the simply connected Lie group G/H. Moreover $\mathbf{L}(G)/H(W)$ is of globality type (Definition VIII.11). Hence the first part of the proof implies that $d\varphi(\mathbf{1})W$ is global in G/H. Consequently W is global in G ([Ne89, III.7]). ■

Remark VIII.13. The property of a Lie algebra to be of globality type is easy to check with the following list of all hermitean simple Lie algebras. The algebras of tubular type ($l = 0$) are:

$$\mathrm{su}(n,n),\ \mathrm{so}(n+2,n),\ \mathrm{sp}(n,\mathrm{I\!R}),\ \mathrm{so}^*(4n), \qquad \forall n \in \mathrm{I\!N}$$

and $e_{7(-25)}$ (cf. [Pa81]). Note that $\mathrm{su}(1,1) \cong \mathrm{so}(2,1) \cong \mathrm{sl}(2,\mathrm{I\!R})$ and $\mathrm{so}(2,2) \cong \mathrm{sl}(2,\mathrm{I\!R})^2$.

These are the Dynkin diagrams of real hermitean simple Lie algebras. The circle $\circ$ marks the unique non-compact base root. Note that this has to be a root with coefficient 1 in the representation of the highest root.

$\mathrm{su}(p, n+1-p), 1 \leq p \leq n,$

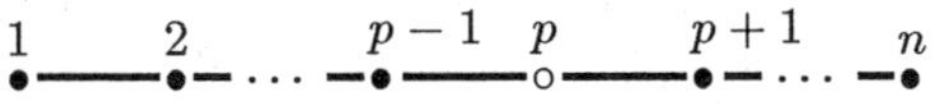

$\mathrm{so}(2n-1, 2), n \geq 2,$

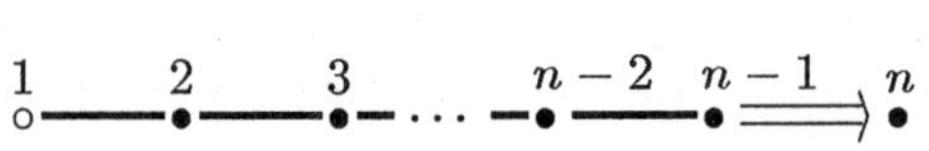

$\mathrm{sp}(n, \mathrm{I\!R}), n \geq 1,$

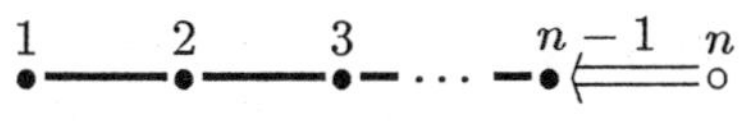

$\mathrm{so}(2n-2, 2), n \geq 2,$

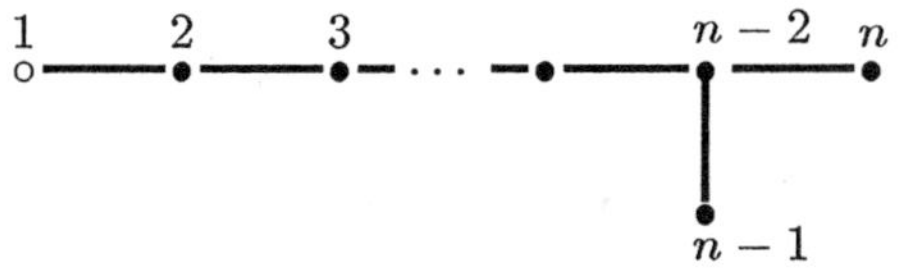

$\mathrm{so}^*(2n), n \geq 3,$

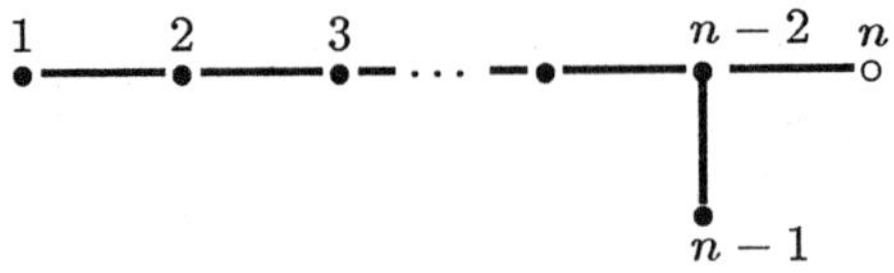

$e_{6(-14)}$

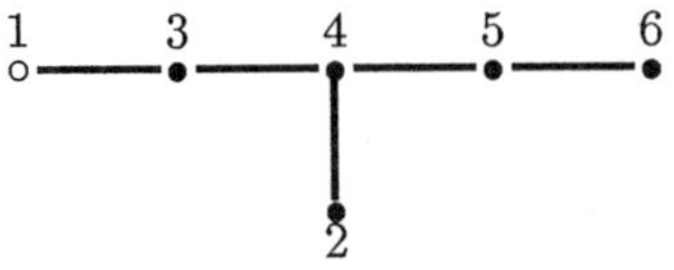

$e_{7(-25)}$

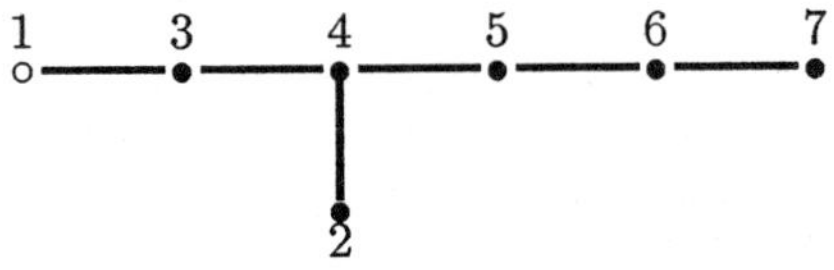

■

The following result was already proved by the author in [Ne88] and in the paper [Gi89] by Gichev.

Corollary VIII.14. *Let G be a solvable simply connected Lie group. Then all invariant wedges $W \subseteq \mathbf{L}(G)$ are global in G.*

Proof. First we note that the normal subgroup $\langle \exp(W - W) \rangle$ is closed in G and simply connected ([Ho65, p.137]). Hence we may assume that W is generating. Now the assertion follows from the Main Globality Theorem (Theorem VIII.12). ■

Example VIII.15. Let $\mathfrak{g} = \mathrm{su}(2,1)$. This is a hermitean Lie algebra that is not of tubular type, i.e., $l \neq 0$. We will see later in Proposition VIII.21 that there are pointed generating invariant cones in $\mathrm{su}(2,1)$ which are not global in the simply connected Lie group $G = \mathrm{SU}(2,1)^\sim$. This example shows that one cannot expect in general that all invariant cones in a simply connected Lie group G are global. We note that this disproves a conjecture of I. E. Segal ([Se76, p.31]). ■

Example VIII.16. Let $\mathfrak{g} = \mathrm{gl}(2,\mathbb{R})$. This is a quasihermitean reductive Lie algebra with strong cone potential. So it contains pointed generating invariant wedges. These wedges are all global in the group

$$\mathrm{Gl}(2,\mathbb{R})^\sim \cong \mathbb{R} \times \mathrm{Sl}(2,\mathbb{R})^\sim$$

since $\mathrm{sl}(2,\mathbb{R}) \cong \mathrm{sp}(1,\mathbb{R})$ is of tubular type. ■

Example VIII.17. If $\mathfrak{g} = \mathfrak{u}(2)$, we also know that this Lie algebra contains pointed generaing invariant wedges W. They are global in the group

$$\mathrm{U}(2)^{\sim} \cong \mathbb{R} \times \mathrm{SU}(2).$$

Since $W_{\mathrm{red}} = \{0\}$ holds for all these wedges, the globality follows from the First Globality Theorem. The same argument works for all compact Lie algebras with non-trivial center. ∎

The Globality theorems we have proved so far consider only the structure a reductive subalgebra and they do not take into account the action of a suitable Levi algebra $\mathfrak{s}$ on the radical $\mathfrak{r}$ and in particular on the space $\mathfrak{r}_{\mathrm{eff}}$ of all solvable root spaces. We have seen in Section III that this action is very nice if the Lie algebra $\mathfrak{g}$ has strong cone potential and in particular if it contains a pointed generating invariant cone (Proposition III.15). This led us to the Structure Theorem for Lie algebras with strong cone potential (Theorem II.38). Now we want to use the information provided by this theorem. It is clear that we first have to consider the proto type of a Lie algebra with strong cone potential. We use the notation of Sections II and III.

Example VIII.18. Let $\mathfrak{g}_J = \mathrm{Lie}\big(\mathrm{sp}(n,\mathbb{R}), \mathbb{R}^{2n}, q, \mathbb{R}\big) = \mathfrak{h}_n \rtimes \mathrm{sp}(n,\mathbb{R})$, where $\mathfrak{h}_n$ denotes the $(2n+1)$-dimensional Heisenberg group, as defined in Example II.37. If $\mathfrak{h}_1$ is the standard compactly embedded Cartan algebra of $\mathrm{sp}(n,\mathbb{R})$ (Example II.37), then $\mathfrak{h} := \{0\} \oplus \mathbb{R} \oplus \mathfrak{h}_1$ is a compactly embedded Cartan algebra of $\mathfrak{g}_J$ (Proposition II.21). We choose a nice positive system Ω^+ such that $\omega(J) > 0$ for all $\omega \in \Omega_S^+$. Then, using Example II.37, we find that

$$C_{\min}(\Omega^+) = \mathbb{R}^+ \oplus C_{\min,\mathfrak{h}_1} \quad \text{and} \quad C_{\max}(\Omega^+) = \mathbb{R} \oplus (C_{\max} \cap \mathfrak{h}_1).$$

Since $C_{\min} \subseteq C_{\max}$ (Proposition III.20), $C_{\min,\mathfrak{h}_1} = C_{\max,\mathfrak{h}_1}$ ([Pa81, p.325]), and $C_{\max} \cap \mathfrak{h}_1 \subseteq C_{\max,\mathfrak{h}_1}$ we conclude that

$$C_{\min,\mathfrak{h}_1} \subseteq C_{\max} \cap \mathfrak{h}_1 \subseteq C_{\max,\mathfrak{h}_1} = C_{\min,\mathfrak{h}_1}.$$

Thus $C_{\max} = \mathbb{R} \oplus C_{\min,\mathfrak{h}_1}$. Let $W_J \subseteq \mathrm{sp}(n,\mathbb{R})$ denote the unique pointed generating invariant cone such that

$$W_J \cap \mathfrak{h}_1 = C_{\min,\mathfrak{h}_1} = C_{\max,\mathfrak{h}_1}.$$

We set $W_{\max,\mathrm{J}} := \mathbb{R}^{2n} \oplus \mathbb{R} \oplus W_J$ and write $W_{\min,\mathrm{J}}$ for the unique pointed generating cone in $\mathfrak{g}_J$ with $W_{\min,\mathrm{J}} \cap \mathfrak{g} = C_{\min}(\Omega^+)$. Then $W_{\max} \cap \mathfrak{h} = C_{\max}(\Omega^+)$. Let $G_J \cong H_n \rtimes \mathrm{Sp}(n,\mathbb{R})^{\sim}$ denote the simply connected group with $\mathbf{L}(G_J) = \mathfrak{g}_J$. The Main Theorem on Globality implies that $W_{\min,\mathrm{J}}$ is global in G_J and the globality of $W_{\max,\mathrm{J}}$ follows from the globality of W_J, the fact that $G_J \cong H_n \rtimes \mathrm{Sp}(n,\mathbb{R})$, and

$$W_{\max,\mathrm{J}} = \mathfrak{h}_n \oplus \mathbf{L}(\overline{\langle \exp W_J \rangle}).$$

Suppose that $\mathfrak{g}$ is a Lie algebra with strong cone potential such that $Z_{\mathfrak{s}}(\mathfrak{r}) = \{0\}$ and $\pi : \mathfrak{g} \to \mathfrak{g}_J$ the homomorphism from the Structure Theorem for Lie algebras with strong cone potential (Theorem II.38) with respect to a nice positive system Ω^+. Then

$$\pi(C_{\min}) \subseteq C_{\min,J} \quad \text{and} \quad \pi(C_{\max}) \subseteq C_{\max,J} \tag{8.3}$$

because $\pi\big(C_{\min} \cap Z(\mathfrak{g})\big) \subseteq \{0\} \oplus \mathbb{R}^+ \oplus \{0\}$,

$$C_{\min} \cap \mathfrak{h}' \subseteq C_{\max} \cap \mathfrak{h}' \subseteq \{h \in \mathfrak{h}' : \omega(h) \geq 0 \text{ for all } \omega \in \Omega_R^+\} = \pi^{-1}(W_J) \cap \mathfrak{h}',$$

and $C_{\max} = Z(\mathfrak{g}) \oplus \mathfrak{h}'$, where $\mathfrak{h}' \subseteq \mathfrak{h}$ is a vector space complement to $Z(\mathfrak{g})$ in $\mathfrak{h}$. ■

Theorem VIII.19. *Let G be a simply connected Lie group, $W \subseteq \mathfrak{g} := \mathbf{L}(G)$ a pointed generating invariant wedge, $\mathfrak{h}$ a compactly embedded Cartan algebra of $\mathfrak{g}$, $\mathfrak{s}$ an $\mathfrak{h}$-invariant Levi algebra of $\mathfrak{g}$ and $Z_{\mathfrak{s}}(\mathfrak{r}) = \{0\}$. Then W is global in G.*

Let $W_{\max,\mathfrak{g}}$ be an invariant wedge in $\mathfrak{g}$ such that

$$W_{\max,\mathfrak{g}} \cap \mathfrak{h} = C_{\max}(\Omega^+) \supseteq W \cap \mathfrak{g}.$$

Then $W_{\max,\mathfrak{g}}$ is global in G.

Proof. We choose a positive system such that

$$C_{\min}(\Omega^+) \subseteq C := W \cap \mathfrak{h} \subseteq C_{\max}(\Omega^+)$$

(Proposition III.15). Let $\omega \in \operatorname{int} W^*$ and set $\alpha := \omega|_{Z(\mathfrak{g})}$. Then $\alpha\big(Q(x)\big) > 0$ for all $x \in \mathfrak{r}_{\text{eff}} \setminus \{0\}$ since $Q(x) \in C_{\min} \setminus \{0\} \subseteq Z(\mathfrak{g})$ and $C_{\min} \subseteq C$ is pointed. Using the Structure Theorem for Lie algebras with strong cone potential (Theorem II.38) we find a homomorphism $\pi : \mathfrak{g} \to \mathfrak{g}_J$, with $\mathfrak{g}_J$ as in Example VIII.18, such that $\ker \pi = \ker \alpha \subseteq Z(\mathfrak{g})$. This homomorphism extends to a homomorphism $\widetilde{\pi}_1 : G \to G_J$ of the corresponding simply connected groups. Then the facts that

$$\pi(C) \subseteq \pi(C_{\max}) \subseteq C_{\max,J} \subseteq W_{\max,J}$$

and

$$W \subseteq \overline{\operatorname{conv}(\operatorname{Inn}_{\mathfrak{g}} \mathfrak{g})(C)}$$

show that $\pi(W) \subseteq W_{\max}$. Now Lemma III.7 in [Ne89] and Example VIII.18 imply that the invariant wedge $W_1 := \pi^{-1}(W_{\max})$ is global in G. The edge of W_1 is exactly $\mathfrak{n}$ since $H(W_{\max}) = \mathfrak{h}_n \oplus \{0\}$. As a consequence of the Reduction Theorem VIII.7 the wedge W, which is contained in W_1, is global in G.

To prove that $W_{\max,\mathfrak{g}}$ is global we note that the Uniquenes of Reconstruction (Proposition III.34) shows that

$$W_{\max,\mathfrak{g}} = \mathfrak{n} + W_{\max,\mathfrak{g}} \cap (\mathfrak{h}_1 + \mathfrak{s}),$$

where $\mathfrak{h}_1$ is a vector space complement to $Z(\mathfrak{g})$ in $\mathfrak{h}$ and $W_{\max,\mathfrak{g}} \cap (\mathfrak{h}_1 + \mathfrak{s})$ is the unique pointed generting invariant wedge in the reductive Lie algebra $\mathfrak{h}_1 + \mathfrak{s}$ wich intersects $\mathfrak{h}_1$ exactly in the pointed generating wedge $C_{\max} \cap \mathfrak{h}_1$. It follows from Theorem III.20 that this wedge is invariant under $\mathcal{W}$ and C. Again we find that $\pi(W_{\max,\mathfrak{g}}) \subseteq W_{\max}$, i.e., $W_{\max,\mathfrak{g}} \subseteq W_1$. But the edges of W_1 and $W_{\max,\mathfrak{g}}$ agree. So Corollary IV.31 shows that $W_{\max,\mathfrak{g}}$ is global. ■

Theorem VIII.20. (The Third Globality Theorem) *Let G be a simply connected Lie group, $W \subseteq \mathfrak{g} := \mathbf{L}(G)$ a pointed generating invariant wedge, $\mathfrak{h}$ a compactly embedded Cartan algebra of $\mathfrak{g}$, $\mathfrak{s}$ an $\mathfrak{h}$-invariant Levi algebra of $\mathfrak{g}$, and $\pi : \mathfrak{g} \to \mathfrak{s}_0$ the projection onto the sum of all hermitean simple ideals of $\mathfrak{s}$ commuting with $\mathfrak{r}$. Then the wedge $\overline{\pi(W)} \subseteq \mathfrak{s}_0$ is pointed generating and invariant, and its globality implies the globality of W.*

Proof. With $[\mathfrak{s}_0, \mathfrak{s}] \subseteq \mathfrak{s}_0$ and $[\mathfrak{s}_0, \mathfrak{r}] = \{0\}$ we see that $\mathfrak{s}_0$ is an ideal of $\mathfrak{g}$. Let $\mathfrak{s} = \mathfrak{s}_1 \oplus \mathfrak{k}_1 \oplus \mathfrak{s}_0$, where $\mathfrak{s}_1$ is the sum of all ideals not commuting with $\mathfrak{r}$, and $\mathfrak{k}_1$ is the sum of all compact ideals commuting with $\mathfrak{r}$. Then $\mathfrak{g}_1 := \mathfrak{r} + \mathfrak{s}_1 \cong \mathfrak{r} \rtimes \mathfrak{s}_1$ is an ideal of $\mathfrak{g}$ with $\mathfrak{g} \cong \mathfrak{g}_1 \oplus \mathfrak{k}_1 \oplus \mathfrak{s}_0$. Hence π is a homomorphism of Lie algebras, the projection onto the third ideal summand, and $\overline{\pi(W)}$ is a generating invariant wedge in $\mathfrak{s}_0$. Using Theorem I.10 we see that the intersection $W \cap (\mathfrak{g}_1 + \mathfrak{s}_0)$ equals the projection of W along $\mathfrak{k}_1$. The group G is simply connected, hence $G \cong G_1 \times \exp \mathfrak{k}_1 \times G_0$ with $G_1 := \langle \exp \mathfrak{g}_1 \rangle$ and $G_0 := \langle \exp \mathfrak{s}_0 \rangle$. It follows from [Ne89, III.5] that W is global in G iff $W + \mathfrak{k}_1$ is global in G, but this is equivalent to the globality of $W \cap (\mathfrak{g}_1 + \mathfrak{s}_0)$ since $W + \mathfrak{k}_1 = \mathfrak{k}_1 + W \cap (\mathfrak{g}_1 + \mathfrak{s}_0)$, and $\exp \mathfrak{k}_1$ is a direct factor of G. Therefore we may assume that $\mathfrak{k}_1 = \{0\}$. We choose a positive system of real roots with respect to a compactly embedded Cartan algebra $\mathfrak{h}$ such that

$$C_{\min}(\Omega^+) \subseteq C := W \cap \mathfrak{h} \subseteq C_{\max}(\Omega^+).$$

Then $C_{\max} = C_{\max} \cap \mathfrak{g}_1 + \overline{C_{\max} \cap \mathfrak{s}_0}$ (Proposition III.20) and $C_{\max} \cap \mathfrak{s}_0$ is pointed. Now $C \subseteq C_{\max}$ and $W \subseteq \overline{\operatorname{conv}(\operatorname{Inn}_{\mathfrak{g}} \mathfrak{g})(C)}$ shows that

$$\overline{\pi(W)} \subseteq W_{\max,\mathfrak{s}_0},$$

where $W_{\max,\mathfrak{s}_0} \subseteq \mathfrak{s}_0$ is the unique pointed generating invariant wedge with $W_{\max,\mathfrak{s}_0} \cap \mathfrak{h} = C_{\max} \cap \mathfrak{s}_0$. This proves that $\overline{\pi(W)}$ is pointed. We set $V := \mathbf{L}(\overline{\langle \exp W \rangle})$ and have to show that $V = W$. The globality of the wedge $\mathfrak{g}_1 + \overline{\pi(W)}$ shows that $H(V) \subseteq \mathfrak{g}_1$.

The Lie algebra $\mathfrak{g}_1$ has strong cone potential and is quasihermitean. Hence Theorem VIII.19 shows that the wedge $W_{\max,\mathfrak{g}_1}$ and therefore $W_{\max,\mathfrak{g}_1} + \mathfrak{s}_0$ is global in G. But this invariant wedge satisfies

$$(W_{\max,\mathfrak{g}_1} + \mathfrak{s}_0) \cap \mathfrak{h} \supseteq C_{\max} \supseteq C,$$

and thus $W \subseteq W_{\max,\mathfrak{g}_1} + \mathfrak{s}_0$. We conclude that

$$H(V) \subseteq H(W_{\max,\mathfrak{g}_1} + \mathfrak{s}_0) = \mathfrak{n} + \mathfrak{s}_0.$$

In view of the preceding argument, we get that $H(V) \subseteq \mathfrak{n}$ is a nilpotent ideal of $\mathfrak{g}$. So the Reduction Theorem applies and shows that W is global in G. ∎

To prove the next Globality Theorem we first have to consider the case of simple Lie algebras. The following result was already proved by Ol'shanskiĭ in [Ol82b] and by the author in [Ne90] as a corollary from the solution of the controllability problem for invariant wedges in simple Lie groups.

Proposition VIII.21. *Let G be a simply connected simple Lie group and $W \subseteq \mathbf{L}(G)$ a pointed generating invariant wedge. Then W is global iff*

$$\widehat{l} \in C^\star.$$

There exists a largest invariant wedge $W_0 \subseteq \mathbf{L}(G)$ such that $l \in C_0^\star$ and $C_{\min} \subseteq C_0$. In particular, the minimal invariant wedge $W_{\min}$ with $W_{\min} \cap \mathfrak{h} = C_{\min}$ is global in G. If $l \neq 0$, then $C_0 \neq C_{\max}$ and $W_{\max}$ is not global in G.

Proof. It follows from the Second Globality Theorem that W is global if $l \in C^\star \subseteq W^*$. If W is global in G, then it is not controllable and therefore Theorem VII.21 shows that $\widehat{l} \in C^\star$. The existence of W_0 is a consequence of Proposition 2.21 in [Ne90], it requires some computations in simple hermitean Lie algebras. The last assertion follows from [Ne90, 3.8]. ∎

Theorem VIII.22. (The Fourth Globality Theorem) *Let G be simply connected Lie group and $W \subseteq \mathfrak{g} := \mathbf{L}(G)$ a pointed generating invariant wedge. Then there exists a pointed generating invariant wedge $V \subseteq W$ which is global in G.*

Proof. Let $\mathfrak{g} = \mathfrak{g}_1 \oplus \mathfrak{s}_0$ be the decomposition of Theorem VIII.20, where $\mathfrak{s}_0$ is the sum of all hermitean simple ideals commuting with $\mathfrak{r}$. We choose a positive system such that

$$C_{\min}(\Omega^+) \subseteq C \subseteq C_{\max}(\Omega^+).$$

We write $W_{\min,\mathfrak{s}_0}$ for the unique pointed generating invariant wedge in $\mathfrak{s}_0$ with $W_{\min,\mathfrak{s}_0} \cap \mathfrak{h} = C_{\min} \cap \mathfrak{s}_0$. According to Proposition VIII.21, this wedge is global in the closed subgroup $G_0 := \langle \exp \mathfrak{s}_0 \rangle$ of G. Note that $G \cong G_1 \times G_0$ with $G_1 := \langle \exp \mathfrak{g}_1 \rangle$. Since the Lie algebra $\mathfrak{g}_1$ has strong cone potential and is quasihermitean there are two cases (Theorem III.36):
Case 1) $\mathfrak{g}_1$ is compact and semisimple: Then $W_{\min,\mathfrak{s}_0} + \mathfrak{g}_1$ is global in G and its intersection V with W is global since $W \cap \mathfrak{g}_1 \subseteq H(W) = \{0\}$ (Theorem I.10) and $W_{\min,\mathfrak{s}_0} + \mathfrak{g}_1$ is global (Corollary IV.31). Now, according to Theorem I.10,

$$\operatorname{algint} C_{\min} \subseteq \operatorname{algint} C = \operatorname{int} W \cap \mathfrak{h}$$

and consequently $\operatorname{int} W \cap V \neq \emptyset$. This proves that $\operatorname{int} V \neq \emptyset$.
Case 2) $\mathfrak{g}_1$ contains a pointed generating invariant wedge W_1 such that

$$C_{\min} \cap \mathfrak{g}_1 \subseteq W_1 \cap \mathfrak{h} \subseteq C_{\max} \cap \mathfrak{g}_1.$$

We set $V := W \cap (W_1 + W_{\min,\mathfrak{s}_0})$. The wedge $W_1 + W_{\min,\mathfrak{s}_0}$ is pointed generating and global, hence V is global, too (Corollary IV.31). Again we have that

$$\operatorname{algint} C_{\min} \subseteq \operatorname{algint} C = \operatorname{int} W \cap \mathfrak{h},$$

and $\operatorname{algint} C_{\min} \subseteq W_1 + W_{\min,\mathfrak{s}_0}$. Thus $\operatorname{int} V \neq \emptyset$. ∎

Corollary VIII.23. *Let G be a simply connected Lie group. Then the following are equivalent:*

(1) $\mathbf{L}(G)$ *contains a pointed generating invariant wedge which is global in* G.

(2) $\mathbf{L}(G)$ *has strong cone potential, is quasihermitean and not compact semisimple.*

Proof. In view of Theorem III.36 we only have to apply Theorem VIII.22 to see that (2) implies (1). ∎

Example VIII.24. To see that there are cases where the Main Theorem on Globality does not help and only the Third Globality Theorem is applicable, we construct such an example. We set

$$\mathfrak{s} := \mathrm{su}(2,1) = \{X \in \mathrm{sl}(3,\mathbb{C}) : X^*D + DX = 0\}, \qquad \text{where} \qquad D = \begin{pmatrix} 1 & 0 & 0 \\ 0 & 1 & 0 \\ 0 & 0 & -1 \end{pmatrix}.$$

The subalgebra $\mathfrak{h}_\mathfrak{s}$ of diagobal matrices in $\mathfrak{s}$ is a compactly embedded Cartan algebra because all these matrices have purely imaginary entries. We let $\mathfrak{s}$ act on $M := \mathbb{C}^3$ as it is defined by the natural matrix representation. Since the hermitean form $(x,y) = \langle x, Dy\rangle$, where $\langle\cdot,\cdot\rangle$ denotes the canonical hermitean product, is invariant under the action of $\mathfrak{s}$ on M, the bilinear form

$$q : M \times M \to \mathbb{R}, \quad (z,z') \mapsto \mathrm{Re}\langle x, iDy\rangle = \mathrm{Im}\langle x, Dy\rangle$$

is invariant under the action of $\mathfrak{s}$, and skew symmetric since $D^* = D$. Therefore the Lie algebra

$$\mathfrak{g} := \mathrm{Lie}(\mathfrak{s}, M, q, \mathbb{R})$$

defined in Definition II.22 has a compactly embedded Cartan algebra (Proposition II.21). We claim that it has strong cone potential. To see this, we define the H_S-invariant complex structure I on $\mathbb{C}^3$ by setting $I := iD$. Then we have that

$$q(Iz, z) = \mathrm{Re}\langle Iz, Iz\rangle > 0 \quad \text{for} \quad z \in M \setminus \{0\}.$$

The positive solvable roots with respect to I and the base

$$h_1 = \begin{pmatrix} i & 0 & 0 \\ 0 & 0 & 0 \\ 0 & 0 & -i \end{pmatrix} \quad \text{and} \quad h_2 = \begin{pmatrix} 0 & 0 & 0 \\ 0 & i & 0 \\ 0 & 0 & -i \end{pmatrix}$$

of $\mathfrak{h}_\mathfrak{s}$ are given by

$$\omega_1(\alpha h_1 + \beta h_2) = \alpha, \quad \omega_2(\alpha h_1 + \beta h_2) = \beta, \quad \text{and} \quad \omega_3(\alpha h_1 + \beta h_2) = \alpha + \beta.$$

This is a positive system because the set $\{\alpha, \beta, \alpha+\beta\}$ is contained is the half space of all linear functionals on $\mathfrak{h}_\mathfrak{s}$ which are positive on $h_1 + h_2$. The Lie algebra $\mathfrak{g}$ is quasihermitean (Proposition III.16) and not compact semisimple. Hence the

Characterization Theorem for Lie algeras with Invariant Cones (Theorem III.36) shows that $\mathfrak{g}$ contains a pointed generating invariant cone W. The globality of W in the associated simply connected group follows from the Third Globality Theorem and the Main Theorem on Globality does not apply because $l \neq 0$ for su$(2,1)$ (Remark VIII.13). To see that the Second Globality Theorem does also apply in this case, we have to show that the projection of W along the nilradical $\mathfrak{n} = M \oplus \mathbb{R}$ onto $\mathfrak{s}$ yields a global cone. This is true because $\mathfrak{h} = Z(\mathfrak{g}) \oplus \mathfrak{h}_\mathfrak{s}$ is a compactly embedded Cartan algebra of $\mathfrak{g}$,

$$H(C_{\max}) = Z(\mathfrak{g}), \quad C_{\max} \cap \mathfrak{s} = C_{\min,\mathfrak{s}}, \quad \text{and} \quad \hat{l} \in C^\star_{\min,\mathfrak{s}}$$

(Proposition VIII.21). To see that $C_{\min,\mathfrak{s}} = C_{\max} \cap \mathfrak{s}$ we use the fact that

$$C_{\min,\mathfrak{s}} = \mathbb{R}^+ h_1 + \mathbb{R}^+ h_2 \quad \text{and} \quad \Omega_S^+ \cap \Omega_P^+ = \{\omega_4, \omega_5\}$$

with

$$\omega_4(\alpha h_1 + \beta h_2) = 2\alpha + \beta, \quad \text{and} \quad \omega_5(\alpha h_1 + \beta h_2) = \alpha + 2\beta.$$

Hence

$$C_{\max} \cap \mathfrak{s} = \bigcap_{i=1}^{5} \omega^{-1}(\mathbb{R}^+) = \omega_1^{-1}(\mathbb{R}^+) \cap \omega_2^{-1}(\mathbb{R}^+) = \mathbb{R}^+ h_1 + \mathbb{R}^+ h_2 = C_{\min,\mathfrak{s}}.$$

One should note at this point that no computation with roots and weights where necessary to apply the Third Globality Theorem. This shows that this theorem is a rather effective tool to check the globality of an invariant wedge in a simply connected Lie group. ■

Problems VIII.25. Finally we show up a class of examples that cannot be treated with the Globality Theorems of this section. Let $\mathfrak{g} = \mathrm{sl}(2,\mathbb{R}) \oplus \mathrm{su}(2,1)$. We fix an invariant cone W_1 in sl$(2,\mathbb{R})$, and we write $W_{\min,2}$ and $W_{\max,2}$ for a minimal and maximal invariant cone in su$(2,1)$. It follows from Proposition VIII.21 that $W_{\min,2} = W_{0,2}$ and that $W_{\max,2}$ is not global in $\mathrm{SU}(2,1)^\sim$. So it is clear that

$$W_{\min} = W_1 + W_{\min,2}$$

is global in $G := \mathrm{Sl}(2,\mathbb{R})^\sim \times \mathrm{SU}(2,1)^\sim$, and that

$$W_{\max} = W_1 + W_{\max,2}$$

is not global in G. Let $l \in \mathrm{su}(2,1)$ be as in Definition VII.19. Then the Second Globality Theorem asserts that an invariant cone $W \subseteq \mathfrak{g}$ containing $W_{\min}$ is global in G whenever $l \in W^*$. This means that the projection of W onto su$(2,1)$ is contained in $W_{\min}$. This holds iff $W = W_{\min}$.

On the other hand it is clear that every invariant cone W containing $W_{\min}$ is not controllable because it is contained in $W_{\max}$ and

$$\langle \exp W_{\max} \rangle = \langle \exp W_1 \rangle \times \mathrm{SU}(2,1)^\sim.$$

So the next idea is to use Theorem VII.17. If $\mathfrak{g} = \mathfrak{k} + \mathfrak{t}$ is an Iwasawa decomposition, this theorem asserts that W is global if the cone

$$V := W^* \cap \mathfrak{t}^\perp \cap [\mathfrak{k}, \mathfrak{k}]^\perp$$

is two dimensional. Note that this cone always contains the ray $W_1^* \cap \big(\mathfrak{t} \cap \mathrm{sl}(2, \mathbb{R})\big)^\perp \subseteq \mathrm{sl}(2, \mathbb{R})\widehat{}$. In fact, this theorem proves the globality of some invariant wedges W different from $W_{\min}$ (cf. Example XII.4). But as an easy geometric construction shows, there still exist invariant cones W in $\mathfrak{g}$ such that V is one-dimensional and $V \cap \mathrm{su}(2, 1) = W_{\min,2}$. For these cones we do not know whether they are global or not. ■

IX. BOHR COMPACTIFICATIONS

The objective of the next sections (Section IX to XII) is to describe the Bohr compactification S^b of an invariant Lie semigroup (S, G). The motivation for this work is the following. First it is interesting to see how such an abstract universal object as the Bohr compactifiction can be described by using the data at hand. We compute the unit group of the Bohr compactification S^b and therefore, since all the subsemigroups eS^b, with e idempotent, are again Bohr compactifications of invariant Lie semigroups, of all the groups $H(e)$ (Section X). The lattice of idempotents can be embedded isomorphically into the lattice of faces of the cone dual to $\mathbf{L}(S) \cap \mathfrak{h}$, where $\mathfrak{h}$ is a compactly embedded Cartan algebra of $\mathbf{L}(G)$ (Section XI). We also describe how the Bohr compactification is built as a union of subsets of invariant Lie semigroups.

Another motivation was to find interesting examples of compact topological semigroups which are not too far from Lie semigroups. That means that the subsemigroup generated by the images of all one-parameter subsemigroups is dense and the set of all one-parameter semigroups is a finite dimensional wedge. We call these semigroups *generalized Lie semigroups.* They show how the nowadays classical theory of compact semigroups and the very recent Lie theory of semigroups combine to yield new interesting results.

Definition IX.1. The *Bohr compactification* of a topological semigroup T is a compact semigroup T^b together with a continuous morphism of topological semigroups $i_T : T \to T^b$ such that the following universal property holds: If $\varphi : T \to T'$ is a morhism into a compact topological semigroup T, then there exists a morphism $\varphi^b : T^b \to T'$ such that $\varphi^b \circ i_T = \varphi$. The Bohr compactification is a functor from the category $\underline{TSg}$ of topological semigroups to the category $\underline{CTSg}$ of compact topological semigroups. It is the adjoint of the forgetful functor from $\underline{CTSg}$ to $\underline{TSg}$. We endow a topological semigroup T with the quasi order

$$s \prec s' \quad \text{if } s' \in sT.$$

We write $E(T)$ for the set of idempotents in a semigroup T. Note that the above order, restricted to the set of idempotents, is opposite to the order usually considered on this set. It seems to us that it is more natural in our context to use this order because we have already the order $\leq_S$ on the group G and the semigroup S and we want to avoid confusion. For an idempotent $e \in E(T^b)$ we define

$$F_e := \{t \in T : ei_S(t) \in H(e)\}. \qquad \blacksquare$$

Proposition IX.2. *Let S be a topological monoid with $S = N(S) := \{s \in S : sS = Ss\}$ and $e \in E(S^\flat)$. Then the following assertions hold:*

(1) $sS^\flat = S^\flat s$ *for every* $s \in S^\flat$.

(2) $E(S^\flat)$ *is central in* $S^\flat$.

(3) $E := E(S^\flat)$ *is a compact topological semilattice, i.e., a compact abelian topological semigroup such that each element is idempotent. The poset* $(E, \prec)$ *is a complete lattice,* $\mathbf{1}$ *is a minimal element and there exists a maximal element* 0 *that is a zero element in* $S^\flat$. *The supremum of two elements is* $\sup(e, e') = ee'$ *and the sets*

$$\uparrow A = \{b \in E : (\exists a \in A) a \prec b\}, \quad \text{and} \quad \downarrow A = \{b \in E : (\exists a \in A) b \prec a\}$$

are closed for every closed subset $A \subseteq E$.

(4) $H(e) = \uparrow e \cap \downarrow e$ *is a compact subgroup of* $S^\flat$, *the unit group of the compact monoid* $eS^\flat$.

(5) *The quasi order* $\prec$ *on* $S^\flat$ *is closed and* $ss' \prec t$ *implies* $s, s' \prec t$.

Proof. (1) (cf. [Rup88, p.326]) We have $sS = Ss$ for every $s \in S$ and therefore

$$i(s)S^\flat = \overline{i(sS)} = \overline{i(Ss)} = S^\flat i(s).$$

For $s, t \in S^\flat$ and $s = \lim s_n$ with $s_n \in i(S)$, we find $t_n \in S^\flat$ with $s_n t = t_n s_n$. We assume that $t' = \lim t_n$, hence

$$st = \lim s_n t = \lim t'_n s_n = t's.$$

We conclude that $sS^\flat \subseteq S^\flat s$. The reverse inclusion follows similarly.
(2) [Rup88, p.326].
(3) [CHK86, p.11].
(4) Using [CHK83, p.18, p.183] we see that the set

$$H(e) = eS^\flat e \cap \{s \in S^\flat : e \in sS^\flat \cap S^\flat s\}$$

is a compact group. With (1) we get

$$H(e) = eS^\flat \cap \{x \in S^\flat : e \in sS^\flat\} = \uparrow e \cap \downarrow e.$$

(5) Let (s_n, t_n) be a net in $S^\flat \times S^\flat$ and $(s, t) = \lim(s_n, t_n)$ with $s_n \prec t_n$. Then we find $\widetilde{s}_n \in S^\flat$ with $s_n \widetilde{s}_n = t_n$. We have to show that $s \prec t$. Firstly we may assume that $\widetilde{s} := \lim \widetilde{s}_n$ exist in $S^\flat$. Then $s\widetilde{s} = \lim s_n \widetilde{s}_n = \lim t_n = t$ and therefore $s \prec t$. The last assertion follows from $N(S) = S$ and the definition of $\prec$. ■

In the following we write $i := i_S : S \to S^\flat$ if no confusion is possible.

Lemma IX.3. $F_e = i^{-1}(\downarrow e)$.

Proof. Let $s \in i^{-1}\downarrow e$, then there exists $t \in S^\flat$ such that $i(s)t = e$, hence $e = ei(s)et \in H(e) = H(eS^\flat)$ and therefore $ei(s) \in H(e)$ because $eS^\flat \setminus H(e)$ is an ideal in $eS^\flat$. Conversely, let $s \in F_e$. Then $ei(s) \in H(e)$ and we find $t \in H(e)$ such that $ei(s)t = i(s)et = e$, hence $i(s) \prec e$. ■

Lemma IX.4. *For every subsemigroup $F \subseteq S$, the semigroup $\overline{i(F)}$ contains a unique maximal idempotent e_F. Moreover, every closed subsemigroup of $S^\flat$ contains a unique maximal idempotent.*

Proof. The semigroup $F' := \overline{i(F)}$ is compact, hence contains a maximal idempotent with respect to $\prec\mid_{F'}$ ([CHK83, p.29]). This idempotent must be unique since $E(F') \subseteq E(S^\flat)$ is abelian. ■

From now on we use the notation S only for a generating invariant Lie semigroup (S, G), where G is a connected Lie group and $W = \mathbf{L}(S)$ for its tangent wedge.

Definition IX.5. For a subsemigroup $F \subseteq S$ we write e_F for the maximal idempotent in $\overline{i(F)}$. For $x \in \mathbf{L}(G)$ we set $e_x := e_{\exp \mathbb{R}^+ x}$. We say that an idempotent $e \in E(S^\flat)$ is *accessible* if $e = e_F$ for a subsemigroup $F \subseteq S$ and *directly accessible* if $e = e_x$ for an $x \in W$. We write $E_a(S^\flat)$ for the set of accessible idempotents and $E_{da}(S^\flat)$ for the set of directly accessible idempotents. ■

Example IX.6. In the same sense as the semigroup $\mathbb{R}^+$ plays a central role in the Lie theory of semigroups, the semigroup $T := (\mathbb{R}^+)^\flat$ plays a central role in the theory of compact generalized Lie semigroups. We give some information on this semigroup which is also called the *universal solenoidal semigroup.* Let $e = \lim i_{\mathbb{R}^+}(x_i)$ and $f = \lim i_{\mathbb{R}^+}(y_j)$ be idempotents in T, where x_i and y_j denote nets in $\mathbb{R}^+$. There are two cases. If $e \neq 0$, it is clear that $y_j \prec e$ for all j and therefore $f \prec e$. We conclude that

$$E(T) = \{0, \omega\}$$

consists of exactly two idempotents, the first one, namely 0, is an identity element, and the second one, we call it ω in the following, is a zero element. The group $H(\omega)$ is isomorphic to $\mathbb{R}^\flat$, the Bohr compactification of $\mathbb{R}$ which may also be obtained by taking the dual $\widehat{\mathbb{R}_d}$ of the locally compact abelian group $\mathbb{R}$ with the discrete topology. This is a consequence of a more general result in Section XI about the structure of the groups $H(e)$ sitting in $S^\flat$.

Since each one-parameter semigroup

$$\gamma_x : \mathbb{R}^+ \to S, \quad t \mapsto \exp(tx)$$

gives rises to a morphism $\gamma_x^\flat : T \to S^\flat$ with $\gamma_x^\flat \circ i_{\mathbb{R}^+} = i_S \circ \gamma_x$ we find that

$$e_x = \gamma_x^\flat(\omega).$$

This means that the directly accessible idempotents are exactly the images of ω under morphisms $T \to S^\flat$ induced by one-parameter semigroups starting in $\mathbf{1}$. ■

Proposition IX.7. *For every $e \in E(S^\flat)$ the set $F_e \subseteq S$ is a compact exposed face and for every compact exposed face $F \in \mathcal{F}_c(S)$ we have $F = F_{e_F}$. In particular $F_1 = H(S)$.*

Proof. (1) Let $\varphi : S \to eS^\flat$ denote the continuous homomorphism $s \mapsto ei(s)$. This is a homomorphism because e is a central idempotent. The rest follows from the definition of compact exposed faces since S is invariant (Proposition VI.11 and IX.2).
(2) Let $F \in \mathcal{F}_c(S)$ be a compact exposed face. We find a homomorphism $\varphi : S \to K$, where K is a compact monoid and $F = \varphi^{-1}(H(K))$. Then there exists a unique homomorphism $\varphi^\flat : S^\flat \to K$ with $\varphi = \varphi^\flat \circ i$. Let $e \in {\varphi^\flat}^{-1}(H(K))$ be the unique maximal idempotent. Then

$$\varphi^\flat(\downarrow e) \subseteq \downarrow \varphi^\flat(e) = \downarrow \mathbf{1} = H(K),$$

hence $\downarrow e = {\varphi^\flat}^{-1}(H(K))$ and by Lemma IX.3 we obtain

$$F_e = i^{-1}(\downarrow e) = i^{-1}\left({\varphi^\flat}^{-1}(H(K))\right) = \varphi^{-1}(H(K)) = F.$$

■

Theorem IX.8. *The mappings*

$$d : (\mathcal{F}(S), \subseteq) \to (E(S^\flat), \prec), \quad F \mapsto e_F$$

and

$$g : (E(S^\flat), \prec) \to (\mathcal{F}(S), \subseteq), \quad e \mapsto F_e$$

are monotone and define a Galois connection, i.e.,

$$F \subseteq F_e = g(e) \quad \Longleftrightarrow \quad d(F) = e_F \prec e.$$

The following assertions hold

(1) *The compact exposed faces are the fixed points of $g \circ d$ in $\mathcal{F}(S)$ and the accessible idempotents are the fixed points of $d \circ g$ in $E(S^\flat)$.*
(2) $F \subseteq F_{e_F}$, $F_e = F_{e_{F_e}}$.
(3) $e_{F_e} \prec e$, $e_{F_{e_F}} = e_F$.
(4) $(\exists F) e = e_F \Longleftrightarrow e_{F_e} = e$, $F = F_{e_F} \Longleftrightarrow (\exists e) F_e = F$
(5) *g preserves infs and d preserves sups.*
(6) *The mapping $d|_{\mathcal{F}_c(S)} : (\mathcal{F}_c(S), \subseteq) \to (E_a(S^\flat), \prec)$ is an order isomorphism of the complete lattice of compact exposed faces onto the lattice of accessible idempotents.*

Proof. The isotony of the above mappings follows directly from the definitions. Let $F \subseteq F_e$. Then $\overline{ei(F)} \subseteq H(e)$ and therefore $ee_F = e$ because $H(e)$ contains only one idempotent. We conclude that $e_F \prec e$. Conversely, assume that $e_F \prec e$, then $\overline{i(F)} \subseteq \downarrow e_F \subseteq \downarrow e$ and therefore

$$e \cdot \overline{i(F)} \subseteq e \cdot \downarrow e \subseteq \uparrow e \cap \downarrow e = H(e).$$

This shows that $F \subseteq F_e$. The rest follows from Proposition IX.7, the definition of the accessible idempotents, [GHK80, pp.18-20] and the fact that $\mathcal{F}_c(S)$ is a complete lattice (Corollary IV.17). ■

Lemma IX.9. *Let G be a connected Lie group, $U \subseteq G$ be an open set, $x \in \mathbf{L}(G) \setminus \{0\}$, $I \subseteq \mathbf{L}(G)$ the smallest ideal containing x and $n = \dim I$. Then we find elements $s_1 = \mathbf{1}, s_2, \ldots, s_n \in U$ such that*

$$I = \operatorname{span}\{\operatorname{Ad}(s_1)x, \ldots, \operatorname{Ad}(s_n)x\}.$$

Proof. Let $U \in \mathcal{U}(\mathbf{1})$ and $J := \operatorname{span}\{\operatorname{Ad}(s)x : x \in U\}$. The mapping $g \mapsto \operatorname{Ad}(g)x$ is analytic on G and $\operatorname{Ad}(g)x \in J$ for $g \in U$, hence $\operatorname{Ad}(g)x \in J$ for all $g \in G$, thus $I = J$. ∎

Lemma IX.10. $\mathbf{L}(F_e) = \{x : e_x \prec e\}$.

Proof. For $x \in \mathbf{L}(G)$ the following assertions are equivalent:

$$\begin{aligned} x \in \mathbf{L}(F_e) &\Leftrightarrow \exp(\mathbb{R}^+ x) \subseteq F_e \\ &\Leftrightarrow ei\big(\exp(\mathbb{R}^+ x)\big) \subseteq H(e) \\ &\Leftrightarrow ee_x = e \\ &\Leftrightarrow e_x \prec e \end{aligned}$$

because $ei\big(\exp(\mathbb{R}^+ x)\big) \subseteq H(e) \subseteq {\downarrow} e$ implies that $ee_x \subseteq {\downarrow} e \cap {\uparrow} e = H(e)$, and conversely, $e_x \prec e$ implies that

$$ei\big(\exp(\mathbb{R}^+ x)\big) \subseteq {\downarrow} ee_x \subseteq {\downarrow} e.$$

Thus $ei\big(\exp(\mathbb{R}^+ x)\big) \subseteq H(e)$. ∎

Lemma IX.11. *Let $x \in \mathbf{L}(S)$. Then the following assertions hold:*

(1) *$e_x = e_F$, where F is the smallest normal exposed face of S which contains $\exp x$.*

(2) *Let $F \in \mathcal{F}_n(S)$ and $x \in \operatorname{algint} \mathbf{L}(F)$. Then $\mathbf{L}(F) = \mathbf{L}(F_{e_x})$.*

(3) *Let $x \in \operatorname{int} \mathbf{L}(S)$. Then e_x is the unique maximal idempotent in $S^\flat$.*

Proof. (1) Let $U \in \mathcal{U}(\mathbf{1})$, I the smallest ideal of $\mathbf{L}(G)$ containing x and $s_1 = \mathbf{1}, s_2, \ldots, s_n \in U \cap \operatorname{int} S$ such that $I = \operatorname{span}\{\operatorname{Ad}(s_1)x, \ldots, \operatorname{Ad}(s_n)x\}$ (Lemma IX.9). Set $e_i := e_{\operatorname{Ad}(s_i)x}$ and $e := e_1 \cdot e_2 \ldots \cdot e_n$. Now Lemma IX.10 shows that $\operatorname{Ad}(s_i)x \subseteq \mathbf{L}(F_e)$ for $i = 1, \ldots, n$. Let $N := \langle \exp I \rangle$ and $F := H(\overline{SN}) \cap S$. This is an normal exposed face of S since

$$F = T_N(S) \cap S \subseteq T_F(S) \cap S \subseteq H\big(L_N(S)\big) \cap S = T_N(S) \cap S = F$$

and $\operatorname{Ad}(s_i)x \in \mathbf{L}(F)$ for every i. We conclude that $e_i \prec e_F$ for every i and therefore that

$$e = e_1 \cdot e_2 \cdot \ldots \cdot e_n \prec e_F^n = e_F.$$

We claim that $e_F \prec e$, i.e., $F \subseteq F_e$ (Theorem IX.8). The one-parameter semigroups $s_i \exp(\mathbb{R}^+ x) s_i^{-1} = \exp\big(\mathbb{R}^+ \operatorname{Ad}(s_i)x\big)$ are contained in $F_e \subseteq S$ and they generate the whole group N, hence

$$\operatorname{Ad}(s_i)x \in \mathbf{L}\left(T_{F_e}(S)\right).$$

Thus $N \subseteq T_{F_e}(S)$ and this implies that

$$F = H(\overline{SN}) \cap S \subseteq H\big(L_{F_e}(S)\big) \cap S = T_{F_e}(S) \cap S = F_e.$$

According to Proposition IX.2 we have $e_x i(s_i) = i(s_i)e_x = e_i i(s_i)$ because $s_i \exp(\mathbb{R}^+ x) = \exp\big(\mathbb{R}^+ \operatorname{Ad}(s_i)x\big)s_i$. Hence

$$e_F i(s_1 \cdot \ldots \cdot s_n) = e_1 i(s_1) \cdot \ldots \cdot e_n i(s_n) = e_x i(s_1 \cdot \ldots \cdot s_n).$$

Using the fact that the elements s_i are arbitrary near to $\mathbf{1}$ and n is a fixed number, we see that $e_F = e_x$. For every other normal exposed face $F' \subseteq S$ with $x \in F'$ we have $e_x = e_F \prec e_{F'}$, hence $F \subseteq F_{e_{F'}} = F'$.
(2) Let $A := \langle \exp \mathbf{L}(F) - \mathbf{L}(F)\rangle$ and $G(A)$ the same group with its Lie group topology. We set $F' := \langle \exp_{G(A)} \mathbf{L}(F)\rangle$. Then $s := \exp x \in \operatorname{int} F'$ and therefore $s \in \big(G(A), \preceq_{F'}\big)$ is an order unit ([Ne91b, 2.11]). We conclude that for every $f \in \mathbf{L}(F)$ we find an $n \in \mathbb{N}$ with $\exp_{G(S)} f \preceq_{F'} s^n$, hence $\exp_G(f) \preceq_S \exp_G(x)^n$. Thus

$$\langle \exp \mathbf{L}(F)\rangle \subseteq i_S^{-1}(\downarrow e_x) = F_{e_x} \subseteq F$$

since $e_x \prec e_F$. Consequently F and F_{e_x} have the same tangent wedge.
(3) Using (1) and Lemma IV.31 we find that $e_x = e_S$ because S is the smallest face of S containing $\exp x$. This proves (3). ∎

Proposition IX.12. *Let $e = e_F$ be an accessible idempotent. Then the unique extension $\varphi_F : L_F(S) \to eS^\flat$ of the mapping $s \mapsto ei(s)$ induces an isomorphism $\varphi_F^\flat : L_F(S)^\flat \to eS^\flat$. Suppose that, in addition, $F \in \mathcal{F}_n(S)$, $I_g^\flat \in \operatorname{Aut}(S^\flat)$ and let $\widetilde{I}_g^\flat \in \operatorname{Aut}\big(L_F(S)^\flat\big)$ denote the unique extensions of $I_g\,|_S : s \mapsto gsg^{-1}$ and $I_g\,|_{L_F(S)}$ for $g \in G$. Then*

$$I_g^\flat \circ \varphi_F^\flat = \varphi_F^\flat \circ \widetilde{I}_g^\flat.$$

$$\begin{array}{ccc} S & \xrightarrow{\;I_g\;} & S \\ \big\downarrow{\scriptstyle i_S} & & \big\downarrow{\scriptstyle i_S} \\ S^\flat & \xrightarrow{\;I_g^\flat\;} & S^\flat \end{array} \qquad \begin{array}{ccc} L_F(S) & \xrightarrow{\;I_g\;} & L_F(S) \\ \big\downarrow{\scriptstyle i_{L_F(S)}} & & \big\downarrow{\scriptstyle i_{L_F(S)}} \\ L_F(S)^\flat & \xrightarrow{\;\widetilde{I}_g^\flat\;} & L_F(S)^\flat \end{array}$$

$$\begin{array}{ccccc} S & \xrightarrow{\;j\;} & L_F(S) & \xrightarrow{\;\varphi_F\;} & eS^\flat \\ \big\downarrow{\scriptstyle i_S} & & \big\downarrow{\scriptstyle i_{L_F(S)}} & & \big\downarrow \\ S^\flat & \xrightarrow{\;j^\flat\;} & L_F(S)^\flat & \xrightarrow{\;\varphi_F^\flat\;} & S^\flat \end{array}$$

$$\begin{array}{ccc} eS^\flat & \xrightarrow{\;I_g^\flat\;} & eS^\flat \\ \big\downarrow & & \big\downarrow \\ L_F(S)^\flat & \xrightarrow{\;\widetilde{I}_g^\flat\;} & L_F(S)^\flat \end{array}$$

Proof. The existence of φ_F follows from Lemma VI.10. Then the universal property of $L_F(S)^\flat$ guarantees the existence of a unique homomorphism $\varphi_F^\flat$:

$L_F(S)^\flat \to eS^\flat$ such that $\varphi_F^\flat \circ i_{L_F(S)} = \varphi_F$. On the other hand, the inclusion mapping $j : S \to L_F(S)$ induces a homomorphism $j^\flat : S^\flat \to L_F(S)^\flat$. Then

$$\varphi_F^\flat \circ j^\flat\big(i_S(s)\big) = \varphi_F^\flat \circ \Big(i_{L_F(S)}\big(j(s)\big)\Big) = \varphi_F\big(j(s)\big) = ei_S(s),$$

and

$$\varphi_F^\flat \circ j^\flat = \lambda_e : s \mapsto es, S^\flat \to eS^\flat.$$

Hence $(\varphi_F \circ j)^\flat|_{eS^\flat} = \mathrm{id}_{eS^\flat}$. In addition, we have

$$j^\flat \circ \varphi_F^\flat\Big(i_{L_F(S)}\big(j(s)\big)\Big) = j^\flat \circ \varphi_F\big(j(s)\big) = j^\flat\big(ei(s)\big) = j^\flat(e)j^\flat\big(i(s)\big) = i_{L_F(S)}\big(j(s)\big)$$

since

$$j^\flat(e) \in j^\flat\big(\overline{i_S(F)}\big) \subseteq \overline{j^\flat \circ i_S(F)} = \overline{i_{L_F(S)}(F)} \subseteq H\big(L_F(S)^\flat\big)$$

and therefore $j^\flat(e) = \mathbf{1}_{L_F(S)^\flat}$. We conclude that

$$j^\flat \circ \varphi_F^\flat = \mathrm{id}_{L_F(S)^\flat}$$

because these two continuous homomorphisms agree on $i_{L_F(S)}\big(j(S)\big)$ and thus on the dense set $i_{L_F(S)}(SF^{-1})$. Hence $\varphi_F^\flat$ is an isomorphism.

To prove the last assertion, it suffices to show that the above homomorphisms agree on the dense subsemigroup $i_{L_F(S)}(SF^{-1})$ of $L_F(S)^\flat$. But $i_{L_F(S)}(F^{-1}) = i_{L_F(S)}(F)^{-1}$ and we only have to check the assertion on $i_{L_F(S)}(S)$. First we note that the invariance of F under I_g implies that $I_g^\flat(e) = e$. Consequently

$$\begin{aligned} I_g^\flat \circ \varphi_F^\flat \circ i_{L_F(S)}(s) &= I_g^\flat \circ \varphi_F(s) \\ &= I_g^\flat\big(ei_S(s)\big) = eI_g^\flat\big(i_S(s)\big) \\ &= ei_S \circ I_g(s) = \varphi_F \circ I_g(s) \\ &= \varphi_F^\flat \circ i_{L_F(S)} \circ I_g(s) \\ &= \varphi_F^\flat \circ \widetilde{I}_g^\flat\big(i_{L_F(S)}(s)\big). \end{aligned}$$

■

One should notice that the unit group of the semigroup $L_F(S)^\flat$ is not necessarily trivial (cf. Section X). Our next objective is to show that all the idempotents in $S^\flat$ are accessible in finitely many steps. Then the fact that all the semigroups $eS^\flat$ are Bohr compactifications of Lie semigroups $S_e \supseteq S$ will permit us an iductive argument. From this we will deduce that all compact faces are normal and that all idempotents are invariant under the automorphisms $I_g^\flat$ for $g \in G$. First we take a closer look at the idempotents of $S^\flat$ and how they are related to the other elements of $S^\flat$. First we need two results to get a reduction to the case where the unit group is trivial.

Lemma IX.13. *Let T be a compact topological semigroup and $H \subseteq T$ a closed subset. Then the following assertions hold:*

(1) $N(H) := \{t \in T : tH = Ht\}$ *is closed.*

(2) *If S is a topological semigroup, $H \subseteq S$ a normal subset, i.e., $N(H) = S$, and $\varphi : S \to T$ a morphism. Then $\overline{\varphi(H)}$ is normal in $\overline{\varphi(S)}$.*

Proof. (1) Let $(t_n)_{n\in D}$ be a net in $N(H)$ with $t_n \to t$ and $h \in H$. Then we find $h_n \in H$ such that $t_n h = h_n t_n$. We may assume that $h_n \to h' \in H$ because H is compact. Then $th = h't \in Ht$ and therefore $tH \subseteq Ht$. The converse inclusion follows similarly. Hence $t \in N(H)$ and thus $N(H)$ is closed.
(2) Let $s \in S$. Then

$$\varphi(s)\overline{\varphi(H)} = \overline{\varphi(s)\varphi(H)} = \overline{\varphi(sH)} = \overline{\varphi(Hs)} = \overline{\varphi(H)}\varphi(s)$$

because $\varphi(s)\overline{\varphi(H)}$ contains the dense subset $\varphi(s)\varphi(H)$ of $\overline{\varphi(s)\varphi(H)}$. Thus $\varphi(S) \subseteq N\big(\overline{\varphi(H)}\big)$ and the closedness of the normalizer of $\overline{\varphi(H)}$ proves (2). ∎

Proposition IX.14. *The mapping $(S,G) \to E(S^\flat)$, $\varphi \to E(\varphi^\flat) := \varphi^\flat|_{E(S^\flat)}$ defines a functor from the category $\underline{ILSg}$ to the category of compact semilattices. Let $\varphi : (S,G) \to (S',G')$ be an $\underline{ILSg}$-morphism such that $\ker\varphi \subseteq H(S)$ and $\varphi(S) = S'$. Then $E(\varphi^\flat)$ is an isomorphism.*

Proof. The first assertion is clear since $\varphi^\flat\big(E(S^\flat)\big) \subseteq E\big(S'^\flat\big)$ for every homomorphism $\varphi^\flat : S^\flat \to S'^\flat$. Suppose that $\ker\varphi \subseteq H(S)$ and $\varphi(S) = S'$. Let $H := \overline{i_S(\ker\varphi)} \subseteq H(S^\flat)$. Then H is a normal subgroup of $H(S^\flat)$ (Lemma IX.13) and therefore the relation $x \sim y$ if $xH = yH$ is a closed congruence on $S^\flat$. We set $\widetilde{S} := S^\flat/H$ and denote the quotient homomorphism $S^\flat \to \widetilde{S}$ with q ([CHK83, p.48]). Then

$$q \circ i_S(\ker\varphi) = \{1\}$$

and we find a homomorphism $q' : S' \to \widetilde{S}$ such that $q' \circ \varphi = q \circ i_S$.

$$\begin{array}{ccccc}
S & \xrightarrow{\quad\varphi\quad} & S' & \xrightarrow{\quad i_{S'}\quad} & S'^\flat \\
\big\downarrow i_S & & \big\downarrow q' & & \big\downarrow q'^\flat \\
S^\flat & \xrightarrow{\quad q\quad} & S^\flat/H = \widetilde{S} & \xrightarrow{\quad \mathrm{id}_{\widetilde{S}}\quad} & \widetilde{S}
\end{array}$$

$E(q)$ is an isomorphism: If $e, f \in E(S^\flat)$ and $q(e) = q(f)$, then $e \in fH$ and therefore $H(e) = H(f)$ which implies that $e = f$. To see that $E(q)$ is surjective, let $e \in E(\widetilde{S})$. Then $q^{-1}(e)$ is a subsemigroup of $S^\flat$ which contains an idempotent (Lemma IX.4).
$E(\varphi)$ is an isomorphism: E is a functor, hence

$$E(q' \circ \varphi) = E(q') \circ E(\varphi) = E(q) \circ E(i_S) = E(q).$$

Thus $E(\varphi)$ is injective. The surjectivity of φ implies, by the same argument as above, that $E(\varphi)$ is surjective. ∎

Definition IX.15. Let (S, G) be a generating invariant Lie semigroup. We set $S_0 := S/H(S)$ and write $\pi : G \to G/H(S)$ for the quotient homomorphism. Moreover we recall the definition

$$\widetilde{\mathrm{comp}}(S) := \pi^{-1}\big(\mathrm{comp}(S)\big) \quad \text{and set} \quad S(\mathbf{1}) := i_S\big(\widetilde{\mathrm{comp}}(S)\big)H(\mathbf{1}).$$

Note that $\widetilde{\mathrm{comp}}(S) = \mathrm{comp}(S)$ iff $H(S)$ is compact (Theorem V.8). In the following we will often need the diagram

$$\begin{array}{ccc} S & \xrightarrow{\quad i_S \quad} & S^\flat \\ \Big\downarrow{\scriptstyle \pi} & & \Big\downarrow{\scriptstyle \pi^\flat} \\ S_0 & \xrightarrow{\quad i_{S_0} \quad} & S_0^\flat. \end{array}$$

The following lemma provides information on the kernel of $\pi^\flat$.

Lemma IX.16. *The morphism $\pi^\flat : S^\flat \to S_0^\flat$ is a quotient homomorphism with*

$$\ker \pi^\flat = H(S^\flat) = \overline{i_S\big(H(S)\big)}.$$

Proof. Using Lemma IX.13 we see that $H := \overline{i_S\big(H(S)\big)}$ is normal in $S^\flat$ because $H(S)$ is normal in S. Then $\widetilde{S} := S^\flat/H$ is a compact topological semigroup ([CHK83, p.48]). Moreover, writing $q : S^\flat \to \widetilde{S}, x \mapsto xH$, we obtain that

$$q \circ i_S\big(sH(S)\big) \subseteq q\big(i_S(s)\big)q(H) = \{q\big(i_S(s)\big)\}.$$

Thus we find a continuous homomorphism $\varphi : S_0 \to \widetilde{S}$ with $\varphi \circ \pi = q \circ i_S$. Hence

$$q \circ i_S = \varphi \circ \pi = \varphi^\flat \circ i_{S_0} \circ \pi = \varphi^\flat \circ \pi^\flat \circ i_S$$

and therefore $q = \varphi^\flat \circ \pi^\flat$. For $\pi^\flat(x) = \pi^\flat(y)$ this leads to $q(x) = \varphi^\flat \circ \pi^\flat(x) = \varphi^\flat \circ \pi^\flat(y) = q(y)$ which means that $y \in Hx$. If, conversely, $h \in H(S)$, then

$$\pi^\flat\big(i_S(h)\big) = i_{S_0}\big(\pi(h)\big) = i_{S_0}(\mathbf{1}) = \mathbf{1}.$$

Consequently $\pi^\flat(H) = \{\mathbf{1}\}$. In view of Corollary V.10 we know that $H(S_0^\flat) = \{\mathbf{1}\}$. Hence $\pi^\flat\big(H(S^\flat)\big) = \{\mathbf{1}\}$. This implies that $H(S^\flat) = \overline{i_S\big(H(S)\big)} = \ker \pi^\flat$. ∎

Lemma IX.17. *Let $s \in S^\flat$. Then*

$$\downarrow s \cap E(S^\flat) = \{e \in E(S^\flat) : es = s\}$$

is a compact subsemigroup of $E(S^\flat)$ which contains a unique maximal element.

Proof. First we note that $e \in \downarrow s \cap E(S^\flat)$ is equivalent to $es = s$. The set on the right side is compact because multiplication is continuous on $S^\flat \times S^\flat$. Moreover, $es = s$ and $fs = s$ implies that $(ef)s = e(fs) = s$. So it is also a semigroup. ∎

Definition IX.18. We define the function $E : S^{\flat} \to E(S^{\flat})$ by $s \mapsto e$, where e is the unique maximal idempotent in $\downarrow s \cap E(S^{\flat})$ (Lemma IX.17). ■

Lemma IX.19. *We have that*

$$\pi^{\flat}\big(S(\mathbf{1})\big) = S_0(\mathbf{1}) \quad \text{and} \quad (\pi^{\flat})^{-1}\big(S_0(\mathbf{1})\big) = S(\mathbf{1}).$$

Proof. It follows from the definitions that

$$\pi^{\flat}\big(S(\mathbf{1})\big) = \pi^{\flat} \circ i_S\big(\widetilde{\mathrm{comp}}(S)\big) = i_{S_0}\big(\mathrm{comp}(S_0)\big) = S_0(\mathbf{1})$$

since $\pi\big(\widetilde{\mathrm{comp}}(S)\big) = \mathrm{comp}(S_0)$. But $S(\mathbf{1}) = S(\mathbf{1})H(\mathbf{1})$ implies that $S(\mathbf{1})$ is $\pi^{\flat}$-saturated (Lemma IX.16). This proves the lemma. ■

Lemma IX.20. $E(S^{\flat}) \cap S(\mathbf{1}) = \{\mathbf{1}\}$.

Proof. Suppose that $e = i_{S_0}(s) \in S_0(\mathbf{1})$ is an idempotent of $S_0^{\flat}$. Then $i_{S_0}(s^2) = i_{S_0}(s) = e$. There are two cases:
Case 1) $s^2 \in \mathrm{comp}(S_0)$: Then $s^2 = s$ because $i_{S_0}|_{\mathrm{comp}(S_0)}$ is injective (Theorem V.9). This proves that $s = \mathbf{1}$ since S_0 contains only one idempotent.
Case 2) $s^2 \notin \mathrm{comp}(S_0)$: Then $e \in i_{S_0}\big(S_0 \setminus \mathrm{comp}(S_0)\big) \cap S_0(\mathbf{1})$ which contradicts Theorem V.9 because there exists a compact semigroup S_1 with 0 and a homomorphism $\varphi : S_0 \to S_1$ with $\varphi\big(S_0 \setminus \mathrm{comp}(S_0)\big) = \{0\}$ and $0 \notin \varphi\big(\mathrm{comp}(S_0)\big)$. Consequently this case is impossible.

So we have proved the lemma for S_0. If $e \in E(S^{\flat}) \cap S(\mathbf{1})$, then, in view of Lemma IX.19,

$$\pi^{\flat}(e) \in S_0(\mathbf{1}) \cap E(S_0^{\flat}) = \{\mathbf{1}\}.$$

Hence $\pi^{\flat}(e) = \mathbf{1}$ and this implies that $e = \mathbf{1}$ (Proposition IX.14). ■

Proposition IX.21. $S(\mathbf{1}) = \{s \in S^{\flat} : E(s) = \mathbf{1}\}$.

Proof. Let $s \in S(\mathbf{1})$ and $e = E(s)$. Then $\pi^{\flat}(e) \prec \pi^{\flat}(s) \in S_0(\mathbf{1})$ (Lemma IX.19) and Lemma IX.20 proves that $\pi^{\flat}(e) = \mathbf{1}$. Hence $e = \mathbf{1}$ by Proposition IX.14 and therefore $E\big(S(\mathbf{1})\big) = \{\mathbf{1}\}$.

If, conversely, $E(s) = \mathbf{1}$ for an element $s \in S^{\flat}$ we have to show that $s \in S(\mathbf{1})$. Suppose that this is false. Then $\pi^{\flat}(s) \notin S_0(\mathbf{1})$ (Lemma IX.19) and, according to Proposition VI.2, we find $x \in \mathbf{L}(S_0)\setminus\{0\}$ such that $i_{S_0}(\exp \mathbb{R}^+ x) \subseteq \downarrow\pi^{\flat}(s)$. Taking $x' \in \mathbf{L}(S)$ with $d\pi(\mathbf{1})x' = x$ we get an element $x' \in \mathbf{L}(S) \setminus H\big(\mathbf{L}(S)\big)$ with

$$\pi^{\flat} \circ i_S(\exp \mathbb{R}^+ x') = i_{S_0}(\exp \mathbb{R}^+ x) \subseteq \downarrow\pi^{\flat}(s) = \pi^{\flat}(\downarrow s)$$

beause the relation $\pi^{\flat}([t,t']) = [\pi^{\flat}(t), \pi^{\flat}(t')]$ follows from Lemma IX.16 and the surjectivity of $\pi^{\flat}$. Now

$$\mathbf{1} \neq e_x \prec s$$

since $x \notin H\big(\mathbf{L}(S)\big)$ (Proposition IX.7). This contradicts the assumption that $E(s) = \mathbf{1}$. ■

Our objective is to describe $S^\flat$ as a union

$$S^\flat = \bigcup_{e \in E(S^\flat)} S_e(\mathbf{1}),$$

where S_e are certain generating invariant Lie semigroup s in G containing S. To do this, we need more information about the idempotents in $S^\flat$, namely that they are invariant under the inner automorphisms $I_g^\flat$ of $S^\flat$ for all $g \in G$, and that $eS^\flat \cong S_e^\flat$ for a suitable generating invariant Lie semigroup S_e. The strategy to obtain these results is to climb up on $S^\flat$ along finite chains of idempotents, where each one is directly accessible from its predecessor. The following lemma will be crucial for the inductive argument.

Lemma IX.22. *For every idempotent $e \in E(S^\flat) \setminus \{\mathbf{1}\}$ there exists a normal exposed face $F \in \mathcal{F}_n(S)$ and $x \in \mathbf{L}(S) \setminus H(\mathbf{L}(S))$ such that*

$$\mathbf{1} \neq e_F = e_x \prec e.$$

Proof. Using Proposition IX.14 we see that $\pi^\flat(e) \neq \mathbf{1}$. Now Lemma IX.20 entails that $\pi^\flat(e) \not\in i_{S_0}(\operatorname{comp}(S_0))$.

So, according to Proposition VI.2, we find $x' \in \mathbf{L}(S_0) \setminus \{0\}$ such that $i_{S_0}(\exp \mathbb{R}^+ x') \subseteq \downarrow \pi^\flat(e)$, i.e., $\mathbf{1} \neq e_{x'} \prec \pi^\flat(e)$ (Lemmas IX.3, IX.10). Taking $x \in \mathbf{L}(S)$ with $d\pi(\mathbf{1})x = x'$ we get that

$$\pi^\flat(e_x) = \pi^\flat \circ \gamma_x^\flat(\omega) = (\pi \circ \gamma_x)^\flat(\omega) = \gamma_{x'}^\flat(\omega) = e_{x'}.$$

Hence $\mathbf{1} \neq e_x \prec e$ (Proposition IX.14). Now an application of Lemma IX.11 completes the proof. ∎

Lemma IX.23. *Let $j : S \to S'$ be an embedding of generating invariant Lie semigroups in G. Suppose that $j^\flat : S^\flat \to S'^\flat$ is an isomorphism. Then $S = S'$.*

Proof. The compact monoid $S^\flat/H(S^\flat)$ (Lemma IX.16) has a neighborhood of $\mathbf{1}$ which is homeomorphic to a neighborhood of 0 in the pointed cone $\mathbf{L}(S)/H(\mathbf{L}(S))$ (Proposition VI.12, Theorem V.9). Therefore we conclude that

$$\dim \mathbf{L}(S)/H(\mathbf{L}(S)) = \dim \mathbf{L}(S')/H(\mathbf{L}(S')).$$

This proves that $H(\mathbf{L}(S)) = H(\mathbf{L}(S'))$ and therefore $H(S) = H(S')$. Let $\pi : G \to G/H(S) = G/H(S')$ denote the quotient homomorphism. Then the embedding $\widetilde{j} : S_0 \to S_0'$ satisfies $\widetilde{j} \circ \pi = \pi' \circ j$ and therefore $\widetilde{j}^\flat \circ \pi^\flat = \pi'^\flat \circ j^\flat$. Since $\pi^\flat$, $\pi'^\flat$ and $j^\flat$ are surjections, it follows that $\widetilde{j}^\flat$ is surjective. Suppose that $\widetilde{j}^\flat(x) = \widetilde{j}^\flat(y)$ and that $x = \pi^\flat(x')$, $y = \pi^\flat(y')$. Then $\pi'^\flat \circ j^\flat(x') = \pi'^\flat \circ j^\flat(y')$ and consequently $j^\flat(y') \in j^\flat(x')H(S'^\flat)$ (Lemma IX.16). Hence we find $z' \in H(S^\flat)$ with

$$j^\flat(y') = j^\flat(x')j^\flat(z') = j^\flat(x'z').$$

We conclude that $y' = x'z'$ because $j^\flat$ is an isomorphism. Hence

$$y = \pi^\flat(y') = \pi^\flat(x'z') = \pi^\flat(x') = x$$

(Lemma IX.16) and thus $\widetilde{j}^\flat$ is injective, hence an isomorphism. So we may assume that $H(S) = H(S') = \{\mathbf{1}\}$. Let $x \in \mathbf{L}(S')$. Then $\gamma_x : \mathbb{R}^+ \to S'$ induces a one-parameter semigroup $i_{S'} \circ \gamma_x$ of $S'^\flat$ and $(j^\flat)^{-1} \circ i_{S'} \circ \gamma_x : \mathbb{R}^+ \to S^\flat$ is also a one-parameter semigroup. Now we use Corollary V.10 to see that

$$(j^\flat)^{-1} \circ i_{S'} \circ \gamma_x = i_S \circ \gamma_y$$

for an $y \in \mathbf{L}(S)$ because i_S maps a neighborhood of $\mathbf{1}$ homeomorphically onto a neighborhood of $\mathbf{1}$ in $S^\flat$. Applying $j^\flat$ to this relation, we obtain

$$i_{S'} \circ \gamma_x = j^\flat \circ i_S \circ \gamma_y = i_{S'} \circ j \circ \gamma_y.$$

Now we use the fact that $i_{S'}$ is a local homeomorphism to conclude that $\gamma_x = j \circ \gamma_y$ and consequently that $x = y \in \mathbf{L}(S)$. So $\mathbf{L}(S) \subseteq \mathbf{L}(S')$ which implies that $S = S'$. ∎

Theorem IX.24. *For every idempotent $e \in E(S^\flat)$ there exists a sequence*

$$e_0 = \mathbf{1} \prec e_1 \prec \ldots \prec e_n = e$$

in $E(S^\flat)$ such that $e_{i+1} \in E_{da}(e_i S^\flat)$ and a generating invariant Lie semigroup $S_e \supseteq S$ such that the following assertions hold:

(1) $I_g^\flat(e) = e$.

(2) $F_e \in \mathcal{F}_n(S)$.

(3) *If* $\operatorname{rank} \pi_1(G) \leq 1$, *then* $\mathbf{L}(S_e)^* \in \mathcal{F}\big(\mathbf{L}(S)^*\big)$.

(4) *There exists a homomorphism $\varphi_e : S_e \to eS^\flat$ such that the inclusion $j_e : S \to S_e$ satisfies:*

  (a) $\varphi_e \circ j_e = \lambda_e \circ i_S$ *(φ_e extends $\lambda_e \circ i_S$),*

  (b) $j_e^\flat \circ \varphi_e^\flat = \mathrm{id}_{S_e^\flat}$ *($\varphi_e^\flat$ is an isomorphism), and*

  (c) $I_g^\flat \circ \varphi_e^\flat = \varphi_e^\flat \circ I_g^\flat$ *for all $g \in G$.*

(5) *The extension φ_e of $\lambda_e \circ i_S$ is unique.*

(6) *The properties* (4a) *and* (4b) *determine S_e uniquely.*

(7) *If e is accessible, then $S_e = L_F(S)$ for $F \in \mathcal{F}_n(S)$ with $e = e_F$.*

Proof. We start with the inductive construction of the idempotents e_i and show that it stops after finitely many steps. We set

$$e_0 := \mathbf{1}_{S^\flat} \quad \text{and} \quad S_{e_0} := S.$$

It is trivial to verify that (1)-(5) hold in this case for e_0 because $F_{e_0} = H(S)$ (Proposition IX.7). Assertion (6) follows from Lemma IX.23 because (4a) and (4b) imply that $\varphi_{e_0}^\flat$ is an isomorphism. Now we assume that

$$e_0 = \mathbf{1} \prec e_1 \prec \ldots e_i \prec e$$

are constructed such that $e_j \in E_{da}(e_{j-1}S^\flat)$ for $j = 1,\ldots,i$ and that (1)-(5) hold for these idempotents. If $e_i = e$ then the theorem is proved. If not, we use Lemma IX.22 to find an element $x \in \mathbf{L}(S_{e_i}) \setminus H\big(\mathbf{L}(S_{e_i})\big)$ such that

$$\varphi^\flat_{e_i}(\mathbf{1}) = e_i \neq \varphi^\flat_{e_i}(e_x) \prec e.$$

Let $F_i \in \mathcal{F}_n(S_{e_i})$ be the smallest normal exposed face containing $\exp x$. Then $e_{F_i} = e_x$ (Lemma IX.11). We define

$$e_{i+1} := \varphi^\flat_{e_i}(e_x) \in E_{da}(e_i S^\flat), \qquad \text{and} \qquad S_{e_{i+1}} := L_{F_i}(S_{e_i}).$$

Now we give the proof of the steps $i \to i+1$ for (1) - (6). We also note that $x \in H\big(\mathbf{L}(S_{e_{i+1}})\big) \setminus H\big(\mathbf{L}(S_{e_i})\big)$. Thus the dimension of the ideal $H\big(\mathbf{L}(S_{e_i})\big)$ is enlarged in every step and the construction eventually has to stop.

(1) We have that

$$\begin{aligned} I^\flat_g(e_{i+1}) &= I^\flat_g \circ \varphi^\flat_{e_i}(e_x) = I^\flat_g \circ \varphi^\flat_{e_i}(e_{F_i}) = \varphi^\flat_{e_i} \circ I^\flat_g(e_{F_i}) \\ &= \varphi^\flat_{e_i}\big(e_{I_g(F_i)}\big) = \varphi^\flat_{e_i}(e_{F_i}) = \varphi^\flat_{e_i}(e_x) = e_{i+1} \end{aligned}$$

since $F_i \in \mathcal{F}_n(S_{e_i})$ is an invariant subsemigroup of G.

(2) This follows from (1) and $I_g(F_e) = F_{I^\flat_g(e)} = F_e$.

(3) Using the Face Theorem VI.18, we obtain for $\operatorname{rank}\pi_1(G) \leq 1$ that

$$\mathbf{L}(S_{e_{i+1}})^* = \mathbf{L}\left(L_{F_i}(S_{e_i})\right)^* \in \mathcal{F}\big(\mathbf{L}(S_{e_i})^*\big) \subseteq \mathcal{F}\big(\mathbf{L}(S)^*\big)$$

because the induction hypothesis shows that $\mathbf{L}(S_{e_i})^* \in \mathcal{F}\big(\mathbf{L}(S)^*\big)$ (Proposition I.5(1)).

(4) Since $F_i \in \mathcal{F}_e(S_{e_i})$ and S_{e_i} is a generating invariant Lie semigroup, the semigroup $S_{e_{i+1}}$ is a generating invariant Lie semigroup (Proposition IV.36). Proposition IX.12 implies that the unique extension

$$\varphi_{F_i} : S_{e_{i+1}} \to e_{F_i} S^\flat_{e_i}$$

induces a homomorphism $\varphi^\flat_{F_i}$ satisfying

$$\text{(9.1)} \quad \varphi_{F_i} \circ j_i = \lambda_{e_{F_i}} \circ i_{S_{e_i}}, \quad j^\flat_i \circ \varphi^\flat_{F_i} = \mathrm{id}_{S^\flat_{e_{i+1}}}, \qquad \text{and} \qquad I^\flat_g \circ \varphi^\flat_{F_i} = \varphi^\flat_{F_i} \circ I^\flat_g,$$

where $j_i : S_{e_i} \to S_{e_{i+1}}$ is the inclusion. We define

$$\varphi_{e_{i+1}} := \varphi^\flat_{e_i} \circ \varphi_{F_i} : S_{e_{i+1}} \to \varphi^\flat_{e_i}(e_{F_i} S^\flat_{e_i}) = e_{i+1} S^\flat.$$

(a) $\varphi_{e_{i+1}}$ extends $\lambda_{e_{i+1}} \circ i_S$:

$$\begin{aligned} \varphi_{e_{i+1}} \circ j_{e_{i+1}} &= \varphi^\flat_{e_i} \circ \varphi_{F_i} \circ j_i \circ j_{e_i} \\ &= \varphi^\flat_{e_i} \circ \lambda_{e_{F_i}} \circ i_{S_{e_i}} \circ j_{e_i} \\ &= \lambda_{e_{i+1}} \circ \varphi^\flat_{e_i} \circ i_{S_{e_i}} \circ j_{e_i} = \lambda_{e_{i+1}} \circ \varphi_{e_i} \circ j_{e_i} \\ &= \lambda_{e_{i+1}} \circ \lambda_{e_i} \circ i_S = \lambda_{e_{i+1}} \circ i_S. \end{aligned}$$

(b)

$$\begin{aligned} j^b_{e_{i+1}} \circ \varphi^b_{e_{i+1}} &= j^b_{e_{i+1}} \circ \varphi^b_{e_i} \circ \varphi^b_{F_i} \\ &= j^b_i \circ j^b_{e_i} \circ \varphi^b_{e_i} \circ \varphi^b_{F_i} = j^b_i \circ \varphi^b_{F_i} = \mathrm{id}_{S^b_{e_{i+1}}}. \end{aligned}$$

(c) Using (9.1) and (4c) for e_i, we obtain

$$I^b_g \circ \varphi^b_{e_{i+1}} = I^b_g \circ \varphi^b_{e_i} \circ \varphi^b_{F_i} = \varphi^b_{e_i} \circ \varphi^b_{F_i} \circ I^b_g = \varphi^b_{e_{i+1}} \circ I^b_g.$$

This completes the proof of (4).

(5) Let $\alpha : S_{e_{i+1}} \to e_{i+1}S^b$ be a morphism with

$$\alpha \circ j_{e_{i+1}} = \lambda_{e_{i+1}} \circ i_S.$$

To see that $\alpha = \varphi_{e_{i+1}}$, we prove by induction over j that

$$\alpha|_{S_{e_j}} = \lambda_{e_{i+1}} \circ \varphi_{e_j}.$$

This is true for $j = 0$ since $\varphi_{e_0} = i_S$. For $j = i+1$ this will show that $\alpha = \lambda_{e_{i+1}} \circ \varphi_{e_{i+1}} = \varphi_{e_{i+1}}$. Suppose that $\alpha|_{S_{e_j}} = \lambda_{e_{i+1}} \circ \varphi_{e_j}$ and that $j \leq i$. Then

$$\alpha(F_j) = e_{i+1}\varphi_{e_j}(F_j) \subseteq e_{i+1} \cdot \downarrow\varphi^b_{e_j}(e_{F_j}) = e_{i+1} \cdot \downarrow e_{j+1} \subseteq H(e_{i+1}).$$

Consequently

$$\begin{aligned} \alpha(sf^{-1}) = \alpha(s)\alpha(f)^{-1} &= \lambda_{e_{i+1}}\big(\varphi_{e_j}(s)\varphi_{e_j}(f)^{-1}\big) \\ &= \lambda_{e_{i+1}}\varphi_{e_{j+1}}(sf^{-1}). \end{aligned}$$

Thus $\alpha|_{S_{e_{j+1}}} = \lambda_{e_{i+1}} \circ \varphi_{e_{j+1}}$ because both homomorphisms agree on the dense subsemigroup $S_{e_j}F_j^{-1}$.

(6) Let $\widetilde{S} \supseteq S$ be a generating invariant Lie semigroup such that there exists a homomorphism $\varphi : \widetilde{S} \to eS^b$ with

$$\varphi \circ \widetilde{j} = \lambda_e \circ i_S \quad \text{and} \quad \widetilde{j}^b \circ \varphi^b = \mathrm{id}_{\widetilde{S}^b},$$

where $\widetilde{j} : S \to \widetilde{S}$ is the inclusion. We prove by induction over $k = 0, \ldots, n$ that $S_{e_k} \subseteq \widetilde{S}$ and that the inclusion $\widetilde{j}_k : S_{e_k} \to \widetilde{S}$ satisfies

$$\varphi \circ \widetilde{j}_k = \lambda_e \circ \varphi_{e_k} \quad \text{and} \quad \widetilde{j}^b_k \circ (\varphi^b_{e_k})^{-1} \circ \varphi^b = \mathrm{id}_{\widetilde{S}^b}. \tag{9.2}$$

For $k = 0$ this holds by assumption since $\varphi_{e_0} = i_S$. Suppose that $k \leq n-1$ and that the assertion holds for $k \leq i$. Then there exists an element $y \in \mathbf{L}(S_{e_k})$ with $\varphi^b_{e_k}(e_y) = e_{k+1}$. We set

$$\widetilde{e}_y := (\widetilde{j}_k \circ \gamma_y)^b(\omega) = \widetilde{j}^b_k(e_y).$$

This is the idempotent of $\widetilde{S}^\flat$ reached by the one-parameter semigroup $\exp \mathbb{R}^+ y \subseteq S_{e_k} \subseteq \widetilde{S}$. Then

$$\varphi^\flat(\widetilde{e}_y) = \varphi^\flat \circ \widetilde{j}_k^\flat(e_y) = \lambda_e \circ \varphi_{e_k}^\flat(e_y) = \lambda_e(e_{k+1}) = e$$

and therefore

$$\begin{aligned} \widetilde{e}_y &= \widetilde{j}_k^\flat \circ (\varphi_{e_k}^\flat)^{-1} \circ \varphi^\flat(\widetilde{e}_y) \\ &= \widetilde{j}_k^\flat \circ (\varphi_{e_k}^\flat)^{-1}(e) \\ &= \widetilde{j}_k^\flat \circ (\varphi_{e_k}^\flat)^{-1} \circ \varphi^\flat(\mathbf{1}) \\ &= \mathbf{1}_{\widetilde{S}^\flat}. \end{aligned}$$

This implies with Proposition IX.7 that $y \in H(\mathbf{L}(\widetilde{S}))$. Thus $H(\widetilde{S}) \cap S_{e_k}$ is a normal exposed face of S_{e_k} containing $\exp y$. So the definition of $S_{e_{k+1}}$ shows that $S_{e_{k+1}} \subseteq \widetilde{S}$ since

$$F_k \subseteq H(\widetilde{S}) \quad \text{and} \quad S_{e_k} \subseteq \widetilde{S}.$$

Then, in view of (9.2), we have for $s \in S_{e_k}$ and $f \in F_k$ that

$$\begin{aligned} \varphi \circ \widetilde{j}_{k+1}(sf^{-1}) &= \varphi\big(\widetilde{j}_{k+1}(s) \cdot \widetilde{j}_{k+1}(f)^{-1}\big) \\ &= \varphi \circ \widetilde{j}_k(s) \cdot \varphi \circ \widetilde{j}_k(f)^{-1} \\ &= e\varphi_{e_k}(s) \cdot \big(e\varphi_{e_k}(f)\big)^{-1}. \end{aligned}$$

For $s \in S_{e_k}$ this leads to

$$\begin{aligned} e\varphi_{e_{k+1}}(s) &= e\varphi_{e_k}^\flat \circ \varphi_{F_k} \circ j_k(s) = e\varphi_{e_k}^\flat\big(e_{F_k} i_{S_{e_k}}(s)\big) \\ &= ee_{k+1}\varphi_{e_k}(s) = e\varphi_{e_k}(s). \end{aligned}$$

Hence

$$\varphi \circ \widetilde{j}_{k+1}(sf^{-1}) = e\varphi_{e_{k+1}}(s)\big(e\varphi_{e_{k+1}}(f)\big)^{-1} = \lambda_e \circ \varphi_{e_{k+1}}(sf^{-1}).$$

Therefore $\varphi \circ \widetilde{j}_{k+1} = \lambda_e \circ \varphi_{e_{k+1}}$. Moreover, in view of (9.1), we have

$$\begin{aligned} \widetilde{j}_{k+1}^\flat \circ \big(\varphi_{e_{k+1}}^\flat\big)^{-1} \circ \varphi^\flat &= \widetilde{j}_{k+1}^\flat \circ (\varphi_{F_k}^\flat)^{-1} \circ (\varphi_{e_k}^\flat)^{-1} \circ \varphi^\flat \\ &= \widetilde{j}_{k+1}^\flat \circ j_k^\flat \circ (\varphi_{e_k}^\flat)^{-1} \circ \varphi^\flat \\ &= (\widetilde{j}_{k+1} \circ j_k)^\flat \circ (\varphi_{e_k}^\flat)^{-1} \circ \varphi^\flat \\ &= \widetilde{j}_k^\flat \circ (\varphi_{e_k}^\flat)^{-1} \circ \varphi^\flat \\ &= \mathrm{id}_{\widetilde{S}^\flat}. \end{aligned}$$

So we have shown that (9.2) holds.

For $k = n$ equation (9.2) shows that $S_e = S_{e_n} \subseteq \widetilde{S}$, that

$$\varphi \circ \widetilde{j}_n = \lambda_e \circ \varphi_{e_n} = \lambda_e \circ \varphi_e = \varphi_e,$$

and $\widetilde{j}_n^\flat \circ (\varphi_{e_n}^\flat)^{-1} \circ \varphi^\flat = \mathrm{id}_{\widetilde{S}^\flat}$. This implies that

$$\varphi^\flat \circ \widetilde{j}_n^\flat = \varphi_e^\flat : S_e^\flat \to eS^\flat$$

and therefore that $\widetilde{j}_n^\flat$ is an isomorphism of $S_e^\flat$ and $\widetilde{S}^\flat$. Now Lemma IX.23 shows that $S_e = \widetilde{S}$.

(7) This is a direct consequence of (9.1). ∎

Theorem IX.25. *Let $e \in E(S^\flat)$, then the following assertions hold:*

(1)
$$sH(e) = H(e)s \quad \text{for all} \quad s \in S^\flat.$$

(2) $F_e \in \mathcal{F}_n(S)$ *and consequently*
$$\mathcal{F}_c(S) = \mathcal{F}_n(S).$$

(3) $F_e \subseteq H(S_e)$.

Proof. (1) For $s \in S$ we conclude with Theorem IX.24 that
$$i_S(s)H(e) = I_s^\flat\big(H(e)\big)i_S(s) = H\big(I_s^\flat(e)\big)i_S(s) = H(e)i_S(s).$$
This leads to $sH(e) = H(e)s$ for every $s \in S^\flat$ since $H(e)$ is compact.
(2) Let $F = F_e$ with $e = e_F$. Then $I_g(F_e) = F_{I_g^\flat(e)} = F_e$ and therefore $F \subseteq G$ is an invariant subsemigroup. Hence $T_F(S)$ is a normal subgroup and $F = T_F(S) \cap S \in \mathcal{F}_n(S)$.
(3) We know that
$$\varphi_e^\flat\big(i_S(F_e)\big) = \varphi_e(F_e) = \lambda_e \circ i_S(F_e) \subseteq \lambda_e(\downarrow e) \subseteq H(e).$$
Therefore
$$i_{S_e}(F_e) \subseteq H(S_e^\flat).$$
Hence $F_e \subseteq i_{S_e}^{-1}(\downarrow 1) = H(S_e)$ (Proposition IX.7). ■

Corollary IX.26. *The mapping*
$$\big(\mathcal{F}_n(S), \subseteq\big) \to \big(E_a(S^\flat), \prec\big), \quad F \mapsto e_F$$
is an order isomorphism which preserves arbitrary sups. For two faces $F_1, F_2 \in \mathcal{F}_n(S)$ we have that
$$F_1 \vee F_2 = T_{F_1 F_2}(S) \cap S$$
and $e_{F_1} e_{F_2} = e_{F_1} \vee e_{F_2}$ is the maximal idempotent in the subsemigroup $\overline{i(F_1 F_2)}$ of $S^\flat$.

Proof. This follows from Theorem IX.8 and Theorem IX.25. ■

Remark IX.27. If $F \in \mathcal{F}_n(S)$ and $x \in \operatorname{algint} \mathbf{L}(F)$ we have seen in Lemma IX.11 that $e_x = e_{F'}$, where F' is the smallest exposed face of S containing $\exp x$. Moreover $\mathbf{L}(F') = \mathbf{L}(F)$. So we have two normal exposed faces $F' \subseteq F$ with the same tangent wedge. The idempotent
$$e_x = e_{F'} \prec e_F$$
is the maximal directly accessible idempotent in $\overline{i(F)}$. So the following questions are equivalent:
(1) $E_{da}(S^\flat) \neq E_a(S^\flat)$?
(2) $\big(\exists F \neq F' \in \mathcal{F}_n(S)\big)\, \mathbf{L}(F) = \mathbf{L}(F')$?

Since we do not know whether such examples exist or not, we cannot show that every accessible idempotent is directly accessible as is true in the abelian case because all faces of a wedge are wedges. ■

Remark IX.28. To see that in general $E(S^\flat) \neq E_a(S^\flat)$, we consider an abelian example. Let $G = \mathbb{R}^n$ and $W = S \subseteq \mathbb{R}^n$ be a pointed generating wedge such that

$$\mathcal{F}(W^*) \neq \mathcal{F}_e(W^*)$$

(cf. Example I.6). We will prove in Section XI (Corollary XI.17) that an idempotent $e \in W^\flat$ is accessible if and only if W_e^* is an exposed face of W^* and that the mapping

$$\delta : E(W^\flat) \to \mathcal{F}(W^*), \quad e \mapsto W_e^*$$

is a bijection. So we have non-accessible idempotents in $W^\flat$ iff all faces of W^* are exposed. Having this in mind, we can easily describe how non-accessible idempotents occur. Let $E \in \mathcal{F}(W^*)$ be non-exposed such that E is an exposed face of $\widetilde{E}$, the exposed face generated by E (Proposition I.5). We set $F := E^\perp \cap W \in \mathcal{F}_e(W)$ and $\widetilde{W} := L_F(W)$. Then $\widetilde{W}^* = W^* \cap F^\perp = \widetilde{E}$ (Proposition I.4). Since E is exposed in $\widetilde{E}$, we find an element $x \in \widetilde{W}$ such that

$$x^\perp \cap \widetilde{E} = E.$$

We contend that $x \notin F$. If $x \in F \subseteq W$, then $x^\perp \cap W^*$ is an exposed face of W^* which contains E and therefore $\widetilde{E}$. This contradicts the assumption that $E \neq \widetilde{E}$. Now we claim that $e_x \in E(W^\flat) \setminus E_a(W^\flat)$. If this is not so, there exists an element $y \in W$ such that $e_x = e_y$. Consequently x and y generate the same exposed face $\widetilde{F}$ of $\widetilde{W}$. We conclude that

$$E = x^\perp \cap \widetilde{E} = \operatorname{op}(x) = \operatorname{op}(y) = y^\perp \cap \widetilde{E}.$$

Again this is contradicts $E \neq \widetilde{E}$ since $y \in W$. ■

Now we have all the information which is necessary to give a very concrete description of $S^\flat$.

Definition IX.29. For an idempotent $e \in E(S^\flat)$ we define

$$S(e) := \varphi_e^\flat\big(S_e(\mathbf{1})\big) = \varphi_e\big(\widetilde{\operatorname{comp}}(S_e)\big)H(e).$$

We write $\operatorname{rank}(e)$ for the length of a Jordan-Hölder series of the quotient Lie algebra $\mathbf{L}\left(H(S_e)\right)/\mathbf{L}\left(H(S)\right)$ or equivalently the dimension of the radical plus the number of simple ideals in a Levi subalgebra. This is the length of a maximal chain of ideals between $\mathbf{L}\left(H(S)\right)$ and $\mathbf{L}\left(H(S_e)\right)$ in $\mathbf{L}(G)$.

Lemma IX.30. *For $e \prec f$ with $e \neq f$ in $E(S^\flat)$ we have that*

$$\operatorname{rank} e < \operatorname{rank} f.$$

Proof. Since $\mathbf{L}\left(H(S_e)\right) \subseteq \mathbf{L}\left(H(S_f)\right)$ we only have to show that $H(S_e) \neq H(S_f)$. In view of Theorem IX.23(3) we may assume that $e = \mathbf{1}$. Now we use Lemma IX.22 to find $x \in \mathbf{L}(S) \setminus H\big(\mathbf{L}(S)\big)$ such that $\mathbf{1} \neq e_x \prec f$. Then $x \in \mathbf{L}(F_f) \subseteq H\big(\mathbf{L}(S_f)\big)$ (Lemma IX.10, Theorem IX.25) satisfies $x \notin H\big(\mathbf{L}(S_e)\big) = H\big(\mathbf{L}(S)\big)$. ■

Corollary IX.31. *Let e_S be the maximal idempotent in $S^\flat$. Then every chain of idempotents has at most length* $\operatorname{rank}(e_S)$, *or equivalently, contains at most* $\operatorname{rank}(e_S)+1$ *elements.*

Proof. This is a consequence of Lemma IX.30. ■

Theorem IX.32. (The Structure Theorem) *Let (S,G) be a generating invariant Lie semigroup and $S^\flat$ its Bohr compactification. Then the following assertions hold:*

(1) $S^\flat = \bigcup_{e\in E(S^\flat)} S(e)$.

(2) $S(e)\cap S(f)=\emptyset$ *for* $e\neq f$.

(3) $S(e)=\{s\in S^\flat : E(s)=e\}$.

Proof. (1) Let $s\in S^\flat$ and $e=E(s)$. Then $s\in eS^\flat\cong S_e^\flat$. Using Proposition IX.21 we find that $s\in\varphi_e^\flat(S_e(\mathbf{1}))=S(e)$ because $E\big((\varphi_e^\flat)^{-1}(s)\big)=\mathbf{1}_{S_0^\flat}$.

(2) Let $s\in S(e)\cap S(f)$ for $e,f\in E(S^\flat)$. Then $ef\in E(S^\flat)$ and $efs=s$. So $e\prec ef\prec s$. Hence $(\varphi_e^\flat)^{-1}(ef)\in E(S_e^\flat)\cap S_e(\mathbf{1})=\{\mathbf{1}\}$ (Lemma IX.20). This shows that $ef=e$ since $\varphi_e^\flat : S_e^\flat\to eS_e^\flat$ is an isomorphism, consequently $f\prec e$. By symmetry we find that $e\prec f$ and therefore $e=f$.

(3) We have already proved in (1) that $E(s)=e$ implies that $s\in S(e)$. Let $s\in S(e)$ and suppose that $E(s)=f$. Then $s\in S(f)\cap S(e)$ and $e=f$ follows with (2). ■

Remark IX.33. If $\mathbf{L}(G)$ is a compact Lie algebra, then it follows from Corollary VI.26 that $\widetilde{\operatorname{comp}}(S_e)=S_e$ for all $e\in E(S^\flat)$. Hence

$$S(e)=\varphi_e^\flat\big(S_e(\mathbf{1})\big)=\varphi_e(S_e)H(e)$$

is a subsemigroup of $eS^\flat$ which contains $ei_S(S)$. Therefore it is dense in $eS^\flat$. This holds in particular in the abelian case ([Rup87, 5.3]).

We use Example VI.24 to see that $S(e)$ is not always dense in $eS^\flat$. In this case $G=\mathcal{O}_1$ is the four-dimensional oscillator group and there exists a homomorphism $\varphi : G\to\mathbb{R}$ such that

$$\operatorname{comp}(S)=\varphi^{-1}([0,\pi[).$$

According to Proposition IV.12 this homomorphism corresponds to the exposed normal face $F:=\ker\varphi\cap S$ of S since $\varphi:(S,G)\to(\mathbb{R}^+,\mathbb{R})$ is a morphism of Lie semigroups. We consider $\varphi^\flat : S^\flat\to(\mathbb{R}^+)^\flat$. Then $\varphi^\flat$ is surjective and

$$\varphi^\flat\big(S(\mathbf{1})\big)=\varphi^\flat\Big(i_S\big(\operatorname{comp}(S)\big)\Big)=\varphi\big(\operatorname{comp}(S)\big)=[0,\pi[$$

is not dense in $\mathbb{R}^+$. This shows that $S(\mathbf{1})$ is not dense in $S^\flat$. ■

Proposition IX.34. *$S(e)$ is open in $eS^\flat$ and*

$$\varphi_e^\flat(S_e \setminus \widetilde{\mathrm{comp}}(S_e)) \cap S(e) = \emptyset.$$

Proof. Since $\varphi_e^\flat : S_e^\flat \to eS^\flat$ is an isomorphism, we may assume that $e = \mathbf{1}$. In view of Lemma IX.19 we also may assume that $H(S) = \{\mathbf{1}\}$. Now we use Theorem V.9 to find a surjective morphism $\varphi : S \to K$ onto a compact monoid such that $\varphi(S \setminus \mathrm{comp}(S))$ is a point and $\varphi(\mathrm{comp}(S))$ is its open complement. Let $\varphi^\flat : S^\flat \to K$ be the induced homomorphism. Then

$$\varphi^\flat(S(\mathbf{1})) = \varphi(\mathrm{comp}(S)).$$

If $s \in S^\flat$ with $\varphi^\flat(s)$ in $\varphi(\mathrm{comp}(S))$, then we find a net $(s_n)_{n\in D}$ in S such that $\lim i_S(s_n) = s$. Now $\varphi^\flat(i_S(s_n)) = \varphi(s_n) \to \varphi^\flat(s)$, so we may assume that $\varphi(s_n) \in \varphi(\mathrm{comp}(S))$ which is an open subset of K. If $s \notin S(\mathbf{1}) = i_S(\mathrm{comp}(S))$, then s_n eventually leaves every compact subset of $\mathrm{comp}(S)$ and therefore $\varphi(s_n)$ tends to $\varphi(t)$ for every element $t \in S \setminus \mathrm{comp}(S)$. This is impossible since $\varphi^\flat(s) \in \varphi(\mathrm{comp}(S))$. Thus $s \in S(\mathbf{1})$ and

$$(\varphi^\flat)^{-1}\Big(\varphi(\mathrm{comp}(S))\Big) = S(\mathbf{1})$$

is open in $S^\flat$. ∎

An interesting consequence of the previous proposition is the fact that the continuous homomorphisms of $E(S^\flat)$ onto the multiplicative idempotent compact semigroup $\{0,1\}$ separate the points.

Proposition IX.35. *For $e \neq f$ in $E(S^\flat)$ there exists a continuous morphism $\alpha : E(S^\flat) \to \{0,1\}$ such that $\alpha(e) \neq \alpha(f)$.*

Proof. If $e \not\prec f$, then $ef \neq f$ and $\lambda_e : E(S^\flat) \to E(S^\flat)$ is a continuous homomorphism. Hence we may assume that $e \prec f$. Since $eS^\flat \cong S_e^\flat$ we may also assume that $e = \mathbf{1}$ and $f \neq \mathbf{1}$. Then $E(S^\flat) \setminus \{\mathbf{1}\}$ is a closed ideal in $E(S^\flat)$ (Proposition IX.34) because $E(S^\flat) \setminus \{\mathbf{1}\} = E(S^\flat) \setminus S(\mathbf{1})$ (Lemma IX.20) and $S(\mathbf{1})$ is open in $S^\flat$. Setting

$$\alpha(e) = \begin{cases} 0, & \text{if } e \neq \mathbf{1} \\ 1, & \text{if } e = \mathbf{1} \end{cases}$$

we obtain a continuous homomorphism $\alpha : E(S^\flat) \to \{0,1\}$ separating $\mathbf{1}$ and f. ∎

The following theorem describes the topology of $S^\flat$. It carries over, almost word for word, from the abelian case. So the proof is nearly the same as in [Rup87, 5.6].

Theorem IX.36. (The Topology of $S^\flat$)

(1) *A net $(s_n)_{n\in D}$ converges to a point $s \in S(e)$ iff for every subnet (s_m) the following conditions hold:*

(a) *The net (es_m) eventually lies in $S(e)$ and converges to s.*

(b) *e is minimal with respect to* (a).

(2) *A subbasis of the topology of $S^\flat$ is given by the sets $\lambda_e^{-1}(U)$, where U is open in $S(e)$, $e \in E(S^\flat)$; together with the complements of the sets $\lambda_e^{-1}(A)$, where A is compact in $S(e)$, $e \in E(S^\flat)$.*

Proof. (1) It is obvious that $\lim s_n = s \in S(e)$ implies (a) for every subnet (s_m) of (s_n) since $S(e)$ is open in $eS^\flat$ (Proposition IX.34). Suppose that for some $e' \in E(S^\flat)$ we have that $\lim e's_n = e's \in S(e')$. Then $e(e's) = e'(es) = e's$, hence $S(e') \cap eS^\flat \neq \emptyset$ and therefore $e \prec e'$. Thus the condition is necessary. Suppose that, conversely (a) and (b) hold for every subnet (s_m) of (s_n). Choose a subnet s_m which converges to an element s', say in $S(e')$. Then (a) and (b) hold with e' instead of e for the net (s_m). Hence $e \prec e'$ and $e' \prec e$. We conclude that $e = e'$ and $s = s'$. Since $S^\flat$ is compact, this implies $\lim s_n = s$.
(2) Since the set $S(e)$ is open in $eS^\flat$ (Proposition IX.34) and since the map λ_e is continuous for every $e \in E(S^\flat)$, it is obvious that the sets $\lambda_e^{-1}(U)$, U open in $S(e)$, are open, and that the sets $\lambda_e^{-1}(A)$ are closed in $S^\flat$. Thus it suffices to show that the sets $\lambda_e^{-1}(U)$ and $S^\flat \setminus \lambda_e^{-1}(A)$ generate a Hausdorff topology (note that $S^\flat$ is compact). Let $x \in S(e)$, $y \in S(f)$ be distinct points in $S^\flat$. There are three cases:

Case 1) $x \neq ey \in S(e)$: Then we find disjoint open sets $U_1, U_2 \subseteq S(e)$ with $x \in U_1$ and $ey \in U_2$. Hence $\lambda_e^{-1}(U_1)$ and $\lambda_e^{-1}(U_2)$ separate x and y.

Case 2) $x \neq ey \notin S(e)$: Let $U \subseteq S(e)$ be a relatively compact open set with $U \cap (eS^\flat \setminus S(e)) = \emptyset$ and $x \in U$. Hence $\lambda_e^{-1}(U)$ and the complement of $\lambda_e^{-1}(\overline{U})$ separate x and y.

Case 3) $x = ey$: Then $ef \prec ey \in S(e)$ and therefore $ef \prec e$. This implies that $f \prec e$. The fact that $y \neq x$ shows that $e \neq f$. So $x \in fS^\flat \setminus S(f)$ and we find a relatively compact open set $U \subseteq S(f)$ with $U \cap (fS^\flat \setminus S(f)) = \emptyset$. Hence $\lambda_f^{-1}(U)$ and the complement of $\lambda_f^{-1}(\overline{U})$ separate x and y. ■

Remark IX.37. (cf. [Rup87, 5.7]) It follows from the above theorem that a subbase for the topology on $E(S^\flat)$ consists of the sets $\downarrow e$ which are closed and open. So $E(S^\flat)$ is totally disconnected and thus zero-dimensional. This implies with [CHK86, p.14] that $E(S^\flat)$ is a Lawson-semilattice and therefore a continuous lattice with the Lawson topology ([CHK86, p.26]). ■

Corollary IX.38. *The set of accessible idempotents $E_a(S^\flat)$ is closed in $E(S^\flat)$.*

Proof. (cf. [Rup87, 5.8]). Let $(e_n)_{n\in D}$ be a net with $e_n \to e$ and $e_n \in E_a(S^\flat)$. According to Remark IX.37 there exists n_0 such that $e_n \prec e$ for $n \geq n_0$. If $e_{n_0} = e$ the assertion follows. If not, we find $n_1 \geq n_0$ with $e_{n_0}e_{n_1} \neq e_{n_0}$. Repeating this argument we find an infinite chain

$$e_{n_0} \prec e_{n_0}e_{n_1} \prec e_{n_0}e_{n_1}e_{n_2} \prec \ldots \prec e$$

of mutually distinct idempotents. But by Lemma IX.30 every chain of idempotents in $S^\flat$ is finite. So e is accessible. ∎

At this stage of our investigation we are left with the following two problems.

(1) What is the structure of $S(e)$? Since we may consider the set $S_e \setminus S(e)$ as known because it sits in the Lie group G containing S, we have to consider the groups $H(e) = H(eS^\flat) \cong H(S_e^\flat)$.

(2) How can we describe the lattice $E(S^\flat)$?

The answers to both questions will be given in Sections X and XI.

X. THE UNIT GROUP OF $S^\flat$

We have seen in the last section that the Bohr compactification $S^\flat$ of a generating invariant Lie semigroup (S, G) is built up as a disjoint union of the subsets $S(e) = S_e(\mathbf{1})$. The equivalence classes

$$[s] := [s, s] = \{t \in S^\flat : s \prec t \prec s\}$$

of the elements in this set are exactly the translates of the groups $H(e)$ (Proposition IX.16). So we are interested in these groups. In view of the isomorphism $S(e) \cong S_e(\mathbf{1})$, it is clear that it suffices to assume that $e = \mathbf{1}$ and to consider $H(\mathbf{1}) = H(S^\flat)$. We know already that $i_S\big(H(S)\big)$ is dense in $H(S^\flat)$ (Proposition IX.16). Hence $H(S^\flat)$ is a homomorphic image of $H(S)^\flat$. As we will see below it may be substantially smaller. We determine the ideal corresponding to the kernel of the mapping $i_S\,|_{H(S)}$ in the Lie algebra $\mathbf{L}(G)$. Then we consider a reduced situation and describe how to get $H(S^\flat)$ from $H(S)$. We start with the group case.

Proposition X.1. *Let G be a connected Lie group and $\mathfrak{g} = \mathbf{L}(G)$ its Lie algebra. Then $\mathfrak{g}$ contains a minimal ideal I such that $\mathfrak{g}/I$ is a compact Lie algebra. Suppose that $N := \overline{\langle \exp I \rangle}$ and $i_G : G \mapsto G^\flat$ is the Bohr compactification of G. Then $N = \ker i_G$ and $(G/N)^\flat \cong G^\flat$. The group G/N is a direct product of a vector group V and a compact group K:*

$$G/N \cong V \times K.$$

Proof. Suppose that $I_1, I_2 \subseteq \mathfrak{g}$ are ideals such that $\mathfrak{g}/I_1$ and $\mathfrak{g}/I_2$ are compact. Then the Lie algebra $\mathfrak{g}_0 := \mathfrak{g}/I_1 \oplus \mathfrak{g}/I_2$ is compact and the kernel of the homomorphism $\varphi : x \mapsto (x + I_1, x + I_2), \mathfrak{g} \mapsto \mathfrak{g}_0$ agrees with $I_1 \cap I_2$. Hence $\mathfrak{g}/(I_1 \cap I_2)$ is compact because subalgebras of compact Lie algebras are compact. We conclude that there exists a minimal ideal $I \subseteq \mathfrak{g}$ such that $\mathfrak{g}/I$ is compact. It is clear that the group $G/\ker i_G$ is injectable into a compact group, hence $G/\ker i_G \cong V \times K$, where V is a vector group and K is compact ([Di64, p.303]). Therefore

$$\mathbf{L}(G/\ker i_G) \cong \mathbf{L}(G)/\mathbf{L}(\ker i_G)$$

is compact, hence $I \subseteq \mathbf{L}(\ker i_G)$ and $N \subseteq \ker i_G$. Set $G_0 := G/N$. Then $\mathbf{L}(G_0) = \mathbf{L}(G)/\mathbf{L}(N)$ is a compact Lie algebra. Let $K_0 \subseteq G_0$ be a maximal compact subgroup. Then $G_0/K_0 \cong \mathbb{R}^n$ ([Ho65, p.180) and the fact that

$$\mathbf{L}(G) = \mathbf{L}(G)' \oplus Z\big(\mathbf{L}(G)\big)$$

permits us to choose an analytic subgroup $V \subseteq Z(G_0)$ such that $\mathbf{L}(G_0) = \mathbf{L}(K_0) \oplus \mathbf{L}(V)$. Hence V is closed because $\overline{\exp \mathrm{I\!R} x} = \exp \mathrm{I\!R} x$ for every $x \in \mathbf{L}(V)$ ([Ho65, p.192]). Consequently $G_0 \cong K_0 \times V$ and G_0 is injectable into a compact group ([Di64, p.303]). This proves that $\ker i_G \subseteq N$. Putting these facts together, we have

$$\ker i_G = N, \quad (G/N)^\flat \cong G^\flat, \quad \text{and} \quad G_0 = G/N \cong K_0 \times V_0.$$

■

Example X.2. Let $G_1 := \mathrm{Sl}(2, \mathrm{I\!R})^\sim$ and $x \in \mathrm{sl}(2, \mathrm{I\!R}) = \mathbf{L}(G_1)$ such that $Z(G_1) \subseteq \exp(\mathrm{I\!R} x)$. Further we take a 2-torus $\mathrm{T\!T}^2$ and a generator $x' \in \mathbf{L}(\mathrm{T\!T}^2)$ of a dense one-parameter sungroup. We set

$$G := (G_1 \times \mathrm{T\!T}^2)/\exp\big(\mathrm{I\!R}(x, -x')\big).$$

Then $\mathbf{L}(G) \cong \mathrm{sl}(2, \mathrm{I\!R}) \oplus \mathrm{I\!R}$, $I = \mathrm{sl}(2, \mathrm{I\!R})$ is the smallest ideal such that $\mathbf{L}(G)/I$ is compact, and

$$N := \overline{\langle \exp I \rangle} = G \neq \langle \exp I \rangle.$$

Therefore $G^\flat \cong \{\mathbf{1}\}$. ■

Lemma X.3. *Let (S, G) be a generating invariant Lie semigroup and suppose that $i_{H(S)}$ is injective. Then*

$$[G, Z(H(S))_0] = \Big\langle \exp \Big[Z\Big(H\big(\mathbf{L}(S)\big)\Big), \mathbf{L}(G)\Big]\Big\rangle \subseteq \ker i_S.$$

Proof. The subgroup $Z := Z\big(H(S)\big) \subseteq H(S)$ is characteristic, hence normal in G. According to [Ho65, p.138] we have that $N_1 := [Z_0, G]$ is the analytic subgroup of G with Lie algebra

$$\mathbf{L}(N_1) = \big[\mathbf{L}(Z), \mathbf{L}(G)\big].$$

We have to show that $i_S \circ \exp([z, x]) = 0$ for $z \in \mathbf{L}(Z)$ and $x \in \mathbf{L}(G)$. We clearly may assume that $x \in \mathbf{L}(S)$ because $\mathbf{L}(S) - \mathbf{L}(S) = \mathbf{L}(G)$. We choose a net $t_n \in \mathrm{I\!R}^+$ such that $\lim i_S \circ \exp(t_n z) = 1 \in H(S^\flat)$ and $\lim t_n = \infty$. We conclude that

$$\begin{aligned}
\mathbf{1} &= \lim i_S\big(\exp -t_n z\big) i_S\big(\exp \frac{x}{t_n}\big) i_S\big(\exp t_n z\big) \\
&= \lim i_S\big(\exp -t_n z\big) i_S\big(I_{\exp \frac{x}{t_n}}(\exp t_n z)\big) i_S\big(\exp \frac{x}{t_n}\big) \\
&= \lim i_S\big(\exp -t_n z\big) i_S\big(\exp(t_n e^{\frac{1}{t_n} \operatorname{ad} x} z)\big) \\
&= \lim i_S\big(\exp(-t_n z + t_n e^{\frac{1}{t_n} \operatorname{ad} x} z)\big) \\
&= \lim i_S\Big(\exp\big((e^{\frac{1}{t_n} \operatorname{ad} x} - \mathbf{1}) t_n z\big)\Big) \\
&= \lim i_S\big(\exp([x, z])\big) = i_S\big(\exp([x, z])\big).
\end{aligned}$$

■

Lemma X.4. *Let H be a Lie group with compact Lie algebra, $K := [H,H]$ its compact semisimple commutator group, $Z := Z(H)_0$, and $i_Z : Z \to Z^\flat$ the Bohr compactification of Z. Then*

$$H^\flat \cong (H \times Z^\flat)/N,$$

where $N := \{(z, i_Z(z)^{-1}) : z \in Z\}$.

Proof. Z is a closed subgroup of H, hence N is a closed central subgroup of $H \times Z^\flat$. Therefore $H_c := (H \times Z^\flat)/N$ is a locally compact group. Let $\varphi : H \times Z^\flat \to H_c$ be the quotient homomorphism and $h \in H$. Then, since $H = KZ$, we find $k \in K$ and $z \in Z$ with $h = kz$. Therefore

$$\varphi(h, \mathbf{1}) = \varphi(k, \mathbf{1})\varphi(z, \mathbf{1}) = \varphi(k, i_Z(z)) \subseteq \varphi(K \times Z^\flat)$$

and $K \times Z^\flat$ is a compact group. Consequently H_c is compact and there exists a surjective homomorphism

$$\psi : H^\flat \to H_c \quad \text{with} \quad \psi(i_H(h)) = \varphi(h, \mathbf{1}) \quad \text{for} \quad h \in H.$$

Let $\widetilde{\alpha} : H \times Z^\flat \to H^\flat$ be defined by $\widetilde{\alpha}(h,z) = i_H(h)j(z)$, where $j : Z^\flat \to H^\flat$ is induced by the injection $Z \to H$. Then

$$\widetilde{\alpha}(z, i_Z(z)^{-1}) = i_H(z) j(i_Z(z))^{-1} = i_H(z) i_H(z)^{-1} = \{\mathbf{1}\}$$

and we find a homomorphism $\alpha : H_c \to H^\flat$ with $\alpha \circ \varphi = \widetilde{\alpha}$. We claim that $\alpha \circ \psi = \mathrm{id}_{H^\flat}$ and $\psi \circ \alpha = \mathrm{id}_{H_c}$.

$$\begin{array}{ccc} H \times Z^\flat & \xrightarrow{\quad\widetilde{\alpha}\quad} & H^\flat \\ \downarrow{\scriptstyle \varphi} & & \downarrow{\scriptstyle \mathrm{id}_{H^\flat}} \\ H_c & \xleftarrow{\quad\psi\quad} & H^\flat \end{array}$$

(1) $\alpha \circ \psi = \mathrm{id}_{H^\flat}$: Let $h \in H$. Then

$$\alpha \circ \psi(i_H(h)) = \alpha \circ \varphi(h, \mathbf{1}) = \widetilde{\alpha}(h, \mathbf{1}) = i_H(h)$$

and therefore $\alpha \circ \psi = \mathrm{id}_{H^\flat}$ because $i_H(H)$ is dense in $H^\flat$.
(2) $\psi \circ \alpha = \mathrm{id}_{H_c}$: Let $\varphi(h, i_Z(z)) \in H_c$ with $h \in H$ and $z \in Z$. Then

$$\begin{aligned} \psi \circ \alpha\Big(\varphi\big(h, i_Z(z)\big)\Big) &= \psi \circ \widetilde{\alpha}\big(h, i_Z(z)\big) = \psi\Big(i_H(h) j\big(i_Z(z)\big)\Big) \\ &= \psi\big(i_H(h) i_H(z)\big) = \psi \circ i_H(hz) \\ &= \varphi(hz, \mathbf{1}) = \varphi\big(h, i_Z(z)\big). \end{aligned}$$

Now we use the density of $\varphi(H \times i_Z(Z))$ in H_c to complete the proof. ∎

Lemma X.5. *Let (S,G) be a generating invariant Lie semigroup such that $i_{H(S)}$ is injective and*

$$Z := Z\big(H(S)\big)_0 \subseteq Z(G).$$

Then

$$H(S^\flat) \cong H(S)^\flat.$$

Proof. Let $i_Z : Z \to Z^\flat$ be the Bohr compactification of Z. We set

$$G_c := (G \times Z^\flat)/N \quad \text{and} \quad \varphi : G \times Z^\flat \to G_c, g \mapsto gN,$$

where $N := \{(z, i_Z(z)^{-1}) : z \in Z\}$. The subgroup N is closed and central in the locally compact group $G \times Z^\flat$, hence

$$S_c := (S \times Z^\flat)/N$$

is a locally compact monoid. According to Lemma X.4 we have

$$H(S_c) = \varphi\big(H(S \times Z^\flat)\big) = \varphi\big(H(S) \times Z^\flat\big) \cong \big(H(S) \times Z^\flat\big)/N \cong H(S)^\flat.$$

We notice that

$$S/H(S) \cong (S \times Z^\flat)/H(S \times Z^\flat) \cong S_c/H(S_c).$$

Now Theorems V.8 and V.9 apply and show that $\operatorname{comp}(S_c)$ is open because

$$\operatorname{comp}(S_c)/H(S_c) = \operatorname{comp}\big(S_c/H(S_c)\big) = \operatorname{comp}\big(S/H(S)\big).$$

Moreover, we get that $\operatorname{comp}(S_c) \neq \emptyset$, the restriction from i_{S_c} to $\operatorname{comp}(S_c)$ is injective, and $i_{S_c}\big(\operatorname{comp}(S_c)\big)$ is an open neighborhood of $H(S_c^\flat)$ in $S_c^\flat$ such that $S_c^\flat \setminus i_{S_c}\big(\operatorname{comp}(S_c)\big)$ is a closed ideal. Collecting these facts we have a homomorphism

$$\psi : S \to S_c^\flat, \quad s \mapsto i_{S_c}\big(\varphi(s, \mathbf{1})\big)$$

of S into a compact monoid such that $\psi|_{H(S)}$ induces an embedding $H(S)^\flat \to H(S_c^\flat) \cong H(S_c)$. Therefore $i_S|_{H(S)}$ induces an embedding $H(S)^\flat \to H(S^\flat)$ and consequently $H(S^\flat) \cong H(S)^\flat$. ∎

Theorem X.6. (Unit Group Theorem for $S^\flat$) *Let (S,G) be a generating invariant Lie semigroup, $i_{H(S)} : H(S) \to H(S)^\flat$ the Bohr compactification of $H(S)$, $N = \ker i_{H(S)}$, $\varphi : G \to \widetilde{G} := G/N$ the quotient homomorphism and*

$$\widetilde{N} := \varphi^{-1}\Big(\Big[\widetilde{G}, Z\Big(H\big(\varphi(S)\big)\Big)_0\Big]\Big) \subseteq H(S).$$

Then the following assertions hold:

(1) $\widetilde{N} = \ker(i_S|_{H(S)})$.

(2) $H(S^\flat) \cong H(S/\widetilde{N})^\flat$.

(3) $i_{S/\widetilde{N}}|_{H(S/\widetilde{N})}$ *is injective.*

Proof. First we notice that $\ker i_{H(S)} \subseteq \ker i_S|_{H(S)}$. Hence $i_S : S \to S^b$ factors to a mapping $i_{S/N} : S/N \to S^b$ which is the Bohr compactifiaction of S/N and $i_{H(S/N)}$ is injective. Now Lemma X.3 implies that

$$\left[\widetilde{G}, Z\Big(H\big(\varphi(S)\big)\Big)_0\right] \subseteq \ker i_{S/N}$$

and therefore that $\widetilde{N} \subseteq \ker i_S|_{H(S)}$. By factorization we get a homomorphism $i_{S/\widetilde{N}} : S/\widetilde{N} \to S^b$. We have

$$\begin{aligned}[Z(H(S/\widetilde{N}))_0, G/\widetilde{N}] &\subseteq [Z(H(S/N))_0, G/N]/(\widetilde{N}/N)\\ &\subseteq (\widetilde{N}/N)/(\widetilde{N}/N) = \{1\}\end{aligned}$$

because the homomorphism

$$\alpha : G/N \to G/\widetilde{N}$$

maps $Z\big(H(S/N)\big)_0$ onto $Z\big(H(S/\widetilde{N})\big)_0$ and $\ker\alpha \subseteq Z\big(H(S/N)\big)_0$. Consequently

$$Z\big(H(S/\widetilde{N})\big)_0 \subseteq Z(G/\widetilde{N})$$

and Lemma X.5 applies and shows that (2) holds. Now (1) follows from the injectivity of $i_{S/\widetilde{N}}$ on $H(S/\widetilde{N})$. ∎

Note that this theorem reproves in particular that $i_S\big(H(S)\big) \subseteq H(S^b)$ is dense (Lemma IX.16).

Remark X.7. To see how the Bohr compactification of a Lie group with compact Lie algebra looks like, we take such a group G. Then $G = K \times V$, where K is a maximal compact subgroup and V is a central vector group. Then

$$G^b \cong K^b \times V^b \cong K \times (\mathbb{R}^b)^{\dim V}$$

because the Bohr compactification preserves products. In view of Proposition X.1 this gives a picture of all Bohr compactifications of Lie groups as products of compact Lie groups and factors isomorphic to $\mathbb{R}^b$.

To get a clearer picture of $\mathbb{R}^b$ we recall that the Bohr compactification of a locally compact abelian group may be obtained by the mapping

$$i_G : G \to (\widehat{G}_d)\widehat{\ },$$

where $\widehat{G}_d$ is the character group of G endowed with the discrete topology. The character group of this group is compact and isomorphic to G^b. The mapping i_G is give by evaluation:

$$i_G(x)(\chi) = \chi(x) \quad \text{for} \quad x \in G, \chi \in \widehat{G}_d.$$

([HR63, p.430]). Using the duality for locally compact abelian groups the proof is very easy. The continuity of i_G follows from the fact that it is the dual mapping of the inclusion $\widehat{G}_d \to \widehat{G}$. The injectivity of this mapping implies that the image of i_G is dense and the universal property is a consequence of the observation that a morphism of a discrete group $A \to \widehat{G}$ is also continuous as a morphism $A \to \widehat{G}_d$.

In particular this shows that $\mathbb{R}^b \cong \widehat{\mathbb{R}_d}$. Hence the topology of $\mathbb{R}^b$ has no countable base. Moreover

$$\mathrm{Hom}(\mathbb{R}, \mathbb{R}^b) \cong \mathrm{Hom}(\mathbb{R}, \widehat{\mathbb{R}_d}) \cong \mathrm{Hom}(\mathbb{R}_d, \widehat{\mathbb{R}}) \cong \mathrm{Hom}(\mathbb{R}_d, \mathbb{R})$$

implies that the one-parameter groups of $\mathbb{R}^b$ are in one-to-one correspondense with the endomorphisms of the abstract abelian group $\mathbb{R}$. These are exactly the endomorphisms of the $\mathbb{Q}$-vector space $\mathbb{R}$. This shows that in general

$$\mathbf{L}(S^b) := \mathrm{Hom}(\mathbb{R}^+, S^b) \neq \mathbf{L}(S) = \mathrm{Hom}(\mathbb{R}^+, S)$$

for a generating invariant Lie semigroup (S, G) if $H(S) \neq \{\mathbf{1}\}$. ∎

XI. FACES AND IDEMPOTENTS

In this section we consider the lattice $E(S^\flat)$ of idempotents of the Bohr compactification of a generating invariant Lie semigroup (S, G). Since this lattice is the same for $S_0 := S/H(S)$ we assume throughout this section that $H(S) = \{1\}$ (Proposition IX.14). Moreover, we show that, provided $\operatorname{rank} \pi_1(G) \leq 1$, $E(S^\flat)$ can be embedded into the lattice of faces of the wedge $C^\star$, where $C := \mathbf{L}(S) \cap \mathfrak{h}$ is the intersection of $\mathbf{L}(S)$ with a compactly embedded Cartan algebra of $\mathbf{L}(G)$. Such a compactly embedded Cartan algebra exists since $\mathbf{L}(S)$ is a pointed generating invariant cone in $\mathbf{L}(G)$ (Proposition III.15). One should note that this result rests substantially on almost everything that we did in the previous sections. First we need the result on the geometry of cones from Section I, then the Special Reconstruction Theorem from Section III to get the correspondence between the faces of $C^\star$ and the faces of $\mathbf{L}(S)^*$. We also need the crucial Theorem IX.24 on the existence of the semigroup S_e for every idempotent, and its properties. The assumption on the fundamental group of G is needed to apply the Face Theorem from Section VI (cf. Theorem IX.24). Finally the results on globality from Section VIII are needed to characterize those faces of $C^\star$ which correspond to idempotents of $S^\flat$.

Definition XI.1. Let $e \in E(S^\flat)$ be an idempotent and $\varphi_e = \lambda_e \circ i_S : S \to eS^\flat$. Then we define $I(e)$ to be the set

$$\{x \in \mathbf{L}(G) : (\forall x \in \operatorname{int} S)(\exists \varepsilon > 0)(\forall t \in]-\varepsilon, \varepsilon[) : \varphi_e\big(s \exp(tx)\big) \in \varphi_e(s)H(e)\}.$$

∎

We recall the definition of the generating invariant Lie semigroups $S_e \subseteq G$ associated to the idempotents $e \in E(S^\flat)$ (Theorem IX.24).

Lemma XI.2. *Suppose that* $\operatorname{rank} \pi_1(G) \leq 1$. *Then*

$$\mathbf{L}(S_e) = \overline{\mathbf{L}(S) + H\big(L(S_e)\big)}.$$

Proof. According to Theorem IX.24 we know that $\mathbf{L}(S_e)^*$ is a face of $\mathbf{L}(S)^*$. Now the assertion follows from Proposition I.5(5). ∎

Proposition XI.3. *Let $T \subseteq G$ be a generating invariant Lie semigroup containing S and $\alpha : S \to T$ the inclusion. The following assertions hold:*

(1) *If* $\mathbf{L}(T) = \overline{\mathbf{L}(S) - H\big(\mathbf{L}(T)\big)}$, *then there exists a continuation* $\varphi : T \to eS^\flat$ *of* $\varphi_e := \lambda_e \circ i_S : S \to eS^\flat$ *iff*

$$H\big(\mathbf{L}(T)\big) \subseteq I(e).$$

(2) *Let* $e \in E(S^\flat)$. *Then the inclusion* $\alpha : S \to T$ *has an extension* $\widetilde{\alpha} : S_e \to T^\flat$ *with* $i_T \circ \alpha = \widetilde{\alpha} \circ j$, *where* $j : S \to S_e$ *is the inclusion, iff*

$$I(e) \subseteq \mathbf{L}\,\big(H(T)\big).$$

(3) $\mathbf{L}\,\big(H(S_e)\big) = I(e)$.

Proof. (1) "$\Rightarrow$": Let $x \in \mathbf{L}\,\big(H(T)\big), s \in \operatorname{int} S$ and $\varepsilon > 0$ such that $s\exp(tx) \in \operatorname{int} S$ for $|t| < \varepsilon$. Then

$$\begin{aligned}\varphi_e\big(s\exp(tx)\big) &= \varphi_e(s)\varphi_e\big(\exp(tx)\big) \in \varphi_e(s)\varphi_e\big(H(T)\big)\\ &\subseteq \varphi_e(s)H(eS^\flat) = \varphi_e(s)H(e),\end{aligned}$$

thus $x \in I(e)$.

(2) "$\Rightarrow$": Let $x \in I(e)$ and set $H := H(T)$. According to our assumption and Theorem IX.24 we find a mapping $\alpha_1 : eS^\flat \to T^\flat$ such that $\alpha_1 \circ \varphi_e = \widetilde{\alpha}$. We have to prove that $x \in \mathbf{L}(H)$. Let $\pi_G := G \to G/H$ denote the quotient homomorphism $g \mapsto gH$ and $\pi := \pi_G \mid_T : T \mapsto T/H$ the corresponding semigroup homomorphism. Then the universal property of $T^\flat$ guarantees the existence of a continuous homomorphism $\pi^\flat : T^\flat \to (T/H)^\flat$ such that the following diagram commutes:

$$\begin{array}{ccc} T & \xrightarrow{\quad i_T \quad} & T^\flat \\ \big\downarrow{\scriptstyle \pi} & & \big\downarrow{\scriptstyle \pi^\flat} \\ T/H & \xrightarrow{\quad i_{T/H} \quad} & (T/H)^\flat. \end{array}$$

Now $H(T/H) = \{\mathbf{1}\}$ and with Corollary V.10 we find a neighborhood U of $\mathbf{1}$ in T/H such that $i_{T/H}|_U$ is injective. Then we find $t \in \operatorname{int}_{G/H}(U)$ and $s \in S \subseteq T$ with $\pi(s) \in \operatorname{int}_{G/H}(U)$. Choose $\varepsilon > 0$ such that

$$\pi\big(s\exp(tx)\big) \in U \quad \text{and} \quad \varphi_e^\flat\big(s\exp(tx)\big) \in \varphi_e(s)H(e) \quad \text{for} \quad |t| < \varepsilon.$$

Next, in view of

$$\alpha_1(e) = \alpha_1 \circ \varphi_e^\flat(\mathbf{1}) = \widetilde{\alpha}(\mathbf{1}) = i_T \circ \alpha(\mathbf{1}) = \mathbf{1}$$

and $\pi^\flat(H(\mathbf{1}_T)) \subseteq H(T/H(T)) = \{\mathbf{1}\}$, we have that

$$\begin{aligned} i_{T/H} \circ \pi\big(s\exp(tx)\big) &= \pi^\flat \circ i_T\big(s\exp(tx)\big) \\ &= \pi^\flat \circ i_T \circ \alpha\big(s\exp(tx)\big) \\ &= \pi^\flat \circ \widetilde{\alpha} \circ j\big(s\exp(tx)\big) \\ &= \pi^\flat \circ \alpha_1 \circ \varphi_e \circ j\big(s\exp(tx)\big) \\ &= \pi^\flat \circ \alpha_1 \circ \varphi_e\big(s\exp(tx)\big) \\ &\in \pi^\flat \circ \alpha_1 \circ \varphi_e(s) \cdot \pi^\flat \circ \alpha_1\big(H(e)\big) \\ &= \pi^\flat \circ i_T \circ \alpha(s) \cdot \pi^\flat \circ \alpha_1\big(H(e)\big) \\ &= \pi^\flat \circ i_T(s) \cdot \pi^\flat \circ \alpha_1\big(H(e)\big) \\ &\subseteq i_{T/H} \circ \pi(s) \cdot \pi^\flat\Big(H\big(\alpha_1(e)\big)\Big) \\ &\subseteq \{i_{T/H} \circ \pi(s).\} \end{aligned}$$

But $i_{T/H}|_U$ is injective, consequently $\pi\big(s\exp(tx)\big) = \pi(s)$ for $|t| < \varepsilon$. This shows that $d\pi_G(\mathbf{1})x = 0$ and $x \in \mathbf{L}(\ker \pi_G) = \mathbf{L}(H)$.
(3) From (1) we infer that $H\big(\mathbf{L}(S_e)\big) \subseteq I(e)$ and from (2) that

$$I(e) \subseteq \mathbf{L}\left(H(S_e)\right) = H\big(\mathbf{L}(S_e)\big).$$

(1) "$\Leftarrow$": If $H\big(\mathbf{L}(T)\big) \subseteq I(e) = H\big(\mathbf{L}(S_e)\big)$ and $\mathbf{L}(T) = L_{H(\mathbf{L}(T))}\big(\mathbf{L}(S)\big)$, then $\mathbf{L}(T) \subseteq \mathbf{L}(S_e)$ and therefore $T \subseteq S_e$. This guarantees the existence of φ.
2) "$\Leftarrow$": Suppose that $I(e) \subseteq H\big(\mathbf{L}(T)\big)$. Then, in view of (3) and Lemma IX.2, $\mathbf{L}(S_e) \subseteq \mathbf{L}(T)$ and therefore $S_e \subseteq T$. We set $\widetilde{\alpha} := i_T|_{S_e}$. ■

To realize $E(S^\flat)$ as a lattice of faces of $C^\star$ we proceed in two steps. First we map $E(S^\flat)$ into the lattice of invariant faces of $\mathbf{L}(S)^*$. Then, using the tools from Sections I and III, we get a realization in $\mathcal{F}(C^\star)$. For the rest of this section we assume that $\pi_1(G)$ has at most rank 1.

Definition XI.4. Let W be a wedge in a module of a Lie group G. We write $\mathcal{F}_i(W)$ for the set of G-invariant faces of W. We use this mostly for the module $\mathbf{L}(G)$ with the adjoint action and the coadjoint module $\mathbf{L}(G)\widehat{\ }$. ■

Definition XI.5. Let

$$\alpha : \big(E(S^\flat), \prec\big) \to \Big(\mathcal{F}_i\big(\mathbf{L}(S)^*\big), \subseteq\Big), \quad e \mapsto I(e)^\perp \cap \mathbf{L}(S)^* = \mathbf{L}(S_e)^*$$

and

$$\beta : \Big(\mathcal{F}_i\big(\mathbf{L}(S)^*\big), \subseteq\Big) \to \big(E(S^\flat), \prec\big), \quad F \mapsto \sup\{e \in E(S^\flat) : j_F^\flat(e) = \mathbf{1}\},$$

where $S_F := \overline{\langle \exp F^* \rangle}$ and $j_F^\flat : S^\flat \to S_F^\flat$ is the mapping induced by the inclusion $j_F : S \to S_F$. The above supremum exists because $(j_F^\flat)^{-1}(\mathbf{1})$ is a subsemigroup of $S^\flat$ which has a unique maximal idempotent because $E(S^\flat)$ is abelian (Lemma IX.4). ■

Lemma XI.6. *Let K be a compact monoid and $\alpha : S \to K$ a continuous homomorphism. Then α extends to $\widetilde{\alpha} : S_e \to K$ iff $\alpha^\flat(e) = \mathbf{1}$.*

Proof. "$\Leftarrow$": Suppose that $\alpha^\flat(e) = \mathbf{1}$. Then, in view of Theorem IX.24,

$$\alpha^\flat \circ \varphi_e \circ j_e = \alpha^\flat \circ \lambda_e \circ i_S = \lambda_{\alpha^\flat(e)} \circ \alpha^\flat \circ i_S = \alpha^\flat \circ i_S = \alpha,$$

thus $\alpha^\flat \circ \varphi_e : S_e \to K$ is a continuation of α.
"$\Rightarrow$": Assume that $\widetilde{\alpha} : S_e \to K$ exists. Then $\widetilde{\alpha} \circ j_e = \alpha$, where $j_e : S \to S_e$ is the inclusion. Therefore

$$\alpha^\flat(e) = \widetilde{\alpha}^\flat \circ j_e^\flat(e) = \widetilde{\alpha}^\flat(\mathbf{1}) = \mathbf{1}$$

since $j_e^\flat(e) = j_e^\flat \circ \varphi_e^\flat(\mathbf{1}) = \mathrm{id}_{S_e^\flat}(\mathbf{1}) = \mathbf{1}$ (Theorem IX.24). ∎

Lemma XI.7. *For $F \in \mathcal{F}_i(\mathbf{L}(S)^*)$ we have that $S_{\beta(F)} = S_F$.*

Proof. The idempotent $\beta(F)$ was defined to be maximal in $(j_F^\flat)^{-1}(\mathbf{1})$, where $j_F^\flat : S^\flat \to S_F^\flat$ is induced by the inclusion $j_F : S \to S_F$. Therefore $j_F^\flat(\beta(F)) = \mathbf{1}$ and j_F extends to $\widetilde{j}_F : S_{\beta(F)} \to S_F^\flat$ (Lemma XI.6). Hence $I(\beta(F)) \subseteq \mathbf{L}\left(H(S_F)\right)$ (Proposition XI.3) and consequently

$$S_{\beta(F)} = \overline{\langle \exp \mathbf{L}(S) - I(\beta(F)) \rangle} \subseteq S_F.$$

To prove the reverse inclusion we assume that $S_F \neq S_{\beta(F)}$. Then $H(S_F) \not\subseteq S_{\beta(F)}$ and $\mathbf{L}\left(H(S_F)\right) \not\subseteq I(\beta(F))$. We know from Proposition VI.17 that

$$\mathbf{L}(S_F)^* \in \mathcal{F}(F^{**}) = \mathcal{F}(F) \subseteq \mathcal{F}(\mathbf{L}(S)^*)$$

(Proposition I.4) since $F \in \mathcal{F}(\mathbf{L}(S)^*)$. Then Proposition I.4 implies that $\mathbf{L}(S_F)^* \in \mathcal{F}(\mathbf{L}(S_{\beta(F)})^*)$ because $\mathbf{L}(S_F)^* \subseteq \mathbf{L}\left(S_{\beta(F)}\right)^*$. Now $\mathbf{L}(S_F)^* \neq \mathbf{L}\left(S_{\beta(F)}\right)$ and consequently $\mathbf{L}(S_F)^* \cap \mathrm{algint}\, \mathbf{L}(S_{\beta(F)})^* = \emptyset$ (Proposition I.4). Hence there exists

$$x \in \mathbf{L}\left(S_{\beta(F)}\right) \cap \left(\mathbf{L}(S_F)^*\right)^\perp = \mathbf{L}\left(S_{\beta(F)}\right) \cap H\left(\mathbf{L}(S_F)\right)$$

such that $x \notin H\left(\mathbf{L}(S_{\beta(F)})\right)$. Let $t_n \in \mathbb{R}^+$ and $e := e_x \in E(S_{\beta(F)}^\flat)$. Then $e \in \varphi_{\beta(F)}^\flat(S_{\beta(F)}) \subseteq \beta(F) \cdot S^\flat$, i.e., $\beta(F) \prec e$ and $e \neq \beta(F)$ since $x \notin H\left(\mathbf{L}(S_{\beta(F)})\right) = I(\beta(F))$. Furthermore the inclusion $j_1 : S_{\beta(F)} \to S_F$ satisfies (with the notation of Example IX.6)

$$j_1^\flat(e) = j_1^\flat \circ \gamma_x^\flat(\omega) = \gamma_x^\flat(\omega) = \mathbf{1}$$

(Proposition IX.7). Thus, for the inclusion $j_{\beta(F)} : S \to S_{\beta(F)}$, we have that

$$j_F^\flat \circ \varphi_{\beta(F)}^\flat(e) = j_1^\flat \circ j_{\beta(F)}^\flat \circ \varphi_{\beta(F)}^\flat(e) = j_1^\flat(e) = \mathbf{1}$$

and $\beta(F) \prec \varphi_{\beta(F)}^\flat(e) \neq \beta(F)$ (Theorem IX.24). This contradicts the definition of $\beta(F)$ because $\beta(F)$ was the maximal idempotent in $(j_F^\flat)^{-1}(\mathbf{1})$. ∎

Theorem XI.8. *The mappings α and β are antitone and define a Galois connection between $E(S^\flat)$ and $\mathcal{F}_i(\mathbf{L}(S)^*)$, i.e.,*

$$\alpha(e) \subseteq F \iff \beta(F) \prec e.$$

The following assertions hold:

(1) $\beta \circ \alpha = \mathrm{id}_{E(S^\flat)}$.

(2) $\alpha \circ \beta(F) \subseteq F$ *for* $F \in \mathcal{F}_i(\mathbf{L}(S)^*)$.

(3) $\alpha \circ \beta(F) = F$ *iff* F^* *is global in* G.

(4) α, β *map infs into sups.*

(5) $\alpha(e \cdot f) = \alpha(e \vee f) = \mathbf{L}(\overline{S_e S_f})^*$.

Proof. α is antitone: To prove this, we may show that $e \mapsto I(e)$ is monotone. So take $e \prec f$ in $E(S^\flat)$, $x \in I(e)$, $s \in \mathrm{int}(S)$ and $\varepsilon > 0$ such that $s\exp(tx) \in S$ for $|t| < \varepsilon$. Then

$$\begin{aligned}\varphi_f(s\exp(tx)) &= f \cdot i_S(s\exp(tx)) \\ &= f \cdot e \cdot i_S(s\exp(tx)) \\ &= f \cdot \varphi_e(s\exp(tx)) \\ &\in f^2 \varphi_e(s) H(e) = \varphi_f(s)(f \cdot H(e)) \\ &\subseteq \varphi_f(s) H(f)\end{aligned}$$

because $\lambda_f : S^\flat \to S^\flat$ is a semigroup homomorphism with $\lambda_e(f) = ef = f$. Thus $I(e) \subseteq I(f)$ and I is monotone.

β is antitone: Let $F_1 \subseteq F_2 \in \mathcal{F}_i(\mathbf{L}(S)^*)$. Then

$$S \subseteq S_2 := \overline{\langle \exp F_2^* \rangle} \subseteq S_1 := \overline{\langle \exp F_1^* \rangle}$$

are generating invariant Lie semigroups in G. Moreover $j_1 = j_{21} \circ j_2$, where $j_1 : S \to S_1$, $j_2 : S \to S_2$ and $j_{21} : S_2 \to S_1$ are the inclusion homomorphisms. Then $j_1^\flat = j_{21}^\flat \circ j_2^\flat$ and

$$j_1^\flat(\beta(F_2)) = j_{21}^\flat \circ j_2^\flat(\beta(F_2)) = j_{21}^\flat(\mathbf{1}) = \mathbf{1}.$$

Hence $\beta(F_2) \in (j_1^\flat)^{-1}(\mathbf{1})$ and consequently $\beta(F_2) \prec \beta(F_1)$.

α and β define a Galois connection: This follows from the following chain of equivalences:

$$\begin{aligned}\alpha(e) \subseteq F &\Leftrightarrow I(e)^\perp \cap \mathbf{L}(S)^* = \mathbf{L}(S_e)^* \subseteq F && \text{Proposition XI.3(3)} \\ &\Leftrightarrow F^* \subseteq \mathbf{L}(S_e) \\ &\Leftrightarrow S_F := \overline{\langle \exp F^* \rangle} \subseteq S_e \\ &\Leftrightarrow S_{\beta(F)} \subseteq S_e && \text{Lemma XI.7} \\ &\Leftrightarrow \lambda_e \circ i_S \text{ extends to } S_{\beta(F)} && \text{Proposition XI.3(2)} \\ &\Leftrightarrow \lambda_e(\beta(F)) = e && \text{Lemma XI.6} \\ &\Leftrightarrow e\beta(F) = e \\ &\Leftrightarrow \beta(F) \prec e.\end{aligned}$$

(1) First we note that $\beta \circ \alpha(e) \prec e$ because this is equivalent to $\alpha(e) \subseteq \alpha(e)$ which is true. Moreover $\alpha \circ \beta \circ \alpha(e) = \alpha(e)$ and therefore

$$I(e) = H\big(\alpha(e)^*\big) = H\big(\alpha \circ \beta \circ \alpha(e)^*\big) = I\big(\beta \circ \alpha(e)\big).$$

Hence

$$S_e = \overline{\langle \exp L_{I(e)}\, \mathbf{L}(S)\rangle} = \overline{\langle \exp L_{I\left(\beta\circ\alpha(e)\right)}\, \mathbf{L}(S)\rangle} = S_{\beta\circ\alpha(e)}$$

(Theorem IX.24 and Propositions XI.2, XI.3). The inclusion homomorphism $j_e : S \to S_e = S_{\beta\circ\alpha(e)}$ satisfies

$$j_e^\flat \circ \varphi_e^\flat(\mathbf{1}) = j_e^\flat(e) = \mathbf{1}_{S_e^\flat}$$

(Theorem IX.24) and

$$\beta \circ \alpha(e) = \varphi^\flat_{\beta\circ\alpha(e)}(\mathbf{1}) = \varphi^\flat_{\beta\circ\alpha(e)} \circ j_e^\flat(e) = \lambda_{\beta\circ\alpha(e)}(e) = e.$$

Consequently $\beta \circ \alpha(e) = e$ and $\beta \circ \alpha = \mathrm{id}_{E(S^\flat)}$.
(2) The relation $\alpha \circ \beta(F) \subseteq F$ is equivalent to $\beta(F) \prec \beta(F)$ and therefore true.
(3) The fixed points of $\alpha \circ \beta$ are exactly the faces $\alpha(e)$ with an idempotent $e \in E(S^\flat)$. For these faces $\alpha(e)^* = \mathbf{L}(S_e)$ is global. If F^* is global and $F \in \mathcal{F}_i\big(\mathbf{L}(S)^*\big)$, then $S_F = S_{\beta(F)}$ (Lemma XI.7) implies that

$$\mathbf{L}(S_F) = F^* = \mathbf{L}(S_{\beta(F)}) = \alpha \circ \beta(F)^*.$$

Hence $F = \alpha \circ \beta(F)$.
(4) [GHK80, pp.18-20]
(5) It is clear that $e \vee f = e \cdot f \in E(S^\flat)$. In view of (4), we know that $\alpha(e \vee f)$ is the largest face $F \in \mathcal{F}\big(\mathbf{L}(S)^*\big)$ such that $F \subseteq \alpha(e) \cap \alpha(f)$ and F^* is global in G. This is equivalent to $\alpha(e)^* + \alpha(f)^* \subseteq F^*$. We set

$$V := \mathbf{L}\left(\overline{\langle \exp\big(\alpha(e)^* + \alpha(f)^*\big)\rangle}\right).$$

Then, according to Proposition VI.17,

$$V^* \in \mathcal{F}\big(\overline{\alpha(e)^* + \alpha(f)^*}^{\,*}\big) = \mathcal{F}\big(\alpha(e) \cap \alpha(f)\big) \subseteq \mathcal{F}\big(\mathbf{L}(S)^*\big)$$

since $\alpha(e) \cap \alpha(f) \in \mathcal{F}\big(\mathbf{L}(S)\big)$ (Proposition I.5). This proves that $\alpha(ef) = V^*$. ■

Definition XI.9. From now on we fix a compactly embedded Cartan algebra $\mathfrak{h}$ of $\mathbf{L}(G)$. We recall the definition of the Weyl group $\mathcal{W}$ (Definition III.2) and of the wedge $\mathcal{C} \subseteq \mathrm{End}(\mathfrak{h})$ (Definition III.4). We write

$$\mathcal{W}^\frown = \{\widehat{w} : \widehat{\mathfrak{h}} \to \widehat{\mathfrak{h}}, w \in \mathcal{W}\}$$

and

$$\mathcal{C}^\frown = \{\widehat{c} : \widehat{\mathfrak{h}} \to \widehat{\mathfrak{h}}, c \in \mathcal{C}\}.$$

We identify $\widehat{\mathfrak{h}}$ with the subspace of elements in $\mathbf{L}(G)^\frown$ fixed under the action of the compact group $\mathrm{INN}_{\mathfrak{g}}(\mathfrak{h})^\frown$ (Theorem I.10). If $D \subseteq \widehat{\mathfrak{h}}$ is a wedge, we write $\mathcal{F}_i(D)$ for the set of all faces of D which are invariant under $\mathcal{W}^\frown$ and $\mathcal{C}^\frown$. ■

Lemma XI.10. *Set* $C := \mathbf{L}(S) \cap \mathfrak{h}$. *Then the mapping*

$$\gamma : \mathcal{F}_i\big(\mathbf{L}(S)^*\big) \to \mathcal{F}_i(C^\star), \quad F \mapsto F \cap \widehat{\mathfrak{h}}$$

is a monotone intersection preserving bijection.

Proof. Let $F \in \mathcal{F}_i\big(\mathbf{L}(S)^*\big)$. Then $F^* \subseteq \mathbf{L}(G)$ is an invariant wedge and $p(F^*) = (F \cap \mathfrak{h})^\star$ (Theorem I.10). Thus $F \cap \widehat{\mathfrak{h}} \in \mathcal{F}_i(C^\star)$ (Propositions I.12 and III.7). The surjectivity of γ is a consequence of Theorem III.35 and the injectivity follows from Proposition III.34 because $\gamma(F_1) = \gamma(F_2)$ implies that

$$p(F_1^*) = (F_1 \cap \widehat{\mathfrak{h}})^\star = (F_2 \cap \widehat{\mathfrak{h}})^\star = p(F_2^*).$$

∎

Theorem XI.11. *Let* (S, G) *be generating invariant Lie semigroup with* $H(S) = \mathbf{1}$ *and* $\operatorname{rank} \pi_1(G) \leq 1$. *Set* $C = \mathbf{L}(S) \cap H$. *Then the mapping*

$$\delta : E(S^\flat) \to \mathcal{F}_i(C^\star), \quad e \mapsto \alpha(e) \cap \widehat{\mathfrak{h}}$$

is an antitone injection. A face $F \in \mathcal{F}(C^\star)$ *is in* $\delta\big(E(S^\flat)\big)$ *iff it satisfies the following conditions:*

(1) $\mathcal{W}^\curlyvee(F) \subseteq F$.

(2) $C^\curlyvee(F) \subseteq F$.

(3) $\gamma^{-1}(F)^*$ *is global in* G, *where* $\gamma(E) = E \cap \widehat{\mathfrak{h}}$ *for* $E \in \mathcal{F}_i\big(\mathbf{L}(S)^*\big)$.

Proof. The injectivity of $\delta = \gamma \circ \alpha$ follows from Theorem XI.8 and Lemma XI.10. The antitony is a consequence of the antitony of α and the monotony of γ. Knowing that $\delta\big(E(S^\flat)\big) \subseteq \mathcal{F}_i(C^\star) \subseteq \mathcal{F}(C^\star)$, in view of the injectivity of γ, one only has to check that the range of α are exactly the invariant faces F of $\mathbf{L}(S)^*$ for which F^* is global in G. This is exactly the assertion of Theorem XI.8(3). ∎

Corollary XI.12. *Let* (S, G) *be generating invariant Lie semigroup with* $H(S) = \mathbf{1}$ *such that* G *is simply connected and* $\mathbf{L}(G)$ *is of globality type. Set* $C = \mathbf{L}(S) \cap \mathfrak{h}$. *Then the mapping*

$$\delta : E(S^\flat) \to \mathcal{F}_i(C^\star), \quad e \mapsto \alpha(e) \cap \widehat{H}$$

is an antitone bijection.

Proof. This is a direct consequence of Theorem XI.11 since every invariant wedge in a simply connected Lie group whose Lie algebra is of globality type is global (Theorem VIII.12). ∎

Corollary XI.13. *Let (S,G) be generating invariant Lie semigroup with $H(S) = \mathbf{1}$ such that G is simply connected and solvable. Set $C = \mathbf{L}(S) \cap \mathfrak{h}$. Then the mapping*

$$\delta : E(S^\flat) \to \mathcal{F}_i(C^\star), \quad e \mapsto \alpha(e) \cap \widehat{\mathfrak{h}}$$

is an antitone bijection and

$$\mathcal{F}_i(C^\star) = \{F \in \mathcal{F}(C^\star) : C^\curlyvee(F) \subseteq F\}.$$

Proof. This follows from Corollary XI.12 because G is of globality type since it is solvable (Definition VIII.11) and the Weyl group is trivial since $\mathbf{L}(G)$ is solvable (Lemma III.8). ∎

As a last corollary we obtain the known result in the abelian case that the lattice of idempotents in the Bohr compactification of a pointed cone is isomorphic to the face lattice of the dual cone ([Rup87]).

Corollary XI.14. *Let $S \subseteq \mathbb{R}^n$ be a pointed cone in $\mathbb{R}^n$. Then the mapping*

$$\delta : E(S^\flat) \to \mathcal{F}(S^\star), \quad e \mapsto \alpha(e)$$

is an antitone bijection.

Proof. This follows from Corollary XI.13 and the fact that $C = \{0\}$ and $\mathfrak{h} = \mathbf{L}(G)$ whenever $\mathbf{L}(G)$ is abelian (Definition III.4). ∎

Example XI.15. Let $G = \mathcal{O}_n$ and $\mathbf{L}(G) = A_{2n+2}$ the $(2n+2)$-dimensional oscillator algebra. Then $\dim \mathfrak{h} = 2$, $\dim \mathfrak{z} = 1$ and $\Omega_R^+ = \{\omega\}$. We find that $E(S^\flat) = \{\mathbf{1}, e_1, e_2\}$ with

$$\alpha(\mathbf{1}) = \{0\}, \quad \alpha(e_1) = \mathbb{R}^+\omega, \quad \text{and} \quad \alpha(e_2) = C^\star$$

for every pointed generating C-invariant wedge in H. ∎

For the rest of this section we consider the problem of characterizing the faces of $C^\star$ which correspond to the accessible idempotents of $S^\flat$.

Proposition XI.16. *For an idempotent $e \in E(S^\flat)$ the following assertions hold.*

(1) $e \in E_a(S^\flat) \Leftrightarrow \big(\exists F \in \mathcal{F}_n(S)\big)\alpha(e) = \mathbf{L}\left(L_F(S)\right)^*$.

(2) $e \in E_{da}(S^\flat) \Leftrightarrow \Big(\exists E \in \mathcal{F}_{ei}\big(\mathbf{L}(S)\big)\Big)\alpha(e) = \mathbf{L}\left(\overline{\langle \exp L_E\big(\mathbf{L}(S)\big)\rangle}\right)^*$.

Proof. (1) e is accessible if and only if the exists a normal exposed face $F \in \mathcal{F}_n(S)$ with $e = e_F$. Then $S_e = L_F(S)$ (Theorem IX.24(6) and Proposition IX.12). Therefore $\alpha(e) = \mathbf{L}(S_e)^* = \mathbf{L}\left(L_F(S)\right)^*$.

If, conversely, $\alpha(e) = \mathbf{L}(S_e)^* = \mathbf{L}\left(L_F(S)\right)^*$, then $L_F(S) = S_e$ since $L_F(S)$ is a generating invariant Lie semigroup (Proposition IV.36). Hence $S_{e_F} = L_F(S) = S_e$ and thus $\alpha(e_F) = \alpha(e)$ leads to $e_F = e$ (Theorem XI.8).

(2) If $e \in E_{da}(S^\flat)$, then there exists $x \in \mathbf{L}(S)$ with $e = e_x$ and therefore $e = e_F$, where F is the normal exposed face of S generated by $\exp x$ (Lemma IX.11). But $F = H\big(\overline{S\exp(\mathbb{R}^- x)}\big) \cap S$ and therefore

$$F = H\Big(\overline{\langle \exp L_E\big(\mathbf{L}(S)\big)\rangle}\Big) \cap S$$

for $E = \mathbf{L}(F) \in \mathcal{F}_{ei}\big(\mathbf{L}(S)\big)$ (Proposition IV.34). Using Proposition IX.24(7), we conclude that

$$S_e = L_F(S) = \overline{\langle \exp L_E\big(\mathbf{L}(S)\big)\rangle}$$

and therefore $\alpha(e) = \mathbf{L}\left(L_F(S)\right)^*$.

If, conversely, $\alpha(e)^* = \mathbf{L}\Big(\overline{\langle \exp L_E\big(\mathbf{L}(S)\big)\rangle}\Big)$ for $E \in \mathcal{F}_i\big(\mathbf{L}(S)\big)$ and $x \in \operatorname{algint} E$, then $F = H\Big(\overline{\langle \exp L_E\big(\mathbf{L}(S)\big)\rangle}\Big) \cap S$ is the normal exposed face generated by $\exp x$. Therefore

$$\alpha(e_x)^* = \mathbf{L}\left(L_F(S)\right) = \alpha(e)^*.$$

Hence $e = e_x$ since α is injective (Theorem XI.8). ∎

Corollary XI.17. *Suppose that $\mathbf{L}(G)$ is of globality type. Then*

$$e \in E_{da}(S^\flat) \quad \Longleftrightarrow \quad \alpha(e) \in \mathcal{F}_{ei}\big(\mathbf{L}(S)^*\big).$$

Proof. According to our assumption that $\mathbf{L}(G)$ is of globality type Theorem VIII.12 shows that every generating invariant Lie wedge $V \subseteq \mathbf{L}(G)$ is global. Consequently, we have for an invariant Lie semigroup $F \subseteq S$ that

$$L_F(S) = \overline{\langle \exp \mathbf{L}(S) - \mathbf{L}(F)\rangle} \qquad \text{with} \qquad \mathbf{L}\left(L_F(S)\right) = \overline{\mathbf{L}(S) - \mathbf{L}(F)}.$$

In view of Proposition XI.16 the directly accessible idempotents in $S^\flat$ are exactly those with

$$\alpha(e) = \mathbf{L}\Big(\overline{\langle \exp L_E\big(\mathbf{L}(S)\big)\rangle}\Big)^* = L_E\big(\mathbf{L}(S)\big)^* = \mathbf{L}(S)^* \cap E^\perp,$$

where $E \in \mathcal{F}_{ei}\big(\mathbf{L}(S)\big)$. These are exactly the invariant exposed faces of $\mathbf{L}(S)^*$ since op (Proposition I.4) maps normal exposed faces of $\mathbf{L}(S)$ into invariant exposed faces of $\mathbf{L}(S)^*$. ∎

Again we obtain the results in the abelian case by a specialization (cf. [Rup87], [F78a]).

Corollary XI.18. *Let $W \subseteq \mathbb{R}^n$ be a poined generating cone. Then*

$$E_{da}(W^\flat) = E_a(W^\flat) \quad \textit{and} \quad \delta\big(E_a(W^\flat)\big) = \mathcal{F}_e(W^*).$$

Proof. Taking Corollaries XI.14 and XI.17 into account, there remains nothing to prove. ∎

XII. EXAMPLES AND SPECIAL CASES

In this last section we consider interesting examples of Lie semigroups and their compactifications. We use these examples to study finite dimensional representation of G which map S into a relatively compact subset of endomorphism of $\mathbb{R}^n$. Moreover, we show how invariant Lie semigroups arise in the theory of ordered manifolds.

EXAMPLES

Proposition XII.1. *Suppose that G is simple and that (S,G) is a generating invariant Lie semigroup with $S \neq G$. Then $S^\flat \cong S_I$, the one-point compactification of the set* comp(S).

Proof. It is clear that $\mathcal{F}_n(S) = \{\{\mathbf{1}\}, S\}$. Therefore $E(S^\flat) = E_a(S^\flat) = \{\mathbf{1}, e_S\}$ (Lemma IX.22). Moreover, $S_{e_S} = G$ implies that $e_S S^\flat \cong G^\flat = \{e_S\}$ (Proposition X.1). Using the Strucuture Theorem (Theorem IX.32), this proves that

$$S^\flat = S(\mathbf{1}) \cup S(e_S) = i_S\big(\mathrm{comp}(S)\big) \cup \{e_S\}.$$

Now Theorem IX.36 shows that $S^\flat$ is hemeomorphic to the one-point compactification of $S(\mathbf{1}) \cong \mathrm{comp}(S)$ (Corollary V.10). ∎

Example XII.2. (cf. [Rup88]) We recall Example VI.23, the invariant subsemigroup $S = \langle \exp W \rangle$ in $G = \mathrm{Sl}(2,\mathbb{R})^\sim$. We have proved that

$$\mathrm{comp}(S) = \{\exp w : w \in W, k(w) < \pi^2\},$$

where $k(hH + tT + uU) = -h^2 - t^2 + u^2$ with respect to the base H, T, U of $\mathbf{L}(G) = \mathrm{sl}(2,\mathbb{R})$, and

$$W = \{hH + tT + uU : u \geq \sqrt{h^2 + t^2}\}.$$

Thus topologically $S^\flat$ is the one-point compactification of the subset

$$k^{-1}([0, 2\pi[) \cap W$$

of the cone W. ∎

Proposition XII.3. *Let G be a semisimple Lie group and (S, G) a generating invariant Lie semigroup. Then* $\operatorname{card} E(S^\flat) \leq 2^n$*, where n is the number of simple ideals of* $\mathbf{L}(G)$.

Proof. Let $e \in E(S^\flat)$. Then $I(e)$ (Definition XI.1) is an ideal of $\mathbf{L}(G)$ and we recall that the mapping I is injective (Theorem XI.8). Hence there are less idempotents in $E(S^\flat)$ than there are ideals in $\mathbf{L}(G)$. Now the assertion follows from the fact that $\mathbf{L}(G)$ has exactly 2^n ideals. ■

Example XII.4. Let $G = \mathrm{SU}(2,1)^{\sim} \times \mathrm{SU}(2,1)^{\sim}$. Then a compactly embedded Cartan algebra $\mathfrak{h} \subseteq \mathbf{L}(G)$ has dimension 4. We construct a Lie semigroup $S \subseteq G$ with $H(S) = \{\mathbf{1}\}$ and $E(S^\flat) = \{\mathbf{1}, e_F, e_S\}$. This example shows that there may be less than 2^2 idempotents in $E(S^\flat)$. We recall the definition of the element l (Definition VII.19) and denote with l_1 and l_2 these elements in the first and second summand respectively. We also recall the notation of Example VIII.24. So, with respect to the base h_1, h_2 of the compactly embedded Cartan algebra of $\mathrm{su}(2,1)$, we know that the non-compact roots are

$$\omega_1 = (2,1), \qquad \text{and} \qquad \omega_2 = (1,2).$$

The minimal cone is $C_{\min} = \mathbb{R}^+ h_1 + \mathbb{R}^+ h_2$ and the functional $\widehat{l}$ is, up to a constant, given by $\widehat{l} = (0,1)$. The Weyl group interchanges h_1 and h_2 and consists of two elements. We denote the elements of the Cartan algebra of $\mathbf{L}(G) = \mathrm{su}(2,1)^2$ with quadrupels; the first two components belonging to the first summand and the second two components to the other one. Then $\widehat{l_1} + \widehat{l_2} = (0,1,0,1)$. Let $h_0 := (-1,2,2,2)$. Then $h_0 \notin C_{\min} = (\mathbb{R}^+)^4$ and the functional $\widehat{l_1} + \widehat{l_2}$ is non-negative on the whole Weyl group orbit of h_0 which lies in $C_{\max}$. We set

$$C := C_{\min} + \sum_{w \in \mathcal{W}} \mathbb{R}^+ w.h_0.$$

Then there exists a pointed generating invariant wedge $W \subseteq \mathbf{L}(G)$ with $W \cap \mathfrak{h} = C$. Now Theorem VIII.10 shows that W is global in G because $\widehat{l_1} + \widehat{l_2} \in C^\star$. We set $S := \overline{\langle \exp W \rangle}$. The projection of W onto the first summand is not global and therefore it is impossible for a semigroup S_e that $H(S_e)$ is the second factor. Hence $\operatorname{card} E(S^\flat) \leq 3$. Since

$$\mathcal{F}_n(S) = \{\{\mathbf{1}\}, (\mathrm{SU}(2,1)^{\sim} \times \{\mathbf{1}\}) \cap S, S\},$$

we find that

$$E(S^\flat) = E_a(S^\flat) = \{\mathbf{1}, e_F, e_S\},$$

where $F = (\mathrm{SU}(2,1)^{\sim} \times \{\mathbf{1}\})$. From the facts that $(\mathrm{SU}(2,1)^{\sim})^\flat = \{1\}$ (Proposition X.1), we conclude that $H(e_F) = \{e_F\}$ and $H(e_S) = \{\mathbf{1}\}$. Consequently

$$S^\flat = S(\mathbf{1}) \cup S(e_F) \cup \{e_S\}$$

and $e_F S^\flat \cong S^\flat_{e_F}$ is isomorphic to the Bohr compactification of the invariant Lie semigroup S_1 in $\mathrm{SU}(2,1)^{\sim}$ with $\mathbf{L}(S_1) = W_{\min}$. ■

Example XII.5. We recall Example VI.24, where $G = \mathcal{O}_1$ is the four-dimensional oscillator group. In this case

$$\mathrm{comp}(S) = \exp\{w \in \mathbf{L}(S) : d\alpha(\mathbf{1})w \geq \pi\},$$

where $\alpha : G \to \mathbb{R}$ is a homomorphism. Using Example XI.15 we find that $E(S^\flat) = \{\mathbf{1}, e_1, e_2\}$ with $\mathbf{1} \prec e_1 \prec e_2$ and $S_{e_1} = \alpha^{-1}(\mathbb{R}^+)$ is a four-dimensional half space with $H(S_{e_1})$ isomorphic to the three dimensional Heisenberg group. We write $\ker\alpha = \mathfrak{n}_{\text{eff}} \oplus \mathfrak{z}$, where $\mathfrak{z} \subseteq Z(\mathbf{L}(G))$ and every one-parameter group $\exp\mathbb{R}x$ with $d\alpha(\mathbf{1})x \neq 0$ acts as rotations in $\mathfrak{n}_{\text{eff}}$. Now $\mathbf{L}\left(\ker i_{H(S_{e_1})}\right) = \mathfrak{z}$ and in the quotient we find with Corollary X.3 that $H(S_{e_1}) \subseteq \ker i_{S_{e_1}}$. Thus $S_{e_1}^\flat \cong (\mathbb{R}^+)^\flat$. For e_2 we have that $S_{e_2} = G$ and thus that $S_{e_2}^\flat \cong G^\flat \cong \mathbb{R}^\flat$. Putting these facts together we find with $W_1 := \{w \in \mathbf{L}(S) : d\alpha(\mathbf{1})w < \pi\}$ that

$$S^\flat = i_S\big(\exp W_1\big) \cup e_1 \exp(\mathbb{R}^+ x) \cup e_2(\exp\mathbb{R}x)^\flat,$$

where $x \in \mathbf{L}(S)$ with $d\alpha(\mathbf{1})x > 0$. ∎

CONTRACTIVE REPRESENTATIONS

In the last sections we have described the universal compactification of a Lie semigroup. So it is interesting to ask how many morphism into compact finite dimensional matrix semigroups do we have? In the abelian case they even separate the points ([Rup87, 5.4]). It even suffices to take homomorphisms into the complex unit disk $\{z \in \mathbb{C} : |z| \leq 1\}$, where the structure of a compact semigroup is given by multiplication. In general, we are led to the following definition.

Definition XII.6. Let (S,G) be a Lie semigroup. A *compact representation* of S is a morphism

$$\varphi : (S,G) \to \big(\overline{\varphi(S)}, \mathrm{Gl}(n,\mathbb{R})\big)$$

such that the closure of $\varphi(S)$ in $\mathrm{End}(\mathbb{R}^n)$ is compact. ∎

Example XII.7. If $(S,G) = \big((\mathbb{R}^+)^n, \mathbb{R}^n\big)$, then

$$\varphi(x_1,\ldots,x_n) = \begin{pmatrix} e^{-x_1} & 0 & 0 & \ldots & 0 \\ 0 & e^{-x_2} & 0 & \ldots & 0 \\ \cdot & \cdot & \cdot & \cdot & \cdot \\ 0 & 0 & \ldots & 0 & e^{-x_n} \end{pmatrix}$$

is a compact representation. ∎

Lemma XII.8. *Let (S,G) be a generating invariant Lie semigroup with*

$$\widetilde{\operatorname{comp}}(S) \neq S.$$

Then we find for every $x \in \operatorname{int} \mathbf{L}(S)$ a $T > 0$ such that

$$\exp(Tx) \notin \widetilde{\operatorname{comp}}(S)$$

and an idempotent $e \in E(S^\flat) \setminus \{\mathbf{1}\}$ with

$$e \prec i_S(\exp Tx).$$

Proof. Let $s \in S \setminus \widetilde{\operatorname{comp}}(S)$. Since $G = S \cdot \exp(\mathbb{R}^- x)$, we find a $T > 0$ such that

$$\exp Tx \in sS \subseteq S \setminus \widetilde{\operatorname{comp}}(S).$$

Hence Proposition IX.34 implies that $i_S(\exp Tx) \notin S(\mathbf{1})$ and therefore $\mathbf{1} \neq E\big(i_S(\exp Tx)\big) \prec i_S(\exp Tx)$. ∎

Proposition XII.9. *Let (S,G) be a generating invariant Lie semigroup with $\widetilde{\operatorname{comp}}(S) \neq S$ and*

$$\varphi : (S,G) \to \big(\overline{\varphi(S)}, \operatorname{Gl}(n,\mathbb{R})\big)$$

a compact representation. Then there exists $e \in E(S^\flat) \setminus \{\mathbf{1}\}$ with $\varphi^\flat(e) = \mathbf{1}$.

Proof. We denote the closure of $\varphi(S)$ in $\operatorname{End}(\mathbb{R}^n)$ with $\operatorname{cl}\big(\varphi(S)\big)$. Let $\varphi^\flat : S^\flat \to \operatorname{cl}\big(\varphi(S)\big)$ be the induced homomorphism. For $x \in \operatorname{int} \mathbf{L}(S)$ we choose $T > 0$ and $e \prec i_S(\exp Tx)$ as in Lemma XII.8. Then

$$\varphi^\flat(e) \prec \varphi^\flat\big(i_S(\exp Tx)\big) = \varphi(\exp Tx)$$

and therefore we find an element $s \in S^\flat$ with $\varphi^\flat(e)\varphi^\flat(s) = \varphi(\exp Tx)$. Thus $\varphi^\flat(e)$ is a projection which is invertible since $\varphi(\exp Tx) \in \operatorname{Gl}(n,\mathbb{R})$ is invertible. We conclude that $\varphi^\flat(e) = \mathbf{1}$. ∎

Example XII.10. Suppose that G is simple and (S,G) a generating invariant Lie semigroup. Then Propositions XII.1 and XII.9 imply that every compact representation of S is trivial. ∎

Example XII.11. Recall Examples XII.5 and VI.24. Then

$$E(S^\flat) = \{\mathbf{1}, e_1, e_2\} \quad \text{and} \quad \operatorname{comp}(S) \neq S.$$

So Proposition XII.9 shows that $\varphi^\flat(e_1) = \mathbf{1}$ for every compact representation of S. Using Lemma XI.6 we see that φ extends to S_{e_1} by setting $\widetilde{\varphi} := \varphi^\flat \circ \varphi_{e_i}$. Since $\varphi_{e_1}(G') = \{\mathbf{1}\}$ we find that $d\varphi(\mathbf{1})\,\mathbf{L}(G)' = \{0\}$. Thus $\varphi(G') = \{0\}$. ∎

Theorem XII.12. *Let (S,G) be a generating invariant Lie semigroup with $H(S)=\{\mathbf{1}\}$ and φ a compact representation. Then $d\varphi(\mathbf{1})\,\mathbf{L}(G)$ is a compact Lie algebra.*

Proof. Let $\mathfrak{g}:=\mathbf{L}(G)$ and $\mathfrak{h}\subseteq\mathfrak{g}$ a compactly embedded Cartan algebra and Ω^+ a positive system of roots such that

$$C_{\min}\subseteq\mathbf{L}(S)\cap\mathfrak{h}\subseteq C_{\max}$$

(Proposition III.15). We proceed as in the proof of Theorem VI.25. Let $\omega\in\Omega_{\mathfrak{p}}^+$ be a non-compact root and $0\neq x\in\mathfrak{g}^\omega$. Then $Q(x)\in\mathbf{L}(S)$. We distinguish two cases.
Case 1) $\omega\big(Q(x)\big)<0$: Then $\langle x\rangle\cong\mathrm{sl}(2,\mathbb{R})$ and $Q(x)\in V:=\mathbf{L}(S)\cap\langle x\rangle$. We conclude that V is a non-zero invariant cone in $\mathrm{sl}(2,\mathbb{R})$, hence is generating and Example XII.10 shows that $d\varphi(\mathbf{1})\langle x\rangle=\{0\}$.
Case 2) $\omega\big(Q(x)\big)=0$: Then $A(x)$ is isomorphic to the four dimensional Oscillator algebra, where $A(x)=\mathbb{R}h\oplus\langle x\rangle$ and $h\in\operatorname{int}\mathbf{L}(S)\cap\mathfrak{h}$. We conclude that $V:=A(x)\cap\mathbf{L}(S)$ is a generating invariant cone in $A(x)$. Thus Example XII.11 shows that

$$d\varphi(\mathbf{1})[A(x),A(x)]=d\varphi(\mathbf{1})\operatorname{span}\{x,[h,x],Q(x)\}=\{0\}.$$

These two cases together imply that

$$d\varphi(\mathbf{1})\left(\bigoplus_{\omega\in\Omega_{\mathfrak{p}}^+}\mathfrak{g}^\omega\right)=\{0\}.$$

Thus $d\varphi(\mathbf{1})\mathfrak{g}=d\varphi(\mathbf{1})\mathfrak{k}_{\mathfrak{h}}$ is a compact Lie algebra (Proposition II.18). ∎

The preceding theorem proves that the only Lie semigroups that admit faithful compact representation are those in Lie groups with compact Lie algebra. Let (S,G) be such a Lie semigroup and write $G=K\times V$, where K is a maximal compact subgroup and V a central vector group. Then KS is also a Lie subsemigroup of G and therefore we may assume that $K\subseteq H(S)$. Then $S=K\times W$, where $W\subseteq V$ is a wedge. Let $V=H(W)\oplus V_1$, where V_1 is a vector space complement for $H(W)$ in V. To get a faithful representation of S we take a representation φ_1 of K, a representation φ_2 of $H(W)$ which maps it injectively into a sufficiently large torus and a compact representation φ_3 of the pointed cone $W\cap V_1$ as in Example XII.7 (Note that $W\cap V_1$ is contained in a polyhedral cone). Putting these fact together we obtain a faithful compact representation of S.

SEMIGROUPS OF FORWARD DISPLACEMENTS

Definition XII.13. Let (M,Θ) be a conal manifold, i.e., a manifold M endowed with a *cone field* Θ which assigns to each point $p \in M$ a wedge $\Theta(p) \in T_p(M)$. The set of all diffeomorphisms of M which preserve the cone field is denoted $\mathrm{Aut}(M,\Theta)$.

A vector field $\mathcal{X}$ on M is said to be *temporal* if it satisfies

$$\mathcal{X}(p) \in \Theta(p) \quad \text{for all} \quad p \in M.$$

We write $\mathcal{T}(\Theta)$ for the set of temporal vector fields on M. ∎

Lemma XII.14. *$\mathcal{T}(\Theta)$ is a convex cone in $\mathcal{V}(M)$ which is closed in the topology of pointwise convergence and invariant under the group $\mathrm{Aut}(M,\Theta)$.*

Proof. Let $\varphi \in \mathrm{Aut}(M,\Theta)$ and $\mathcal{X} \in \mathcal{T}(\Theta)$. Then

$$d\varphi(\varphi^{-1}(p))\mathcal{X}(\varphi^{-1}(p)) \in d\varphi(\varphi^{-1}(p))\Theta(\varphi^{-1}(p)) = \Theta(p).$$

This proves the invariance condition. The rest follows from the closedness of the convex cones $\Theta(p)$ for $p \in M$. ∎

Proposition XII.15. *Let $G \times M \to M$ a Lie group action on M such that the corresponding homomorphism $\pi : G \to \mathrm{Diff}(M)$ maps G into $\mathrm{Aut}(M,\Theta)$. Then $d\pi^{-1}(\mathcal{T}(\Theta))$ is an invariant wedge in $\mathbf{L}(G)$.*

Proof. That $C := d\pi^{-1}(\mathcal{T}(\Theta))$ is a wedge follows from the continuity of the point evaluations $X \mapsto d\pi(X)(p)$ which is a consequence of the linearity.

Let $\mathcal{X}(p) = d\pi(X)(p) = \frac{d}{dt}\Big|_{t=0} \exp(-tX).p$. Then we have for $g \in G$ that

$$d\pi(X)(\pi(g)^{-1}(p)) = d\pi(g)^{-1}(p)d\pi(\mathrm{Ad}(g)X)(p)$$

and therefore

$$\begin{aligned}
&d\pi(g)(\pi(g)^{-1}(p))d\pi(X)(\pi(g)^{-1}(p))\\
&= d\pi(g)(\pi(g)^{-1}(p))d\pi(g)^{-1}(p)d\pi(\mathrm{Ad}(g)X)(p)\\
&= d\pi(\mathrm{Ad}(g)X)(p).
\end{aligned}$$

This means that $\pi(g).d\pi(X) = d\pi(\mathrm{Ad}(g)X)$. Now the assertion follows from Lemma XII.14. ∎

Example XII.16. Let H be a Hilbert space and $G = \mathcal{U}(H)$ the unitary group with the strong operator topology. Then G is a topological group and

$$\mathbf{L}(G) := \mathrm{Hom}(\mathbb{R}, G)$$

may be identified with the set skew-adjoint operators on H. We write $W \subseteq \mathbf{L}(G)$ for the subset of all those elements X for which iX is negative semidefinite. Then $W \subseteq \mathbf{L}(G)$ is an invariant cone. Therefore W defines a "cone field" Θ on G which is invariant under left and right translations with elements of G. In this sense $\mathcal{U}(H)$ is an "infinite dimensional conal manifold". A one-parameter semigroup $\gamma : \mathbb{R} \to \mathcal{U}(H)$ consists of future displacements if and only if its infinitesimal generator lies in W. ∎

Let M be a homogeneous conal manifold which is globally orderable (cf. [La89], [Ne91b]), $x_0 \in M$, and $\pi : G \to M, g \mapsto g.x_0$ the corresponding projection. Then $S_0 := \{g \in G : x_0 \preceq g.x_0\}$ is a closed subsemigroup of G and

$$S := \bigcap_{g \in G} gS_0g^{-1}$$

is the *semigroup of future displacements* of M. Its tangent wedge consists of those elements X of $\mathfrak{g}$ for which the vector field

$$\mathcal{X}(p) := \frac{d}{dt}\bigg|_{t=0} \exp(tX).p$$

on M is *temporal*, i.e., it satisfies $\mathcal{X}(p) \in \theta(p)$ for all $p \in M$.

These realizations are used in [Pa81] for the actions of the universal covering groups of simple hermitean Lie groups on the Shilov boundaries of tube domains to prove globality of the invariant cones in this case.

Example XII.17. (a) Let M be a hyperbolic adjoint orbit in $\mathrm{sl}(2, \mathbb{R})$, a hyperboloid, and $G := \mathrm{Sl}(2, \mathbb{R})$. There exist 4 cone fields on M which are G-invariant. Two are globally orderable, two are not. Take one of those which are not. We lift the action of G on M to an action of $\widetilde{G}$ on $\widetilde{M}$. Then the semigroup of future displacements agrees with one of the invariant Lie semigroups in $\widetilde{G}$.
(b) Let $G := \mathrm{Sl}(2, \mathbb{R})$ and set $M := G/AN$, where $G = KAN$ is an Iwasawa decomposition. Then M is diffeomorphic to the circle and G acts orientation preserving, i.e., there exists a G-invariant cone field. We lift the action to an action of $\widetilde{G}$ on $\widetilde{M} \cong \mathbb{R}$. Then the semigroup of future displacements is one of the invariant Lie semigroups in $\widetilde{G}$. ∎

Let G be a hermitean simple Lie group, $\widetilde{G}$ its universal covering, and $G = KAN$ an Iwasawa decomposition. Again we lift the action of G on $M := G/AN \cong K$ to an action of $\widetilde{G}$ on $\widetilde{K} \cong \mathbb{R} \times \widetilde{K}'$.

Theorem XII.18. *There exists a causal structure on $\widetilde{K}$ with the following property: an invariant cone $W \subseteq \mathbf{L}(\widetilde{G})$ is global in $\widetilde{G}$ iff every vector field associated to an element in W (or $-W$), is temporal on $\widetilde{K}$.*

Proof. We take an invariant half space in $\mathfrak{k} = \mathbf{L}(K)$ which is bounded by the compact semisimple Lie algebra $\mathfrak{k}'$. This works since the center of $\mathfrak{k}$ is one-dimensional. Now the results in [Ne90] imply that the cone W_0 of temporal vector fields is one of the two maximal invariant global cones in $\mathbf{L}(G)$. In view of Proposition VIII.21 an invariant cone W is global iff $W \subseteq W_0$ or $W \subseteq -W_0$. ∎

Example XII.19. $\mathrm{SU}(n,n)$ acting on $U(n)$ (cf. [Pa81, p.348] and [Se76, p.37]). ∎

Example XII.20. Let $M = \mathbb{R}^4$ (Minkowski space), $G \cong M \rtimes \mathrm{SO}(3,1)_0$ is the *Poincaré group*, and $\mathrm{SO}(3,1)_0$ the *Lorentz group*. Then the light cones in $\mathbb{R}^4$ are invariant cones in $\mathbf{L}(G)$ which correspond to the future (or past) displacements in G. ∎

Example XII.21. In contrast to this example, the *Gallilean group* is

$$G = \mathbb{R}^4 \rtimes \left(\mathbb{R}^3 \rtimes \mathrm{SO}(3)\right),$$

where $\mathbb{R}^3 \rtimes \mathrm{SO}(3)$ acts on $\mathbb{R}^4 \cong \mathbb{R}^3 \times \mathbb{R}$ by

$$(v, R).(x, t) := (Rx + tv, t).$$

This group preserves the constant field of half spaces $\{t \geq t_0\}$ on $\mathbb{R}^4$. ■

REFERENCES

[BtD85] Bröcker, T., and T. tom Dieck,"Representations of Compact Lie Groups", Springer-Verlag, Berlin Heidelberg New York Tokyo, 1985.

[Bou71] Bourbaki, N., "Groupes et algèbres de Lie, Chap. III", Hermann, Paris, 1971.

[Bou75] —, "Groupes et algèbres de Lie, Chap. VII, VIII", Hermann, Paris, 1975.

[CHK83] Carruth, J. N., Hildebrand, J. A., and R. J. Koch, "The Theory of Topological Semigroups", Marcel Dekker, New York, 1983.

[CHK86] —, "The Theory of Topological Semigroups, Vol. II", Marcel Dekker, New York, 1986.

[CL89] Chon, I., and J. D. Lawson, *Attainable sets and one-parameter semigroup of sets*, submitted.

[Di64] Dixmier, J., "Les C^*-algèbres et leurs représentations", Gauthier-Villars, Paris, 1964.

[Dö92] Dörr N., *A note on the Ocillator Group*, Seminar Sophus Lie **2:1**(1992).

[Fa87] Faraut, J., *Algèbres de Volterra et Transformation de Laplace Sphérique sur certains espaces symétriques ordonnés*, Symp. - Math. **29**(1987),183–196.

[Fa91] Faraut, J., *Espaces symétriques ordonnés et algèbre de Volterra*, J. Math. Soc. Japan **43**(1991), to appear.

[Fo89] Folland, G. B., "Harmonic Analysis in Phase Space", Princeton University Press, Princeton, New Jersey, 1989.

[F78a] Friedberg, M., *Compactifications of finite dimensional cones*, Semigroup Forum **15**(1978), 199–228.

[F78b] —, *On the universal compactification of a cone*, Rocky Mtn. J. **8**(1978), 503–526.

[GG77] Gel'fand, M., and S.G. Gindikin, *Complex Manifolds whose Skeletons are real Lie Groups, and Analytic Discrete Series of Representations*, Funct. Anal. and Appl. **11**(1977), 19–27.

[GHK80] Gierz, G. et al, "A Compendium of Continuous Lattices", Springer, New York, 1980.

[Gi89] Gichev, V. M., *Invariant Orderings in Solvable Lie Groups*, Sib. Mat. Zhurnal **30**(1989), 57–69.

[Gu76] Guts, A. K., *Invariant Orders on Three-Dimensional Lie Groups*, Sib. Mat. Zhurnal **17**(1976), 986–992.

[GL84] Guts, A. K., and A. V. Levichev, *On the Foundations of Relativity Theory*, Sov. Math. Dokl. **30**, 253–257(1984).

[HE73] Hawking, S. W., and G. F. R. Ellis, "The Large Scale Structure of Space-time", Cambride University Press, Cambridge, 1973.

[Hel78] Helgason, S., "Differential Geometry, Lie Groups, and Symmetric Spaces", Acad. Press, London, 1978.

[Hel84] —, "Groups and Geometric Analysis", Acad. Press, London, 1984.

[HR63] Hewitt, E., and K. A. Ross, "Abstract Harmonic Analysis", Springer, 1963.

[Hi89] Hilgert, J., *A note on Howe's oscillator semigroup*, Annales de l'institut Fourier **39**(1989), 663–688.

[HiHo89] Hilgert, J., and K. H. Hofmann, *Compactly embedded Cartan Algebras and Invariant Cones in Lie Algebras*, Advances in Math. **75**(1989), 168–201.

[HiHo90] Hilgert, J., and K. H. Hofmann, *On the causal structure of homogeneous manifolds*, Math. Scand. **67**(1990), 119-144.

[HHL89] Hilgert, J., K. H. Hofmann, and J. D. Lawson, *Lie Groups, Convex Cones, and Semigroups*, Oxford University Press, 1989.

[HiOl90] Hilgert, J., and G. Olafsson, *Analytic continuations of representations, the solvable case*, Jap. Journal of Math., to appear.

[HOØ91] Hilgert, J., Olafsson, G., and B. Ørsted, *Hardy Spaces on Affine Symmetric Spaces*, J. reine angew. Math. **415**(1991), 189–218.

[Ho65] Hochschild, G., "The Structure of Lie Groups", Holden Day, San Franzisco, 1965.

[Ho90a] Hofmann, K. H., *Hyperplane subalgebra of real Lie algebras*, Geometriae Dedicata **36**(1990), 207–224.

[Ho90b] Hofmann, K. H., *A memo on the exponential function and regular points*, Preprint 1349, TH Darmstadt, 1990.

[HoLa83] Hofmann, K. H., and J. D. Lawson, *Foundations of Lie Semigroups*, Springer Verlag, Lecture Notes Math. 998, 1983.

[HLP90] Hofmann, K. H., Lawson, J. D., and J. S. Pym, Eds., "The Analytic and Topological Theory of Semigroups – Trends and Developments", de Gruyter, 1990.

[HoRu91] Hofmann, K. H., and W. A. F. Ruppert, *On the Interior of Subsemigroup of Lie Groups*, Transaction of the AMS **324**(1991), 169–179.

[Hum72] Humphreys, J. E., "Introduction to Lie Algebras and Representation Theory", Springer-Verlag, Berlin Heidelberg New York Tokyo, 1972.

[La87] Lawson, J. D., *Maximal subsemigroups of Lie groups that are total*, Proceedings of the Edinburgh Math. Soc. **87**(1987), 497–501.

[La89] —, *Ordered Manifolds, Invariant Cone Fields, and Semigroups* , Forum Mathematicum **1**(1989), 273–308.

[La91] —, *Polar and Ol'shanskiĭ decompositions*, Seminar Sophus Lie **1:2**(1991), ??–??.

[Le85] Levichev, A., *Sufficient conditions for the nonexistence of closed causal curves in homogeneous space-times*, Izvestia Phys. **10** (1985), 118–119.

[Lyu88] Lyubich, Y.I., "Introduction to the Theory of Banach Representations of Groups", Birkhäuser Verlag, Boston, Berlin,1988.

[MN92] Mittenhuber, D., and K.–H. Neeb, *On the exponential function of ordered manifolds with affine connection*, submitted.

[Ne88] Neeb, K.– H. "Globalität von Lie-Keilen", Diploma thesis, Technische Hochschule Darmstadt, 1988.

[Ne89] —, *The Duality Between Subsemigroups of Lie Groups and Monotone Functions* , Transactions of the Amer. Math. Soc., to appear.

[Ne90] Neeb, K.–H., *Globality in Semisimple Lie Groups*, Annales de l'Institut Fourier **40** (1991), 493–536.

[Ne91a] —, *Semigroups in the Universal Covering Group of Sl(2)*, Semigroup Forum **43**(1991), 33-43.

[Ne91b] —, *Conal orders on homogeneous spaces*, Inventiones math. **104** (1991), 467–496.

[Ne91c] —, *Objects dual to subsemigroups of Lie groups*, Monatshefte für Mathematik **112**(1991), 303–321.

[Ne91d] —, *On the foundations of Lie Semigroups*, submitted.

[Ne91e] —, *Invariant orders on Lie groups and coverings of ordered homogeneous spaces*, submitted.

[Ol82a] Ol'shanskiĭ, G. I., *Invariant cones in Lie algebras, Lie semigroups, and the holomorphic discrete series*, Funct. Anal. and Appl. **15** (1982), 275–285.

[Ol82b] —, *Invariant orderings in simple Lie groups. The solution to* E. B. VINBERG*'s problem*, Funct. Anal. and Appl. **16**(1982), 311–313.

[Ol82c] —, *Complex Lie semigroups, Hardy spaces and the Gelfand Gindikin program*, in Russian, Conference Report, 1982.

[Pal57] Palais, R.S., "A global formulation of the Lie theory of Transformation groups", Mem. of the AMS, 22m, 1957.

[Pa81] Paneitz, S., *Invariant convex cones and causality in semisimple Lie algebras and groups*, J. Funct. Anal. **43**(1981), 313–359.

[Pa84] —, *Determination of invariant convec cones in simple Lie algebras*, Arkiv för Mat. **21**(1984), 217–228.

[Po90] Poguntke, D., *Invariant Cones in Solvable Lie Algebras*, Preprint, 1990.

[Rup87] Ruppert, W. A. F., *A Geometric Approach to the Bohr Compactification of Cones*, Math. Z. **199**(1988), 209–232.

[Rup88] —, *Bohr Compactifications of Non-abelian Lie groups*, Semigroup Forum (1988), 325–342.

[Se76] Segal, I. E., "Mathematical Cosmology and Extragalactic Astronomy", Acad. Press, New York, San Francisco, London, 1976.

[Sp88] Spindler, K., "Invariante Kegel in Liealgebren", Mitt. aus dem mathematischen Sem. Gießen, Heft 188, 1988.

[Sp89] —, *Some remarks on Levi complements and roots in Lie algebras with cone potential*, submitted.

[St51] Steenrod, N., "The Topology of Fibre Bundles", Princeton University Press, 1951.

[Su72] Sussmann, H. J., *The "Bang-bang" Problem for Certain Control Systems in* $\mathrm{Gl}(n, \mathbb{R})$, SIAM J. Control **10**(1972), 470–476.

[Vin80] Vinberg, E. B., *Invariant cones and orderings in Lie groups*, Funct. Anal. and Appl. **14**(1980), 1–13.

Karl-Hermann Neeb
Fachbereich Mathematik
Technische Hochschule Darmstadt
Schlossgartenstrasse 7
W-6100 Darmstadt
Germany

Editorial Information

To be published in the *Memoirs*, a paper must be correct, new, nontrivial, and significant. Further, it must be well written and of interest to a substantial number of mathematicians. Piecemeal results, such as an inconclusive step toward an unproved major theorem or a minor variation on a known result, are in general not acceptable for publication. *Transactions* Editors shall solicit and encourage publication of worthy papers. Papers appearing in *Memoirs* are generally longer than those appearing in *Transactions* with which it shares an editorial committee.

As of May 6, 1993, the backlog for this journal was approximately 8 volumes. This estimate is the result of dividing the number of manuscripts for this journal in the Providence office that have not yet gone to the printer on the above date by the average number of monographs per volume over the previous twelve months, reduced by the number of issues published in four months (the time necessary for preparing an issue for the printer). (There are 6 volumes per year, each containing at least 4 numbers.)

A Copyright Transfer Agreement is required before a paper will be published in this journal. By submitting a paper to this journal, authors certify that the manuscript has not been submitted to nor is it under consideration for publication by another journal, conference proceedings, or similar publication.

Information for Authors and Editors

Memoirs are printed by photo-offset from camera copy fully prepared by the author. This means that the finished book will look exactly like the copy submitted.

The paper must contain a *descriptive title* and an *abstract* that summarizes the article in language suitable for workers in the general field (algebra, analysis, etc.). The *descriptive title* should be short, but informative; useless or vague phrases such as "some remarks about" or "concerning" should be avoided. The *abstract* should be at least one complete sentence, and at most 300 words. Included with the footnotes to the paper, there should be the 1991 *Mathematics Subject Classification* representing the primary and secondary subjects of the article. This may be followed by a list of *key words and phrases* describing the subject matter of the article and taken from it. A list of the numbers may be found in the annual index of *Mathematical Reviews*, published with the December issue starting in 1990, as well as from the electronic service e-MATH [**telnet e-MATH.ams.org** (or **telnet 130.44.1.100**). Login and password are **e-math**]. For journal abbreviations used in bibliographies, see the list of serials in the latest *Mathematical Reviews* annual index. When the manuscript is submitted, authors should supply the editor with electronic addresses if available. These will be printed after the postal address at the end of each article.

Electronically prepared manuscripts. The AMS encourages submission of electronically prepared manuscripts in $\mathcal{A}_{\mathcal{M}}\mathcal{S}$-TeX or $\mathcal{A}_{\mathcal{M}}\mathcal{S}$-LaTeX because properly prepared electronic manuscripts save the author proofreading time and move more quickly through the production process. To this end, the Society has prepared "preprint" style files, specifically the amsppt style of $\mathcal{A}_{\mathcal{M}}\mathcal{S}$-TeX and the amsart style of $\mathcal{A}_{\mathcal{M}}\mathcal{S}$-LaTeX, which will simplify the work of authors and of the production staff. Those authors who make use of these style files from the beginning of the writing process will further reduce their own effort. Electronically submitted manuscripts prepared in plain TEX or LATEX do not mesh properly with the AMS production systems and cannot, therefore, realize the same kind of expedited processing. Users of plain TEX should have little difficulty learning AMS-TEX, and LATEX users will find that AMS-LATEX is the same as LATEX with additional commands to simplify the typesetting of mathematics.

Guidelines for Preparing Electronic Manuscripts provides additional assistance and is available for use with either AMS-TEX or AMS-LATEX. Authors with FTP access may obtain *Guidelines* from the Society's Internet node e-MATH@math.ams.org (130.44.1.100). For those without FTP access *Guidelines* can be obtained free of charge from the e-mail address guide-elec@math.ams.org (Internet) or from the Publications Department, American Mathematical Society, P.O. Box 6248, Providence, RI 02940-6248. When requesting *Guidelines*, please specify which version you want.

At the time of submission, authors should indicate if the paper has been prepared using AMS-TEX or AMS-LATEX. The *Manual for Authors of Mathematical Papers* should be consulted for symbols and style conventions. The *Manual* may be obtained free of charge from the e-mail address cust-serv@math.ams.org or from the Customer Services Department, American Mathematical Society, P.O. Box 6248, Providence, RI 02940-6248. The Providence office should be supplied with a manuscript that corresponds to the electronic file being submitted.

Electronic manuscripts should be sent to the Providence office immediately after the paper has been accepted for publication. They can be sent via e-mail to pub-submit@math.ams.org (Internet) or on diskettes to the Publications Department address listed above. When submitting electronic manuscripts please be sure to include a message indicating in which publication the paper has been accepted. No corrections will be accepted electronically. Authors must mark their changes on their proof copies and return them to the Providence office. Authors and editors are encouraged to make the necessary submissions of electronically prepared manuscripts and proof copies in a timely fashion.

Two copies of the paper should be sent directly to the appropriate Editor and the author should keep one copy. The *Guide for Authors of Memoirs* gives detailed information on preparing papers for *Memoirs* and may be obtained free of charge from AMS, Editorial Department, P. O. Box 6248, Providence, RI 02940-6248. For papers not prepared electronically, model paper may also be obtained free of charge from the Editorial Department.

Any inquiries concerning a paper that has been accepted for publication should be sent directly to the Editorial Department, American Mathematical Society, P. O. Box 6248, Providence, RI 02940-6248.

Recent Titles in This Series

(Continued from the front of this publication)

466 **Shari A. Prevost,** Vertex algebras and integral bases for the enveloping algebras of affine Lie algebras, 1992
465 **Steven Zelditch,** Selberg trace formulae and equidistribution theorems for closed geodesics and Laplace eigenfunctions: finite area surfaces, 1992
464 **John Fay,** Kernel functions, analytic torsion, and moduli spaces, 1992
463 **Bruce Reznick,** Sums of even powers of real linear forms, 1992
462 **Toshiyuki Kobayashi,** Singular unitary representations and discrete series for indefinite Stiefel manifolds $U(p,q;\mathbb{F})/U(p-m,q;\mathbb{F})$, 1992
461 **Andrew Kustin and Bernd Ulrich,** A family of complexes associated to an almost alternating map, with application to residual intersections, 1992
460 **Victor Reiner,** Quotients of coxeter complexes and P-partitions, 1992
459 **Jonathan Arazy and Yaakov Friedman,** Contractive projections in C_p, 1992
458 **Charles A. Akemann and Joel Anderson,** Lyapunov theorems for operator algebras, 1991
457 **Norihiko Minami,** Multiplicative homology operations and transfer, 1991
456 **Michał Misiurewicz and Zbigniew Nitecki,** Combinatorial patterns for maps of the interval, 1991
455 **Mark G. Davidson, Thomas J. Enright and Ronald J. Stanke,** Differential operators and highest weight representations, 1991
454 **Donald A. Dawson and Edwin A. Perkins,** Historical processes, 1991
453 **Alfred S. Cavaretta, Wolfgang Dahmen, and Charles A. Micchelli,** Stationary subdivision, 1991
452 **Brian S. Thomson,** Derivates of interval functions, 1991
451 **Rolf Schön,** Effective algebraic topology, 1991
450 **Ernst Dieterich,** Solution of a non-domestic tame classification problem from integral representation theory of finite groups ($\Lambda = RC_3, v(3) = 4$), 1991
449 **Michael Slack,** A classification theorem for homotopy commutative H-spaces with finitely generated mod 2 cohomology rings, 1991
448 **Norman Levenberg and Hiroshi Yamaguchi,** The metric induced by the Robin function, 1991
447 **Joseph Zaks,** No nine neighborly tetrahedra exist, 1991
446 **Gary R. Lawlor,** A sufficient criterion for a cone to be area-minimizing, 1991
445 **S. Argyros, M. Lambrou, and W. E. Longstaff,** Atomic Boolean subspace lattices and applications to the theory of bases, 1991
444 **Haruo Tsukada,** String path integral realization of vertex operator algebras, 1991
443 **D. J. Benson and F. R. Cohen,** Mapping class groups of low genus and their cohomology, 1991
442 **Rodolfo H. Torres,** Boundedness results for operators with singular kernels on distribution spaces, 1991
441 **Gary M. Seitz,** Maximal subgroups of exceptional algebraic groups, 1991
440 **Bjorn Jawerth and Mario Milman,** Extrapolation theory with applications, 1991
439 **Brian Parshall and Jian-pan Wang,** Quantum linear groups, 1991
438 **Angelo Felice Lopez,** Noether-Lefschetz theory and the Picard group of projective surfaces, 1991
437 **Dennis A. Hejhal,** Regular b-groups, degenerating Riemann surfaces, and spectral theory, 1990
436 **J. E. Marsden, R. Montgomery, and T. Ratiu,** Reduction, symmetry, and phase mechanics, 1990

(See the AMS catalog for earlier titles)